Diptera Diversity: Status, Challenges and Tools

grafikhuset
HOUSE OF GRAPHICS

Cover montage based on photograph by Philippe Verdon.
Back cover photograph by Albert Szappanos.
Layout and cover design by Grafikhuset (House of Graphics),
Denmark, www.grafikhuset.net

Diptera Diversity:
Status, Challenges and Tools

Edited by

Thomas Pape
Daniel Bickel
Rudolf Meier

BRILL

LEIDEN • BOSTON
2009

This book is printed on acid-free paper.

ISBN   978 90 04 14897 0

TABLE OF CONTENTS

CHAPTER SEVEN

ORIENTAL DIPTERA, A CHALLENGE IN DIVERSITY AND TAXONOMY 197

Patrick Grootaert

SECTION II: DIPTERA BIODIVERSITY: CASE STUDIES, ECOLOGICAL APPROACHES AND ESTIMATION

SECTION III: BIOINFORMATICS AND DIPTERAN DIVERSITY

FOREWORD

QUENTIN D. WHEELER

*International Institute for Species Exploration, School of Life
Sciences, Arizona State University, Tempe, USA*

Dipterists have always been at the vanguard of taxonomic progress. The *Manual of Nearctic Diptera* was one of the few comprehensive guides to a hyper-diverse taxon that was at once authoritative, user accessible, and of a uniformly high aesthetic value. Dipterists were among the first entomological communities to complete comprehensive regional catalogs, such as *A catalog of the Diptera of America north of Mexico*, interactive digital keys to families (http://www.lucidcentral.org/keys/viewKeyDetails. aspx?id=346), and an interactive digital anatomical atlas (http://www. ento.csiro.au/biology/fly/flyGlossary.html). Studies of *Drosophila* not only gave to modern genetics and evo-devo their most important 'guinea pig,' they illustrated in Hawaii one of the most compelling examples of evolutionary radiations. And of course the greatest theoretical advances in taxonomy since Linnaeus and Darwin were led by a dipterist, Willi Hennig. Well, they have done it again.

Much has been written over the past twenty years about the so-called 'biodiversity crisis' and what ecologists have come to refer to as the 'taxonomic impediment.' The former refers to mounting and irrefutable evidence that many species face a threat of extinction; the latter to our inability to reliably identify or tell you anything about most of the species encountered during field work in most places on Earth. The depth of our ignorance of the world's species is at once awesome and inexcusable. Perhaps the single greatest factor in the persistence of this ignorance is society's lack of appreciation and support for taxonomy. In the midst of great angst and defeatism there have been few positive suggestions for ways to move forward, or indeed efforts to provide substantive summaries of what we do know. This volume represents in important ways both.

Diptera Diversity has many more far-reaching implications for students of biodiversity than the title might at first suggest. Biogeographers, ecologists, ecosystem scientists, conservation biologists, comparative biologists generally, and especially taxonomists — without regard to taxon speciality — are well advised to read and contemplate this landmark volume. The sweeping breadth of topics and issues raised is quite unusual for any single volume. The fact that these are raised in the context of an empirically-rich tome based on a hyper-diverse taxon of worldwide importance is simply unprecedented.

Even the one chapter that I found myself in deep disagreement with was importantly thought provoking. While I personally disagree with Dr. Bickel's pessimistic conclusions about the prospect of describing all species, his arguments forced me to reassess and justify to myself my opposing belief that taxonomy and taxonomists are indeed up to the challenge. If we are to succeed in gaining the necessary support for taxonomists to fully explore the diversity of species on Earth, then we must address the constraints — scientific, practical, and human — that he has identified. At the heart of his arguments for me was the explicit recognition of what makes taxonomists tick... it is the love for particular groups of species and the insatiable drive and desire to know them and their characters fully. We must heed his implied warning that taxonomy will never be done completely or well if it is merely a mercenary march through countless thousands of species. It must be done *con amore* by self-selected specialists who will strive for the kind of excellence that has always characterized taxonomy done with passion. This to me is no reason to shy from the challenge of mapping the species of our planet but rather a challenge to develop a system that recognizes, nurtures and supports those admittedly rare individuals who do have the capacity to fall in love with a taxon and devote their lives to its exploration and mastery. I know from more than twenty years of teaching entomology at Cornell University that there is a small number of beady-eyed, passionate, driven, 'naturals' in every generation, indeed in every classroom. We as universities have spent decades discouraging them, telling them that they are looking for love in the wrong place and that they will never be employed. What if instead we told them, yes you can become the world's leading authority on an obscure group of flies that make your heart race. You can indulge your impulse to study them and to search tirelessly for their unknown species. You will have a job and be welcomed into an international community of dipterists and be funded to conduct research, teach and mentor the next generation of taxon specialists. Bickel does a great service to the success of taxonomy's mission in forcing us to think, justify, and explain our ability to overcome what he has not alone seen as insurmountable obstacles.

Dikow *et al.* expand upon what I regard as a seminal paper that they published in 2004 on the potential use of revisions and monographs in estimating what we do not yet know about species diversity. Revisions and monographs have always been the gold standards of good taxonomy. Such comprehensive works represent efficient tests of all species described since 1758 and efficient 'tools' for recognizing and describing new species. This has always been known and appreciated by working taxonomists who use such comprehensive studies daily. Dikow *et al.* extend the impact and importance of such documents to everyone interested in biodiversity and its conservation. Analyzing metadata is an inherently risky business. Metadata is only as good as the data it incorporates. While museum data mobilized in recent decades provides exciting opportunities to draw meta-conclusions about species distributions and ecology, the reality is that much (for many insect taxa, most) of this information is outdated or unreliable for various reasons, all ultimately attributable to the neglect of basic taxonomy in recent decades. It is a truism that taxonomic information is as reliable as it will be for years when it is taken from the most recent revision or monograph. Dikow *et al.* show us how to use the best existing basic data to construct the most reliable metadata. Only a fool or someone in a position to profit financially could disagree with such a strategy. Dikow and his co-authors are neither.

An extremely valuable empirical component of the book is the inclusion of one chapter for each biogeographic region. These regional treatments include a summary of what is known about the respective faunas of the world, including numbers of species, and identifies the main gaps in our knowledge of the faunas. Such an attempt to go beyond a statement of the known and to specify exactly where more work is needed is laudable and should be mirrored in all such comprehensive reviews of major taxa.

While this volume summarizes, updates and refines what we have learned to date, it is also prescriptive for what we must do to make good progress in the future. In one of the most expansive and important applications of molecular evidence to questions of species status, Meier and Zhang analyze 1,001 species of Diptera. They are fair and balanced in their assessment of the positive aspects of molecular data. More importantly, they are wisely cautionary in pointing to its limitations. We live in a cynical age and an age preoccupied by technological wizardry, immediate gratification, and modernity. As they conclude, integrative taxonomy is the only logically and practically defensible way forward for taxonomy. DNA provides an important new source of data but is only component of a scientifically defensible modern taxonomy and no replacement for thoughtful, informed scholarship about morphology and other complex characters.

By the way, there is much here for the dipterist, too. Updated and authoritative synopses of what we know about the flies of all the major biogeographic regions in itself makes this volume indispensible for the student of flies and a true benchmark in the progress of entomology. For those with special interest in island faunas and all that implies in regard to speciation and geographic and ecological distributions will find much food for thought in treatments of the Hawaiian Islands and of the Galápagos. The latter chapter is particularly timely in this the year of Darwin.

I congratulate the authors and editors in producing a magistrate volume. I regard *Diptera Diversity* as a singularly important empirical and theoretical achievement. This will stand as an essential reference for dipterists for decades to come, especially those with interests that include biogeography and ecology and conservation. It also has a much broader immediacy in the sense that it offers sound, encouraging, and empowering advice to those of us deeply concerned about species exploration, taxonomy and biodiversity in putting into context the importance of taxonomic revisions and monographs, molecular data, and meta-analyses of all we know of species and their distributions in geographic and ecological space.

CONTRIBUTORS

Dalton de Souza Amorim, Departamento de Biologia, Faculdade de Filosofia, Ciências e Letras de Ribeirão Preto, Universidade de São Paulo, Av. Bandeirantes 14.040-901 Ribeirão Preto SP, Brazil. E-mail: dsamorim@usp.br

Daniel Bickel, Australian Museum, 6 College Street, Sydney, NSW 2010 Australia. E-mail: Dan.Bickel@austmus.gov.au

Don H. Colless, Australian National Insect Collection, CSIRO Entomology, GPO Box 1700 Canberra, ACT 2601 Australia. E-mail: don.colless@csiro.au

Torsten Dikow, Cornell University, Department of Entomology, Comstock Hall, Ithaca, NY 14853, USA and American Museum of Natural History, Division of Invertebrate Zoology, Central Park West at 79th Street, New York, NY 10024, USA. E-mail: torsten@tdvia.de

Neal L. Evenhuis, Hawaii Biological Survey, Bishop Museum, 1525 Bernice Street, Honolulu, Hawai'i 96817-2704, USA. E-mail: neale@bishopmuseum.org

Patrick Grootaert, Department of Entomology, Royal Belgian Institute of Natural Sciences, Vautierstreet 29, B — 1000 Brussels, Belgium. E-mail: Patrick.Grootaert@naturalsciences.be

Michael E. Irwin, Illinois Natural History Survey, Institute of Natural Resource Sustainability, University of Illinois at Urbana-Champaign, 1816 South Oak St., Champaign, IL 61820, USA. E-mail: meirwin@illinois.edu

Gail E. Kampmeier, Illinois Natural History Survey, Institute of Natural Resource Sustainability, University of Illinois at Urbana-Champaign, 1816 South Oak St., Champaign, IL 61820, USA. E-mail: gkamp@illinois.edu

Ashley H. Kirk-Spriggs, Department of Entomology & Arachnology, Albany Museum, Somerset Street, Grahamstown 6139, South Africa. E-mail: a.kirk-spriggs@ru.ac.za

Jason G. H. Londt, Natal Museum, Private Bag 9070, Pietermaritzburg, 3200, South Africa and School of Biological and Conservation Sciences, University of KwaZulu-Natal, Pietermaritzburg, 3200, South Africa. E-mail: londtja@telkomsa.net

David K. McAlpine, Australian Museum, 6 College Street, Sydney, NSW 2010 Australia.

Rudolf Meier, National University of Singapore, Department of Biological Sciences and University Scholars Programme, 14 Science Dr 4, Singapore 117543, Singapore. E-mail: dbsmr@nus.edu.sg

Thomas Pape, Natural History Museum of Denmark, Zoological Museum, Universitetsparken 15, 2100 Copenhagen, Denmark. E-mail: tpape@snm.ku.dk

Marc Pollet, Department of Entomology, Royal Belgian Institute of Natural Sciences, Vautierstraat 27, B — 1000 Brussels, Belgium and Research Group Terrestrial Ecology, University of Ghent, K.L. Ledeganckstraat 35, B — 9000 Ghent, Belgium. E-mail: MP@iwt.be

Bradley J. Sinclair, Entomology, Ontario Plant Laboratories, Canadian Food Inspection Agency, K.W. Neatby Bldg., C.E.F., 960 Carling Ave., Ottawa, ON, Canada K1A 0C6. E-mail: Bradley.Sinclair@inspection.gc.ca

Brian R. Stuckenberg, Natal Museum, Private Bag 9070, Pietermaritzburg 3200, South Africa. E-mail: bstucken@nmsa.org.za

F. Christian Thompson, Systematic Entomology Laboratory, PSI, Agricultural Research Service, U.S. Department of Agriculture, NHB-0169, Smithsonian Institution, Washington, D. C. 20013-7012, USA. E-mail: chris.thompson@ars.usda.gov

Gaurav G. Vaidya, National University of Singapore, Department of Biological Sciences, 14 Science Dr 4, Singapore 117543, Republic of Singapore. E-mail: gaurav@ggvaidya.com

Quentin D. Wheeler, International Institute for Species Exploration, School of Life Sciences, Arizona State University, PO Box 876505, Tempe, AZ 85287-6505, USA. E-mail: Quentin.Wheeler@asu.edu

Shaun L. Winterton, Entomology, Queensland Department of Primary industries & Fisheries, Indooroopilly, Queensland 4068, Australia and School of Integrative Biology, University of Queensland, St. Lucia, Queensland 4072, Australia. E-mail: wintertonshaun@gmail.com

David K. Yeates, Australian National Insect Collection, CSIRO Entomology, GPO Box 1700 Canberra, ACT 2601 Australia. E-mail: David.Yeates@csiro.au

Guanyang Zhang, Department of Entomology, University of California Riverside, Riverside, CA 92521, USA. E-mail: guanyang.zhang@email.ucr.edu

DIPTERA DIVERSITY: STATUS, CHALLENGES AND TOOLS
(EDS T. PAPE, D. BICKEL & R. MEIER). © 2009 KONINKLIJKE BRILL NV.

SECTION I:

REGIONAL DIVERSITY OF DIPTERA

CHAPTER ONE

NEARCTIC DIPTERA: TWENTY YEARS LATER

F. Christian Thompson

*Systematic Entomology Laboratory, PSI, Agricultural
Research Service, Washington, DC, USA*

INTRODUCTION

Flies are found abundantly almost everywhere; they are only rare in oceanic and extreme arctic and antarctic areas. More than 150,000 extant species are now documented (Evenhuis *et al.* 2008). So, given this great diversity, understanding is aided by dividing the whole into pieces. Sclater (1858) proposed a series of regions for the better understanding of biotic diversity. Those areas were based on common shared distribution of bird species and now are understood to reflect the evolution and dispersal/vicariance of species since the mid-Mesozoic era. While the biotic regions defined by Sclater (1858) have been accepted by most zoologists, the precise definition used here follows the standards of the *BioSystematic Database of World Diptera* [BDWD] (Thompson 1999a). Biotic regions are statistical concepts that try to maximize the common (unique to one area only) elements and minimize the shared elements (Darlington 1957, Thompson 1972). For pragmatic reasons, the BDWD has taken the traditional definitions of the biotic regions and normalized them so that they follow political boundaries, which make the assignment of data easier (Thompson 1999a). Earlier authors (Osten Sacken 1858, 1878; Aldrich 1905) divided the New World into a northern and southern component. So their catalogs covered all the species of North America, that is, the Americas north of Colombia. Unfortunately, most subsequent authors decided to re-define both North America and the Nearctic Region as the area north of Mexico (most recently, Poole 1996 & Adler *et al.* 2004). Griffiths (1980) for his *Flies of the Nearctic Region* has adopted the classic definition of

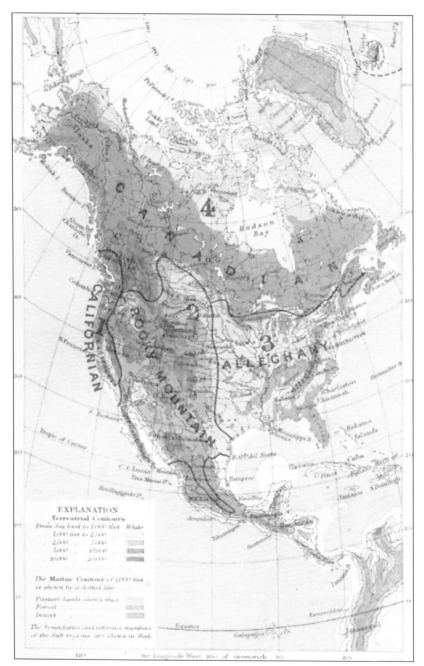

Map 1.1. Nearctic Region
(as illustrated by Wallace 1876).

Wallace (1876). So, users must always be aware of the definition of the words, Nearctic and North America.

The Nearctic Region was defined as essentially the non-tropical areas of North America (Wallace 1876, Map 1.1), a definition now modified slightly to follow the political boundaries of various Mexican states (Map 1.2). While Wallace divided the Nearctic into four subregions and subsequent workers more finely divided these subdivisions, current workers (Heywood 1995, Groombridge 1992) have abandoned this effort and view subdivision of the biogeographic regions as a series of ecological divisions or biomes. The Nearctic has six biomes: Arctic tundra, northern coniferous forest, temperate forest, temperate grassland, Mediterranean vegetation/chaparral and desert (Map 1.3).

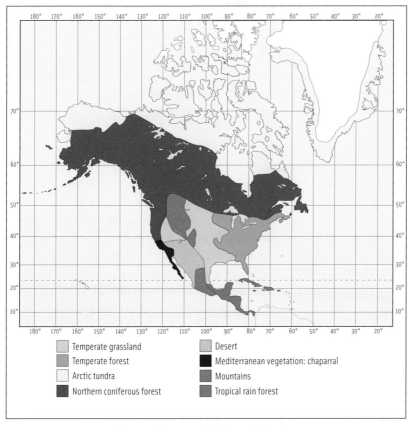

Map 1.2. Biomes of the Nearctic Region
(from Cox & Moore 2005).

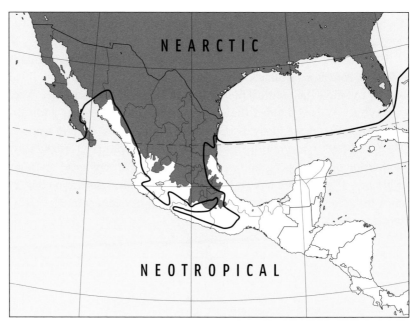

Map 1.3. Mexico and boundary of Nearctic Region
(from Thompson 1999a).

The Nearctic insect fauna was reviewed in the late 1980s (Kosztarab & Schaefer 1990; Diptera by Thompson 1990; also see Ross 1953). Then some 19,500 species of flies were described out of an estimated total of 30,000 species, but less than one per cent of them had been treated comprehensively in monographs and less than a quarter had been thoroughly revised. In the past twenty years, little has changed except that there are fewer workers today (some 250 authors published new species in the last twenty years versus 330 in the preceding 20 years, a 25% decrease). Only some 1,350 new species have been added (some 3,000 added from 1968–1987), and few new monographic works and revisions have been published. All this suggests that the prospects for comprehensive biodiversity inventories of little known groups, such as flies, are abysmal. Promises of new technologies to increase the rate of progress remain only that, as the necessary support for the people to use them is not available. Also, there is a reluctance of workers to abandon the ancient techniques they have used for centuries.

The data for this report are derived from the *BioSystematic Database of World Diptera* that we are building in Washington (Evenhuis *et al.* 2008).

Diptera Diversity: Status, Challenges and Tools
(eds T. Pape, D. Bickel & R. Meier). © 2009 Koninklijke Brill NV.

This database has been built from the earlier regional Diptera catalogs, augmented from the *Zoological Record*, and checked against World-family-level catalogs as they have appeared. Full documentation on the status and sources of the BDWD can be found online (Evenhuis *et al.* 2008). Currently we have nomenclatural and distributional data on all the flies of the world. Also, a data file is maintained on people who work on Diptera. While preparing this summary, we queried several specialists about various questions. However, this report is largely an update of the previous one (Thompson 1990). As a historical footnote, mention needs to be made here of an important but formally unpublished dataset on Nearctic flies. During the late 1980s an effort was started to develop a revised catalog of Nearctic Diptera to up-date the classification and taxonomy in the then current Diptera catalog (Stone *et al.* 1965). This effort was lead by myself and was computerized. All the specialists on Nearctic Diptera contributed and the dataset was completed. Unfortunately, no support was found to publish a revised catalog. The dataset (except Tachinidae), however, was available and used subsequently by some. The dataset was the basis of the figures in my 1990 review. They were made available to the National Oceanographic Data Center (NODC) and appeared in their Taxonomic Code (Hardy 1993). From there, these data records were passed onto the Integrated Taxonomic Information System (ITIS) and eventually to the Species2000 and Global Biodiversity Information Facility (GBIF). Many online sources have copied these data. While this dataset was of the highest quality, there were errors in it. One error, originally made by a data entry clerk in 1987, for example, was discovered recently, and was found to have been duplicated on about a dozen different Internet sites (Animal-Diversity web, ZipCodeZoo, Wikipedia, etc.)!

1. Past

Our knowledge of the taxonomy of Nearctic Diptera began with Linnaeus in 1758, the designated starting date for zoological nomenclature. What needs to be stressed here is not how little Linnaeus knew of Nearctic Diptera, but that we began with a comprehensive summation of all that was known then. *Systema Naturae* (Linnaeus 1758) includes keys and diagnoses, current and correct nomenclature, and synopses of the literature and biology for all taxa. *Systema Naturae* was the last fully comprehensive work published. Works since that time have become ever more restricted

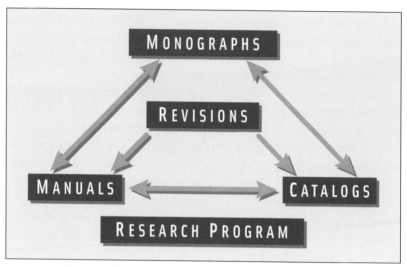

Figure 1.1. Research Program for North American Dipterology,
see text for discussion.

either taxonomically or geographically or both. After Linnaeus, Fabricius continued to try to produce comprehensive works on Insecta (*sensu lato*) (Fabricius 1775). The task, however, became more difficult as others began to adopt the Linnaean method, and more geographic areas were discovered and explored. Fabricius spent his life traveling widely in Europe to maintain contact with all insect systematists and to synthesize their work with his own. Near the end of his life, Fabricius did complete *Systema Antliatorum* (Fabricius 1805), his statement of what was then known about flies. Unfortunately, whereas Linneaus' work was comprehensive by definition, such status cannot be ascribed to the Fabrician *Systema*, which did not include all the discoveries made about flies since 1758.

After Fabricius, systematists specialized more, working either on a single order or particular region. For North America, some Europeans (Macquart, Wiedemann, Walker, *et al.*) specialized on 'exotic' flies, that is, those that did not occur in Europe. During this period there was only one American, Thomas Say, who worked on all insects. Thus, by the middle of the 19th century our knowledge of Nearctic Diptera was in chaos: no comprehensive works, just descriptions scattered through the literature. Fortunately for us, there was a new development in Washington: the Smithsonian Institution. This new organization saw the need for a biotic survey and began sponsoring inventories of our biota. For Diptera, fortunately,

Period	Dates	Species	Rate	Years	Total Species	Total Names	names	%valid
Linnaean	to 1775	87	6,2	14	87	92	92	94,6%
Fabrician	to 1805	163	5,4	30	250	300	208	78,4%
Wiedemann	to 1858	1778	33,5	53	2028	2909	2609	68,1%
Osten Sacken	to 1879	1697	80,8	21	3725	5071	2162	78,5%
Williston	to 1908	3010	100,3	30	6735	9075	4004	75,2%
Alexander, Curran	to 1965	10370	220,6	47	17105	22222	13147	78,9%
McAlpine, Sabrosky	to 1989	3646	145,8	25	20751	26170	3948	92,4%
'US'	to today	1214	71,4	17	21965	27405	1235	98,3%

Figure 1.2. History of Nearctic Diptera Fauna. The columns are: **Period**, named after a prominent dipterist characterizing the period; **Dates**, give the inclusive year; **Species**, gives the total species described within the period; **Rate**, is the total species described divided by the number of years within the period for an average rate of description.

there was a leader to take up the task. Carl Robert Romanovich, Baron von der Osten Sacken (Smith 1977), a Russian diplomat, by example and with the support of the Smithsonian, defined and started the current research program for North American Dipterology. First, Osten Sacken (1858) produced a list of the species already described from North America. Next, he organized people to collect flies, arranged to have the accumulated material studied by the best available specialist (Herman Loew), and arranged eventually to have material deposited in a public museum. Finally, he started a series of monographs (Loew 1862, 1864, 1873; Osten Sacken 1869). Osten Sacken concluded his work on the North American Diptera with a comprehensive synoptic catalog (1878). Samuel Wendell Williston, apparently seeing a weakness in the Osten Sacken program, introduced manuals (Williston 1888) that included keys to the families and genera. This improvement facilitated revisionary work, as the size of the taxonomic unit to be studied could then be as small as a genus. With the master research plan set (Fig. 1.1), the next hundred or so years (1888–1988) saw an alternation between descriptions (and revisions), catalogs (Aldrich 1905, Stone et al. 1965, 1983; Thompson 1988, Poole 1996, also see Arnett 2000), and manuals (Williston 1896, 1908; Curran 1934, 1965; McAlpine 1981, 1987, 1989), with a few monographs being done (Carpenter & La-Casse 1955; Hardy 1943, 1945; Webb 1984, Hogue 1987). This century saw

Diptera Diversity: Status, Challenges and Tools
(eds T. Pape, D. Bickel & R. Meier). © 2009 Koninklijke Brill NV.

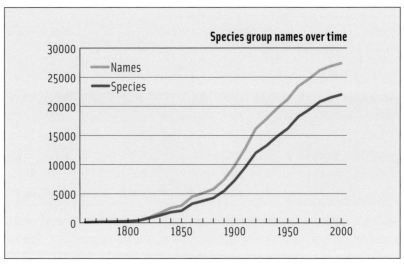

Figure 1.3. Growth of knowledge for Nearctic Diptera, based on increase of species and species-group names over time.

the introduction of regional monographic series (such as Insects of Connecticut, Ohio, Illinois, California, Virginia, Florida, Canada & Alaska), but the coverage of Diptera in them has been limited. With the passing of Williston and the 19th century, a number of highly productive dipterists (Alexander, Felt, Malloch, Melander, Curran and Van Duzee) arrived to build on the foundation of Williston's last manual and Aldrich's catalog. In a short 40 years or so, more flies were described than in the first 150 years and at a rate never since exceeded (Fig. 1.3). The late 20th Century (1960s onward) saw a resurgence starting with a new catalog (Stone *et al.* 1965, 1983) followed by a new manual (McAlpine 1981, 1987, 1989) and then the start of a monographic series, the *Flies of the Nearctic Region* (Griffiths [1980] and others). Unfortunately, the century ended in a decline that continues due to the loss of support and resources. Also, the diminished ranks of specialists on the Nearctic Diptera fauna were distracted by the new and exciting efforts to understand the Neotropics, especially Costa Rica (Brown 2005). In the last 20 years, some 4,500 new species have been described from the Neotropics (almost 700 from Costa Rica alone) compared to only 1,350 for the Nearctic. Also, other exotic survey projects have likewise been distractions (see Evenhuis 2007). So, in summary, the history of Nearctic Dipterology can be viewed as a series of eight periods (Fig. 1.2); a more detailed history of Nearctic dipterology has been written by Stone (1980) (also see Aldrich 1930, Coquillett 1904).

In my prior treatment of the Nearctic Diptera fauna (Thompson 1990), the history was divided into eight periods with dates rounded off to the nearest decade. For this work, the exact year of major defining events have been used. These events are as follows: the Linnaean period runs until the first publication of Fabricius (1775); the Fabrician period runs until his last Diptera publication (1805); the Wiedemann period runs until the first publication of Osten Sacken on North American flies (Osten Sacken 1858); the Osten Sacken period runs to the first publication of Williston (1879) (Osten Sacken's last major North American work was his catalog [1878]); the Williston period runs until his last publication (1908); the Alexander-Curran period runs until publication of the new Diptera catalog (Stone *et al.* 1965, 1983); the McAlpine period runs from the date of that catalog until the publication of the last volume of the Nearctic Diptera Manual (McAlpine 1989).

Given how our knowledge of Nearctic Diptera has developed, the next questions are: 'what do we know and what do we not know?' In considering these questions, we can divide the answers into the description of the problem (fauna), the resources (literature, collections, and human) available or needed to solve the problems, and the approach to solving the problem (research program).

2. Fauna

The Nearctic Diptera fauna is largely a transitional one. The northern two-thirds of the Nearctic region have a Diptera fauna that is largely shared with the Palaearctic region, and the southern third has many elements shared with the Neotropics. There is only one really distinctive clade endemic to the Nearctic area, the Apystomyiidae (Plate 1.1). This group is restricted to California and is probably the sister-group to all higher Diptera (Nagatomi & Liu 1994; Wiegmann, unpubl.). The Oreoleptidae (Zloty *et al.* 2005) are also endemic and restricted to the Rocky Mountains, although this group is probably nothing more than an athericid that has lost a synapomorphy.

The Nearctic Region essentially consists of three major countries, Canada, Mexico and the United States, which now form an economic unit, The North American Free Trade Alliance (NAFTA). Unfortunately, these

Plate 1.1. *Apystomyia elinguis* Melander. Dorsal habitus (above)
and head in profile (below) of adult female.
(Illustration: Marie Metz.)

countries have taken different approaches to their biodiversity. The Convention on Biological Diversity (CBD 1994) defines a standard for the nations of the World, but the USA has not ratified the convention. Hence, there is little official concern about biodiversity within the USA. Canada and Mexico, on the other hand, have joined the convention. Canada established a 'Biological Survey for Terrestrial Arthropods' even before the formation of the CBD. This effort has produced a number of major works on the origin of the North American fauna (Downes & Kavanaugh 1988), the Canadian insect fauna (Danks 1979, 1993), the Diptera fauna (McAlpine 1979) and the changes in it (Downes 1981), the arctic arthropods (Danks 1981a, b), the Yukon insects (Danks & Downes 1997), and arthropods of special habitats, such as springs (Williams & Danks 1991), peatland (Finnamore & Marshall 1994) and marshes (Rosenberg & Danks 1987). Most recently, the Survey has started a new online *Canadian Journal of Arthropod Identification* (CJAI, see Kits *et al.* 2008, for example). Mexico has also established a biodiversity program, the Comisión Nacional Para el Conocimiento y uso de la Biodiversidad (CONABIO), and a couple of works have resulted from the program that cover some groups of flies (Llorente-Bousquets *et al.* 1996). The countries with minor possessions in the Nearctic, Denmark and the United Kingdom, also have had or have recent programs to assess their Diptera or broader, their arthropod fauna. For Bermuda, Woodley & Hilburn (1994) have produced a modern review; for Greenland, a 'Greenart' project is working on an identification handbook of the insects and arachnids of the island (Böcher & Kristensen in prep.).

2.1 Faunal statistics

Where we are today is best summarized by statistics on the fauna (Table 1.1), as well as some statistics on the human, collection, and literature resources. Trend curves plotted for species-group names (Fig. 1.4) show no leveling off; hence, the curves are of little predictive value, merely indicating clearly that the fauna is not fully described (Steyskal 1965, but see also White 1975, 1979, Frank & Curtis 1979, and O'Brien & Wibmer 1979). The percentage of the fauna estimated to be known (49%, Thompson 1990) is probably too low, as Gagné estimates that there are some 14,000 undescribed species of gall midges in the Nearctic Region (1,247 species currently described), an estimate based on the assumption that gall midges are host specific (monophagous) (see Gagné 1983: 9–11, 1989: 2, 34–37).

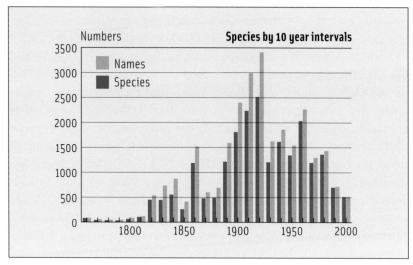

Figure 1.4. Growth of knowledge for Nearctic Diptera, based on increase of species and species-group names by 10 year intervals.

If the gall midge estimate is reduced on the assumption of broader host specificity (polyphagous, using 50% known, instead of 8% known), then the estimate of percentage-fauna-known increases to about 70%. The percentage of species known from only one sex is not estimated, as the statistic is trivial. For many taxa species recognition is based on charac-

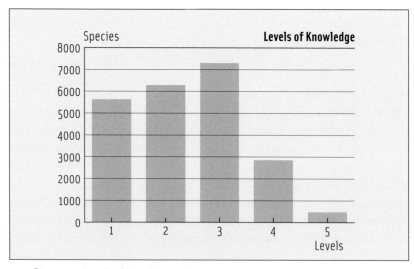

Figure 1.5. Levels of knowledge of Nearctic Diptera, see text for explanation.

DIPTERA DIVERSITY: STATUS, CHALLENGES AND TOOLS
(EDS T. PAPE, D. BICKEL & R. MEIER). © 2009 KONINKLIJKE BRILL NV.

ters of the male genitalia or secondary sexual characters. Hence, in these situations the percentage known from only one sex is by definition 100%. However, females are not unknown, as female specimens are recognized as belonging to higher taxa such as species groups, and these females do provide characters for our classifications. The taxonomy of flies is based on the holomorph. When material has been available, characters have been found in all stages (eggs, larvae, pupae, adult male, and female). Our knowledge of immature stages of Diptera was last reviewed by Hennig (1948, 1950, 1952), and for those of Cyclorrhapha by Ferrar (1987). About 98% of all families need revision. Only five families have been been treated in comprehensive monographs. To produce a more meaningful measure of the status of our knowledge of Diptera of America north of Mexico, I have defined five levels of taxonomic knowledge based on comprehensiveness and quality of publications.

Level 0 — Species descriptions only. Recent examples: Byers & Rossmann (2008), Grogan & Philips (2008), Robinson & Knowles (2008).

Level 1 — Keys to few (about 25% or less) species. Keys usually unreliable as they are based on characters subsequently shown to be variable (such as color) and they are not supported by illustrations. Fortunately, there are no recent examples of poor quality keys, but in many taxa the only keys available are older ones, such as Camras (1945) or Telford (1970).

Level 2 — Keys to some (about 50%) species. Keys reliable, based on non-variable characters (such as male genitalia) and usually illustrated. Examples: Spencer & Stegmaier (1973), Spencer (1981), Spencer & Steyskal (1986).

Level 3 — Keys to most (about 75% or more) species. Keys of high quality, supported by illustrations of essential characters. Usually only adults are treated, and only some species described. Nomenclature and types frequently revised. Examples: Kits *et al.* (2008; regional), Pratt & Pratt (1980), Thompson (1981), Vockeroth (1986).

Level 4 — Revisions. Taxon revised, with keys to most or all adults; all species redescribed; nomenclature, types and literature revised. Examples: Brown (1987), Griffiths (1982–2004), Hall & Evenhuis (1980–2004), Lonsdale & Marshall (2007), Mathis (1982), Michelsen (1988), Thompson (1980).

Level 5 — Monographs. Same as revisions, but immature stages also covered. Examples: Adler *et al.* (2004), Courtney (1990, 1994), Feijen (1989), Hogue (1973, 1987).

Level 4 and 5 are very similar, but differ only in comprehensiveness. The work of Griffiths (1982–2004) in his *Flies of the Nearctic Region* may be considered by some as being level 5, but is here considered level 4 as Griffiths has not treated the immature stages even though they are known for many of the taxa he has covered. Also, mosquitoes represent another special case. The last comprehensive treatment of them was by Carpenter & LaCasse (1955), but even this was not monographic in the sense that it lacks nomenclatural details such as information on types and synonyms. However, the work did present descriptions and keys to all stages. Since then there has been a series of identification guides with distribution data (latest, Darsie & Ward 2005), which keep our knowledge of mosquitoes up to date and make the group the best known Diptera taxon in the Nearctic Region.

When our knowledge of Diptera of America north of Mexico is viewed in terms of these levels (Fig. 1.5), the true magnitude of work remaining to be done is evident. While we may have described two-thirds of the species that exist, we have not properly synthesized these descriptions into comprehensive revisions or monographs. Only five families of North American flies have been effectively treated: black flies, mosquitoes and net-winged midges (Simuliidae, Culicidae, Blephariceridae, Deuterophlebiidae and Nymphomyiidae)!

The above assessment deals only with the extant fauna. While knowledge of the past is always limited, the Nearctic Region has a number of sites that provide exceptional information on the past Diptera faunas. These have been recently summarized: Virginia where there are late Triassic (220 Mya) fossil beds (Blagoderov *et al.* 2007); New Jersey where there are Cretaceous (90 Mya) amber deposits (Grimaldi 2000; Grimaldi & Cumming 1999) and the Rocky Mountains, mainly Florissant and others, where there are late Eocene (34 Mya) shale fossils (Meyer 2003). All together, some 74 families, 229 genera and 516 species are known from fossils in the Nearctic Region. Some represent clades that are unknown from the extant fauna but appear to be endemic elsewhere, such as the tsetse (*Glossina*, Glossinidae, now only known from subsaharan Africa).

Our knowledge of Diptera phylogeny is good: The sister group of Diptera is almost certainly a mecopteran, probably phenetically and cladistically related to Nannochoristidae (Wood & Borkent 1989, but see Whiting 2005 for review). The major monophyletic groups of flies have been blocked out; within the grade 'Nematocera,' the relationships among the

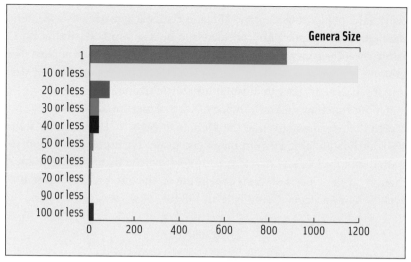

Figure 1.6. Size of Nearctic genera of Diptera.
There is a strong dominance of small genera with either one species
(877 genera) or 2–10 species (1192 genera).
Only two genera have more than one hundred species:
Dolichopus Latreille with 316 spp. and
Tipula Linnaeus with 525 spp.

family group taxa have been largely deciphered (although the contents and position of the Tipulomorpha remain uncertain), and within the grade 'Brachycera,' the major monophyletic clusters have been identified (See Yeates & Wiegmann 2005 for overall review). Much, however, needs to be done to define and objectively rank families; among the cyclorrhaphous flies, monophyletic families have been defined by greatly restricting the scope of these taxa, and much still needs to be discovered to cluster these 'microfamilies'. While the classification of the Nearctic Diptera has been fairly stable in recent times due to the conservative nature of dipterists, this classification does not reflect our progress in the knowledge of Diptera. The current families of Diptera neither conform to cladistic, nor phenetic or 'evolutionary' [*sensu* Mayr] classification conventions. Consider the contradictory treatment of the Phoridae and pupiparous Diptera (Maa & Peterson 1987, Peterson 1987, Peterson & Wenzel 1987, Wenzel & Peterson 1987). Under phenetic or 'evolutionary' conventions, the Phoridae should be treated as a cluster of families equivalent to the present concept of the pupiparous Diptera. Under a cladistic approach as used here[1] the pupiparous Diptera are considered to be one family (Griffiths 1972). Similarly at the generic level, no consistent standard has been applied. For example, the genus *Tipula*, in relation to diversity, age of origin, and size, is more than equivalent to most families of Schizophora (Fig. 1.6)!

On morphology and terminology, North American dipterists have accepted the treatment given in the *Manual of Nearctic Diptera* (McAlpine 1981) as the standard, with one major exception. Terminology for the male genitalia of cyclorrhaphous flies is, unfortunately, theory-laden. Hence, there are different sets of terms depending on the interpretation of genital evolution one accepts (Griffiths 1981, 1984).

3. Resources

3.1 Literature

The current literature resources for Diptera are excellent. Our research program has been and is based on three interrelated core publications (Fig. 1.1): catalogs, manuals, and monographs. Catalogs are the indexes to the diffuse literature of keys, descriptions, and biological data (Thompson & Knutson 1987; but see Steyskal 1988); manuals are the keys to the smallest operational taxonomic group (that is, the genus); and monographs are the ultimate species-level syntheses of all that is known about a taxonomic group, usually a family or subfamily. Today, these categories of publications are represented by the Catalog of the Diptera of America north of Mexico (Stone *et al.* 1965), the *Manual of Nearctic Diptera* (McAlpine 1981, 1987, 1989), and the *Flies of the Nearctic Region* series. The black fly monograph (Adler *et al.* 2004) and the fascicle on Blephariceridae (Hogue 1987) in the *Flies of the Nearctic Region* series, are without doubt the best examples of a monographic treatment of Nearctic insect groups. Similarly, the *Manual of Nearctic Diptera* represents the best ordinal treatment

---

1. The cladistic approach of Hennig and here implemented requires that all families be monophyletic and at least Cretaceous in the age of origin. Age is documented by fossils or inferred by phylogenetic sequence of subordination. Cladistic data are derived principally from Wood & Borkent (1989), Woodley (1989), Griffiths (1972, 1987) and Yeates & Wiegmann (2005); other sources are Hennig (1972), Oosterbroek (1986), Krivosheina (1969, 1978, 1986, 1988), Matile (1990), Chvála (1981, 1983), D.K. McAlpine (1985), and Rotheray & Gilbert (2008). The conventions used follow Wiley (1981) and Griffiths (1972).

of any insect group for any region of the world. However, the *BioSystematic Database of World Diptera* is not as comprehensive as the *Catalog of Hymenoptera in America North of Mexico* (Krombein *et al.* 1979), which is the best example of any systematic catalog ever done. Other types of literature resources are: 1) Handbooks for general users, such as Gagné (1989); 2) comprehensive character surveys (Ferrar 1987); 3) identification aids (Darsie & Ward 2005); 4) regional treatments (Wood *et al.* 1979); 5) type collection listings (Arnaud 1979); 6) annotated bibliographies (West & Peters 1973); 7) parasite and host indexes (Arnaud 1978); and 8) biographies (Shor 1971) with technical summaries (Arnaud & Owen 1981). These examples are the best of their genre; comprehensive listings of literature resources for Nearctic Diptera are found in the Manual and the Catalog and online in the BDWD.

The Internet provides access to digital material for all who have computers and network access. Now some groups have begun to scan and digitize copies of the critical literature and to make those copies accessible via the Internet. For example, more than 80% of the taxonomic literature on mosquitoes is now available online (http://www.wrbu.org/mosqlit.html), and a group of museums and herbaria have joined together to make all the literature on biosystematics of organisms available online (see the *Biodiversity Heritage Library* program at http://www.biodiversitylibrary.org).

The area where dipterists lag behind their colleagues is in ordinal societies and journals. We do not have as many national and international societies as the coleopterists or lepidopterists do, and hence, there are few special journals devoted exclusively to Diptera. For flies, we have *Studia Dipterologica* and its *Supplements*, and there are some specialized journals which deal with groups of flies, such as mosquitoes (*Mosquito News, Mosquito Systematics*) and flower flies (*Volucella*). Unfortunately, the Diptera community is also plagued by personal journals produced by individuals who cannot get their work published in regular, peer-reviewed scientific journals (*Fragmenta Dipterologica, Dipteron, Journal of Dipterological Research*). We also have very few international newsletters as compared to, e.g., the hymenopterists (Bullock 1988). However, we are improving. Since 1988, every four years dipterists hold an International Congress of Dipterology to share our developing knowledge of flies.

What of the future? Literature has always been one of the major stumbling blocks for taxonomy, as the *International Code of Zoological Nomenclature* (ICZN 1999) has enshrined priority and usage as its basic operating

DIPTERA DIVERSITY: STATUS, CHALLENGES AND TOOLS
(EDS T. PAPE, D. BICKEL & R. MEIER). © 2009 KONINKLIJKE BRILL NV.

principles based on printed publications. Early attempts at modernization of the Code failed (ICZN 1989, IUBS 1989, Ride & Yones 1986), but the future is bright as this stumbling block will be removed forever by advances in technology and changes in our code. Already *ZooBank* has been proposed as a universal registration system for names of animals (Polaszek *et al.* 2005a,b, 2008). Technology and projects such as the *Biodiversity Heritage Library* now allows anyone to have an exact copy of any original publication. In building our various regional Diptera catalogs, we have also built our working libraries. So, the sponsors of the various Diptera catalogs may be able to provide copies if one cannot obtain them locally. Technologies that allow rapid computer access to large volumes of information (Internet) as well as archival storage, such as compact disks (CD-ROM, DVD), mean that future publications will be inexpensive and easy to use. For example, *Die Fliegen der Palaearktischen Region* (Lindner 1924–1993), which runs to over 16,000 pages, would cost approximately a million dollars to be printed at today's publication costs, and sells for about $4,000 for a complete set. For the selling price alone, we could produce 1,000 copies on CD-ROM reducing four shelf-feet of books to a single 5 inch disk!! The *Diptera Data Dissemination Disk* is one publication that used CD-ROM technology. Already new publication ventures, such as *Zootaxa* (http://www.mapress.com/zootaxa/) and *ZooKeys* (http://pensoftonline.net/ZooKeys/index.php/journal) provide immediate publication and dissemination via Internet as well as paper copies distributed to libraries. The digital version (Adobe pdf file format) can be readily downloaded and stored on disks. The only hope for completing an inventory of our biota is to use new technologies!

3.2 Collections

Detailed statistics are not available for the holdings of Nearctic Diptera in various collections. This information, however, is part of the *BioSystematic Database of World Diptera* (Evenhuis *et al.* 2008). Preliminary analysis [based on a sample of 15,686 species-group names out of a total of 26,789 names] suggests that, for types, the major depositories [acronyms follow those of *Flies of Nearctic Region* series (Griffiths 1980: viii–xiii)] are: 1) the United States National Collection (USNM: 8,081); 2) the Natural History Museum, London (BMNH: 1,264); 3) the Canadian National Collection (CNC: 1,228; Cooper 1991; Cooper & Cumming 1993, 2000; Cooper & O'Hara 1996); 4) the Museum of Comparative Zoology (MCZ: 810), and 5) the California Academy of Sciences (CAS: 710; Arnaud 1979). After these collections, the following have large holdings of types: American Museum of Natural History (AMNH: 363); Academy of Natural Sciences (ANSP: 341); Illinois Natural History Survey (INHS: 123), University of Kansas (UKaL: 267) and Cornell University (CU: 69). Many foreign museums, especially those in Paris (MNHN: 211), Copenhagen (UZMC: 92), Vienna (NMW: 160), Berlin (ZMHU: 150), Stockholm (NRS: 112) and Lund (ZIL: 291), have a large number of types of American flies. Finally, and surprisingly, virtually all North American collections have at least a few types of Diptera. Only a few collections have adopted a policy of not retaining primary type material and of depositing such material in major collections. For general Diptera material, the Canadian National Collection at the Biosystematics Research Centre clearly has the largest and most diverse holdings of flies from Nearctic America. Once the Museum of Comparative Zoology (MCZ) had the honor. One hundred years ago, the MCZ had the best fly collections, but today it retains status only as a major museum because of the types it has. Some 80 years ago the collection at Washington surpassed that of Cambridge due to the strong programs of the U.S. Department of Agriculture (USDA) and later the Smithsonian Institution (SI), but the building phase of the USNM Diptera Collection petered out some fifty years ago as interests shifted to exotic areas (SI) or programs became more applied in emphasis (USDA). Some forty years ago the Canadian National Collection began its collection building phase, but, at least for flies, that phase has now peaked as there are few dipterists on the staff today. Excellent accumulations of regional material are available in the California Academy of Sciences, Bishop Museum, University of California (Berkeley, Davis and Riverside), University of Guelph, University of Kansas, Kansas State University, University of Minnesota, Florida State Collection of Arthropods, etc. A number of dipterists were queried as to the comprehensiveness of the existing collection resources. The responses to date suggest that the collections provide an adequate sample of adults for most groups of flies. That is, there is now far more material waiting to be studied than there are specialists available to study it! However, in some groups, those with specialized habits or whose taxonomy is based on special characters, such as gall midges, there is a paucity of appropriately collected material.

Today many collections have made lists of their holdings, especially their types, available online, and there is a growing trend to making digital images also available online. The Museum of Comparative Zoology

Specialist	Valid	Names	% Valid
Alexander	1210	1311	92%
Loew	1048	1310	80%
Coquillett	918	1090	84%
Felt	791	1086	73%
Malloch	629	885	71%
Melander	784	866	91%
Walker	437	733	60%
Curran	438	684	64%
Van Duzee	527	649	81%
Osten Sacken	446	517	86%

Figure 1.7. Leading specialists on Nearctic Diptera measured as taxonomic output.

has started this trend (see http://mcz-28168.oeb.harvard.edu/mcztypedb. htm) and other museums are following suit.

While the resources available in collections are adequate to begin the revisionary work which needs to be done, more material will be needed to finish the job. Material is needed of immatures and from certain geographic areas, such as Alaska (Nome Peninsula and Aleutian Islands), the Ozarks, the Red Hills in Alabama, and Nearctic Mexico. Unfortunately, given the history of declining support for surveys and museum programs, the prospects of obtaining the necessary material seem dim.

3.3 Human

Some 1,028 people have contributed to our knowledge of the taxonomy of Nearctic Diptera. The major contributors are listed in Figs. 1.7–1.8. Today, we know some 281 people working on Nearctic Diptera during the past twenty years (out of a data file on some 1,536 workers world-wide for the same period). To characterize these people better, we have grouped them on the basis of their primary occupation, as this gives an indication of the amount of time available for research.

> Volunteers or amateurs, whose occupations are not related to entomology and who do systematics in their leisure time (8) or who are retired (46).

> Entomologists, who are not employed to do systematic work (12) or are consultants (3).

Specialist	Valid	Names	% Valid
Hall	230	246	93%
Griffiths	210	215	98%
Sæther	164	186	88%
Evenhuis	166	181	92%
Gagné	160	164	98%
Marshall	134	135	99%
Sublette	102	119	86%
McAlpine	111	111	100%
Robinson	106	111	95%
Grogan	69	69	100%

Figure 1.8. Top ten living specialist on Nearctic Diptera measured as taxonomic output.

University-based systematists, who may also be required to teach, do extensive research work, and/or curate (19).

Museum-based systematists, who also may be required to curate and do identifications (30).

Then there were others (14) who are now deceased, and students (8) who have left the field after publishing their work.

For the remainder (111), insufficient data were available to classify them in one of the above groups.

What is interesting about these numbers is that the number of university-based systematists has dropped by half (19 now, 37 previously), but the number working in museums has increased slightly (30 now, 24 previously), and previously I did not tally those who were retired as there were so few of them. Retired workers now make up the largest component.

Unfortunately, no data are available on the amount of time spent on (taxonomic) research. An estimate has not been made as there are too many variables involved, and the statistic is not really relevant. Time relates to productivity, that is, the amount of research done per unit time. Productivity varies widely among systematists (Figs 1.7, 1.8); for example, how many 'Alexanders' have there been? While Alexander managed to describe more than 10,000 species in a life-time (Byers 1982), most workers have described only one or two! So, the measure of man-years will not translate to what we really want to know, which is how much research is being done. The amount of research being done is best measured by quan-

tity of research results: the number of genera, species, and names published (Figs 1.3, 1.4, 1.7, 1.8). These data clearly show that while there are
still many species to be described, our rate of description has significantly
declined. The decline is probably directly attributable to the decrease in
number of active systematists, but see Evenhuis (2007) for other reasons.
However, while quantity of taxa described has decreased, the quality of
the work has increased, that is, the percentage of taxa that are valid. Obviously both number of taxa described and the validity of them are only
surrogate measures selected as they are easily obtained from databases.
One only needs to compare the descriptions of Fabricius to one of today's
specialists to see the great improvement in quality, from numerous illustrations to increased number of characters used.

The future for human resources in Diptera remains poor. When this
review was last done, we bemoaned the retirement of those key teach-
ers, such as Alexander of Massachusetts, Berg of Cornell, Byers of Kansas, Cook of Minnesota, Schlinger of California, and Hardy of Hawai'i,
who had trained this generation of dipterists, and noted that only Steve
Marshall of the University of Guelph, Monty Wood of the Biosystematics Research Centre in Ottawa, and the Maryland Center for Systematics
(MCSE) utilizing the dipterists in Washington, had active programs for
training dipterists. Today, Guelph and Ottawa retain active programs, as
the Maryland program has become inactive. Fortunately, a couple of programs are filling this void: new cooperative programs between the American Museum of Natural History and Cornell (Grimaldi), a new molecular
phylogenetic program at North Carolina State University (Wiegmann),
and a revived program at Iowa State University (Courtney). And in between the first report and this, there was an active program at the University of Illinois that trained 7 students (Gaimari, Hauser, Hill, Holston,
Metz, Winterton and Yang).

No short courses are offered in Diptera systematics as, for example,
those provided by the North American hymenopterists (Bee Course, Parasitic Hymenoptera).

4. Research Program

4.1 Approach
The research program established by Osten Sacken and Williston is sufficient for the task. What is needed is the adoption of new technologies

to improve research productivity and distribution of results. We should be using more automated tools in our research: for example, simple word processing to sophisticated data analysis (MacClade, PAUP, TNT, etc.) and presentation (DELTA, Fact Sheet Fusion, Linnaeus II, Lucid, etc.) software (see Thompson *in* Knutson *et al.* 1987; Winterton 2009). We should also not forget who supports our research and should therefore provide our results in user-friendly, interactive expert systems so all can obtain biosystematic information directly. At the Systematic Entomology Laboratory, a prototype Biosystematic Information Database and Expert System for fruit flies (Thompson 1999b) was developed to demonstrate the increased productivity for scientists and greater accessibility for users that the integration of these new technologies will bring. Unfortunately, while the message of the need to recognize user-needs and to increase productivity is accepted, systematists continue to waste resources either by re-inventing proven technologies (as for example, EDIT [Lane 2008; Scoble 2008]) or simply sprucing-up the old (HTML keys as in CJAI).

4.2 Priorities

What taxa should be studied first and what taxa should be left for later? Most families of flies require urgent priority work, as only a few aquatic groups (Culicidae, Blephariceridae, Deuterophlebiidae, Nymphomyiidae, Simuliidae) are truly well known! Why? Because complete knowledge of our biota is, as Aristotle (see Osten Sacken 1869: iii) and E. O. Wilson (1985a,b, 1986, 1987a, b, 1988) stated, an essential humanistic goal, and the time remaining to complete this task is short due to the rapid deterioration of the environment. To set priorities, one needs criteria. Given that the only appropriate goal is a comprehensive knowledge of our entire biota, the criterion for deciding which taxon deserves the highest priority for revision is which is most threatened by extinction. Unfortunately, we do not know enough to apply such a criterion, nor could such a criterion work at a higher taxonomic level as a family group taxon, the usual level of revisionary work. Obviously, given different priority criteria other answers are possible. For example, I work for USDA, and our priorities rank Tephritidae, Cecidomyiidae, and Agromyzidae high for the plant-feeding pests they include, and Tachinidae, Syrphidae, Pipunculidae, *et cetera*, high for the potential biological control agents they include. Obviously, the Department of Defense considers mosquitoes (Culicidae), of the highest priority due to the numerous human disease vectors found among them. The Environmental Protection Agency should rank midges (Chironomidae) of high priority, because of their value as indicators of water quality. Other funding agencies will have different criteria, hence, different priority groups. And, our evaluation of the criteria will vary depending on our knowledge of the taxon. So, I believe the time for 'triage' on the basis of taxon is never: we need to know about all flies!

4.3 Environmental effects

Is there any evidence that flies are affected by acid rain or other air pollutants? Will climate-change/global warming affect fly diversity? Are there endangered habitats that if eliminated would cause the extinction of one or more species of flies? Are there endangered species of flies? These questions cannot be answered readily and in sufficient detail because our knowledge of Diptera is so poor. The general answer is clearly yes, as we do know that Diptera are a major component of all non-marine ecosystems. So, given that some ecosystems are affected by acid rain and aerial pollutants or climate-change, then Diptera are affected. A recent study of pollinators in England and the Netherlands documented that climate-change has affected the ranges of flower flies (Biesmeijer *et al.* 2006). So flies with restricted ranges, such as alpine endemics, will surely disappear along with the polar bears. Given that some specialized habitats are eliminated, then some flies will be too. Many phytophagous flies have narrow host ranges, with most gall midges and leaf miners apparently being species specific (Gagné 1989). So, given endangered plants, there must be endangered phytophages. Evolution is an on-going process; numerous flies have evolved and gone to extinction in the 200 million years that flies have existed on Earth. Obviously, the process is continuing today, so there must be some endangered species of flies somewhere! The problem is the difficulty of separating the real examples of declining and endangered populations from those that appear to be because of a lack of knowledge. For example, there is only one US federally listed endangered species of fly, the Delhi Sands flower-loving fly (*Rhaphiomidas terminatus abdominalis* Cazier). This species was placed on the list as its habitat has been greatly reduced and its nominate subspecies was believed to be extinct. However, further research has revealed that the nominate subspecies is alive and well elsewhere (George & Mattoni 2006). California has a number of other species with very restricted habitats, such as Wilbur Springs Shore Fly (*Paracoenia calida* Mathis), found only at the spring but very abundant there (Mathis 1975), or a couple of robber flies associated with Antioch Dunes (*Efferia antiochi* Wilcox (Wilcox 1966), *Cophura hurdi* Hull (Hull 1960), and *Metapogon hurdi* Wilcox (Wilcox 1964)). These species are endangered in the sense that their habitat is, but as long as the habitat is preserved these species will also be preserved. One recent species of Nearctic fly has been officially listed by the International Union for Conservation of Nature (IUCN 2007) as now extinct, *Stonemyia velutina* (Bigot), and that species was also a narrow California endemic, known from central California (Madera & Mariposa counties (Middlekauff & Lane 1980)) but this status is questioned by the local specialist (J. Burger, pers. comm.) as no scientific survey has been undertaken to access the true status.

Summary

Today, the study of Nearctic Diptera remains stagnant, as in the past twenty years little taxonomic progress was made despite the great promise of technology. An assessment and basic synthesis of our knowledge of flies has been completed. What we know about flies is embodied in the *BioSystematic Database of World Diptera* and the *Manual of Nearctic Diptera*. About two-thirds of all the flies estimated to occur in the Nearctic Region have now been named. Unfortunately, less than one percent of these flies are treated comprehensively in monographs and less than a quarter have been thoroughly revised. To complete the task, a full and comprehensive inventory of the flies of America north of Mexico will require the utilization of new technologies, the training of new dipterists, and the securing of permanent positions for them. Given better tools, which are being developed, we need 30 full-time 'Wirths' (1,200 scientific years) or eight 'Alexanders' (560 SYS) to finish the job of just naming the flies of Nearctic America!

Acknowledgements

I would like to thank all that have contributed to the *BioSystematic Database of World Diptera*, especially the editorial team, Neal Evenhuis, Adrian Pont and Thomas Pape. Alma Solis, Al Norrbom and Michael Gates of SEL, USDA, Washington, are thanked for their critical review of the manuscript.

Table 1.1. Nearctic Diptera Fauna – statistics by families.

Taxon	Genera		Species		% Known		#Specialists		Status of Knowledge	Estimated species
	Valid	Names	Valid	Names	Species	Immatures	World	Nearctic		
NEMATOCERA										
Tipulomorpha										
Cylindrotomidae	4	4	8	15	90	50	4	1	4	9
Limoniidae	105	133	927	1034	80	5	4	1	2	1159
Pediciidae	13	16	149	157	80	8	4	1	2	186
Tipulidae	34	50	620	757	80	8	4	1	2	775
Psychodomorpha										
Canthyloscelidae	3	5	3	3	75	33	2	0	4	4
Psychodidae	33	42	123	162	50	20	4	0	3	246
Scatopsidae	19	22	77	84	80	5	2	0	3	96
Trichoceridae	3	6	30	36	90	20	2	0	3	33
Ptychopteromorpha										
Ptychopteridae	3	5	18	19	80	20	1	0	3	23
Tanyderidae	2	4	4	4	80	50	1	0	3	5
Culicimorpha										
Ceratopogonidae	78	111	614	692	65	5	60	2	3	945
Chaoboridae	4	9	13	29	90	70	2	1	4	14
Chironomidae	234	309	1112	1288	40	30	2	1	3	2780
Corethrellidae	2	3	5	5	70	20	2	1	3	7
Culicidae	81	244	182	311	95	100	60	2	5	192

Diptera Diversity: Status, Challenges and Tools
(eds T. Pape, D. Bickel & R. Meier). © 2009 Koninklijke Brill NV.

Table 1.1. Nearctic Diptera Fauna – statistics by families.

Taxon	Genera		Species		% Known		#Specialists		Status of Knowledge	Estimated species
	Valid	Names	Valid	Names	Species	Immatures	World	Nearctic		
Dixidae	3	4	45	57	90	20	2	1	4	50
Simuliidae	25	50	242	311	80	100	6	3	5	303
Thaumaleidae	3	3	25	27	70	90	3	1	2	36
Blephariceromorpha										
Blephariceridae	6	8	33	42	80	100	2	1	5	41
Deuterophlebiidae	1	1	6	6	80	100	1	1	5	8
Nymphomyiidae	1	3	2	2	100	100	1	1	5	2
Bibionomorpha										
Anisopodidae	7	13	9	13	75	40	3	2	3	12
Bibionidae	5	13	86	133	75	10	3	2	3	115
Bolitophilidae	2	4	20	20	75	10	3	1	3	27
Cecidomyiidae	177	262	1247	1626	8	20	4	1	1	15588
Diadocidiidae	2	3	3	3	66	25	3	1	3	5
Ditomyiidae	3	4	6	8	80	20	2	1	4	8
Hesperinidae	1	2	1	1	100	100	2	1	4	1
Keroplatidae	15	23	85	90	60	5	4	1	3	142
Lygistorrhinidae	1	1	1	1	20	0	4	1	4	5
Mycetophilidae	87	122	672	780	40	10	3	2	2	1680
Pachyneuridae	1	1	1	1	100	100	2	1	4	1
Sciaridae	24	32	172	193	30	6	6	6	1	573

DIPTERA DIVERSITY: STATUS, CHALLENGES AND TOOLS
(EDS T. PAPE, D. BICKEL & R. MEIER). © 2009 KONINKLIJKE BRILL NV.

Table 1.1. Nearctic Diptera Fauna – statistics by families.

Taxon	Genera		Species		% Known		#Specialists		Status of Knowledge	Estimated species
	Valid	Names	Valid	Names	Species	Immatures	World	Nearctic		
Axymyiomorpha										
Axymyidae	1	1	1	1	60	100	2	1	4	2
BRACHYCERA										
Stratiomyomorpha										
Stratiomyidae	46	117	312	516	90	15	4	1	2	347
Xylomyidae	2	21	12	15	90	40	2	1	3	13
Tabanomorpha										
Acroceridae	9	18	62	82	75	5	2	1	2	83
Athericidae	2	2	4	4	100	25	2	1	4	4
Nemestrinidae	3	9	8	11	90	40	2	1	3	9
Oreoleptidae	1	1	1	1	100	100	3	3	4	1
Rhagionidae	10	15	105	143	75	5	3	1	4	140
Spaniidae	3	3	13	16	75	0	3	1	4	17
Tabanidae	39	65	394	624	85	33	10	3	3	464
Xylophagidae	5	26	28	49	80	20	3	2	4	35
Vermileonomorpha										
Vermileonidae	1	1	3	3	100	33	1	0	3	3
Asiloidea										
Apioceridae	2	3	64	66	95	0	1	1	4	67
Apsilocephalidae	1	1	1	1	100	0	1	1	3	1

Table 1.1. Nearctic Diptera Fauna – statistics by families.

Taxon	Genera		Species		% Known		#Specialists		Status of Knowledge	Estimated species
	Valid	Names	Valid	Names	Species	Immatures	World	Nearctic		
Apystomyiidae	1	1	1	1	100	0	0	0	3	1
Asilidae	109	152	1073	1253	80	2	8	2	3	1341
Bombyliidae	66	114	991	1194	70	2	6	1	3	1416
Hilarimorphidae	1	2	27	27	90	0	1	1	4	30
Mydidae	11	18	75	87	95	2	3	1	3	79
Mythicomyiidae	8	15	183	207	70	0	2	1	3	261
Scenopinidae	10	15	148	151	90	1	2	0	4	164
Therevidae	30	39	164	199	90	10	4	2	4	182
Empidoidea										
Atelestidae	1	2	2	2	50	0	3	3	2	4
Brachystomatidae	10	13	17	18	75	0	6	2	2	23
Dolichopodidae	54	102	1383	1558	80	1	6	2	1	1729
Empididae	40	63	468	511	75	3	6	2	2	624
Hybotidae	29	54	316	333	75	1	6	2	2	421
Iteaphila group	2	2	18	23	75	1	6	2	2	24
Oreogetonidae	1	1	8	8	75	1	6	2	2	11
CYCLORRHAPHA										
Aschiza										
Lonchopteridae	1	2	5	7	100	25	1	0	3	5
Phoridae	64	118	421	485	50	10	3	1	3	842

Table 1.1. Nearctic Diptera Fauna – statistics by families.

Taxon	Genera		Species		% Known		#Specialists		Status of Knowledge	Estimated species
	Valid	Names	Valid	Names	Species	Immatures	World	Nearctic		
Pipunculidae	17	26	158	198	70	5	5	1	3	226
Platypezidae	15	29	77	86	90	10	1	0	3	86
Syrphidae	191	357	818	1420	90	10	15	2	2	909
Calyptratae										
Anthomyiidae	40	123	691	881	80	10	5	1	4	864
Calliphoridae	18	39	103	163	90	50	3	1	4	114
Fanniidae	3	8	111	144	90	10	3	0	4	123
Hippoboscidae	27	53	43	63	95	15	3	1	3	45
Muscidae	43	118	632	920	90	10	6	1	2	702
Oestridae	8	27	59	70	70	95	3	1	3	84
Rhiniidae	1	3	1	2	100	100	3	0	4	1
Rhinophoridae	2	6	4	5	100	100	2	0	3	4
Sarcophagidae	80	201	451	624	85	30	5	2	2	531
Scathophagidae	38	62	151	215	75	15	1	1	2	201
Tachinidae	374	921	1440	1843	75	10	6	2	1	1920
Acalyptratae Neriodea										
Cypselosomatidae	2	2	3	3	66	0	1	0	3	5
Micropezidae	12	25	37	50	90	10	3	1	3	41
Neriidae	2	3	2	3	66	50	1	0	3	3

Table 1.1. Nearctic Diptera Fauna – statistics by families.

Taxon	Genera		Species		% Known		#Specialists		Status of Knowledge	Estimated species
	Valid	Names	Valid	Names	Species	Immatures	World	Nearctic		
Diopsoidea										
Diopsidae	1	2	2	2	100	100	1	0	5	2
Psilidae	7	9	32	39	90	10	1	0	2	36
Strongylophthalmyidae	1	2	1	2	100	50	2	1	4	1
Tanypezidae	1	1	2	3	100	50	1	1	3	2
Conopoidea										
Conopidae	11	17	74	154	90	10	6	3	3	82
Tephritoidea										
Lonchaeidae	8	13	136	141	60	10	2	0	2	227
Pallopteridae	4	4	9	9	70	0	1	0	3	13
Piophilidae	10	17	37	49	70	25	1	0	4	53
Platystomatidae	7	13	44	49	80	0	2	0	2	55
Pyrgotidae	6	7	11	20	90	25	1	0	3	12
Richardiidae	6	10	8	10	75	10	1	0	3	11
Tephritidae	53	101	372	545	85	33	8	2	4	438
Ulidiidae	47	75	139	175	90	5	2	0	2	154
Lauxanioidea										
Chamaemyiidae	8	9	80	90	70	30	2	1	2	114
Lauxaniidae	30	35	157	179	80	10	2	1	2	196

Diptera Diversity: Status, Challenges and Tools
(eds T. Pape, D. Bickel & R. Meier). © 2009 Koninklijke Brill NV.

Table 1.1. Nearctic Diptera Fauna – statistics by families.

Taxon	Genera		Species		% Known		#Specialists		Status of Knowledge	Estimated species
	Valid	Names	Valid	Names	Species	Immatures	World	Nearctic		
Sciomyzoidea										
Coelopidae	4	5	5	6	100	20	2	1	4	5
Dryomyzidae	4	6	8	15	85	33	2	1	4	9
Helcomyzidae	1	3	1	1	100	0	1	1	4	1
Heterocheilidae	1	8	1	2	100	0	1	1	4	1
Ropalomeridae	2	2	2	2	80	0	1	1	3	3
Sciomyzidae	22	47	196	243	95	50	5	1	3	206
Sepsidae	9	16	34	54	85	20	2	0	2	40
Opomyzoidea										
Agromyzidae	29	44	763	858	80	33	10	2	4	954
Anthomyzidae	5	6	21	24	66	10	2	1	1	32
Asteiidae	5	6	17	19	66	0	1	1	3	26
Aulacigastridae	1	1	3	3	70	20	2	2	3	4
Clusiidae	6	11	41	51	90	10	3	2	4	46
Fergusoninidae	1	1	1	1	100	100	1	0	4	1
Odiniidae	3	4	11	12	80	30	3	2	4	14
Opomyzidae	3	5	11	13	80	25	1	1	4	14
Periscelididae	3	5	7	7	80	66	2	2	4	9

Table 1.1. Nearctic Diptera Fauna – statistics by families.

Taxon	Genera		Species		% Known		#Specialists		Status of Knowledge	Estimated species
	Valid	Names	Valid	Names	Species	Immatures	World	Nearctic		
Carnoidea										
Acartophthalmidae	1	1	2	2	100	0	1	0	3	2
Braulidae	1	1	1	2	100	100	1	1	3	1
Canacidae	5	5	12	13	90	20	1	1	3	13
Carnidae	4	5	20	25	65	20	2	2	3	31
Chloropidae	54	82	302	380	60	5	2	1	1	503
Cryptochetidae	1	2	1	1	100	100	0	0	3	1
Milichiidae	12	18	43	51	60	10	1	1	4	72
Tethinidae	6	9	28	32	70	0	1	2	3	40
Sphaeroceroidea										
Chyromyidae	3	3	9	11	60	0	1	0	2	15
Heleomyzidae	32	56	152	186	85	10	3	0	3	179
Sphaeroceridae	46	71	283	307	70	5	6	3	3	404
Ephydroidea										
Camillidae	1	2	4	4	100	0	1	0	3	4
Curtonotidae	1	3	1	1	50	0	0	0	3	2
Diastatidae	1	5	8	12	60	0	2	1	4	13
Drosophilidae	33	74	248	315	95	33	8	1	3	261
Ephydridae	70	97	484	554	80	25	10	3	3	605
			21454							44175

REFERENCES

Adler, P.H., Currie, D.C. & Wood, D.M. (2004) *The black flies (Simuliidae) of North America*. Cornell University Press, Ithaca, xv + 941 pp.

Aldrich, J.M. (1905) A catalogue of North American Diptera. *Smithsonian Miscellaneous Collections* 46(2), 680 pp.

Aldrich, J.M. (1930) [Early years in Dipterology]. *Journal of the Washington Academy of Sciences* 20: 495–498.

Arnaud, P.H., Jr. (1978) A host-parasite catalog of North American Tachinidae (Diptera). *Miscellaneous Publication, U. S. Department of Agriculture*, 1319, ii + 860 pp.

Arnaud, P.H., Jr. (1979) A catalog of the types of Diptera in the collection of the California Academy of Sciences. *Myia* 1, 505 pp.

Arnaud, P.H., Jr. & Owen, T.C. (1981) Charles Howard Curran (1894–1972). *Myia* 2, 393 pp.

Arnett, R.H. (2000) *American Insects. A handbook of the insects of America north of Mexico*. 2nd Edition. CRC Press, Boca Raton, xi + 1003 pp. [1st edition 1985.]

Biesmeijer, J.C., Roberts, S.P.M., Reemer, M., Ohlemüller, R., Edwards, M., Peeters, T., Schaffers, A.P., Potts, S.G., Kleukers, R., Thomas, C.D., Settele, J. & Kunin, W.E. (2006) Parallel declines in pollinators and insect-pollinated plants in Britain and the Netherlands. *Science* 313: 351–354.

Blagoderov, V., Grimaldi, D.A. & Fraser, N.C. (2007) How time flies for flies: Diverse Diptera from the Triassic of Virginia and early radiation of the Order. *American Museum Novitates* 3572, 39 pp.

Böcher, J. & Kristensen, N.P. (eds) *The Greenland Entomofauna. An identification manual of the insects, arachnids and myriapods of Greenland*. Zoological Museum, University of Copenhagen, Copenhagen [In preparation.]

Brown, B.V. (1987) Revision of the *Gymnophora* (Diptera: Phoridae) of the Holarctic Region: classification, reconstructed phylogeny and geographic history. *Systematic Entomology* 12: 271–304.

Brown, B.V. (2005) Malaise trap catches and the crisis in Neotropical Dipterology. *American Entomologist* 51: 180–183.

Bullock, J.A. (1988) Entomological Newsletter: Update. *Antenna* 12: 152–152.

Byers. G.W. (1982) In Memoriam. Charles P. Alexander, 1889–1981. *Journal of the Kansas Entomological Society* 55: 409–417.

Byers, G.W. & Rossman, D.A. (2008) A new species of *Dicranomyia* (*Idiopyga*) from Wisconsin, U. S. A. (Diptera: Tipulidae). *Journal of the Kansas Entomological Society* 81: 12–14.

Camras, S. (1945) A study of the genus *Occemyia* in North America (Diptera: Conopidae). *Annals of the Entomological Society of America* 38: 216–222.

Carpenter, S.J. & LaCasse, W. (1955) *Mosquitoes of North America (north of Mexico)*. Berkeley & Los Angeles, University of California Press, vi + 360 pp.

Chvála, M. (1981) Classification and phylogeny of Empididae, with a presumed origin of Dolichopodidae (Diptera). *Entomologica Scandinavica Supplement* 15: 225–236.

Chvála, M. (1983) The Empidoidea of Fennoscandia and Demark. II. General Part. The families Hybotidae, Atelestidae and Microphoridae. *Fauna Entomologica Scandinavica* 12, 279 pp.

Convention on Biological Diversity (CBD) (1994) *Convention on biological diversity.* Text and Annexes. Geneva, UNEP/CBD/94/1, 34 pp.

Cooper, B.E. (1991) Diptera types in the Canadian National Collection of Insects. Part 1 Nematocera. *Research Branch, Agriculture Canada, Publication* 1845/B, iii + 113 pp.

Cooper, B.E. & Cumming, J.M. (1993) Diptera types in the Canadian National Collection of Insects. Part 2 Brachycera (exclusive of Schizophora). *Research Branch, Agriculture Canada, Publication* 1896/B, iii + 105 pp.

Cooper, B.E. & Cumming, J.M. (2000) Diptera types in the Canadian National Collection of Insects. Part 3 Schizophora (exclusive of Tachinidae). *Research Branch, Agriculture and Agri-Food Canada, Publication* [A42-81/3-1999E-IN], iv + 132 pp.

Cooper, B.E. & O'Hara, J.E. (1996) Diptera types in the Canadian National Collection of Insects. Part 4 Tachinidae. *Research Branch, Agriculture and Agri-Food Canada, Publication* 1918/B, iv + 94 pp.

Coquillett, D.W. (1904) A brief history of North American dipterology. *Proceedings of the Entomological Society of Washington* 6: 53–58.

Courtney, G.W. (1990) Revision of Nearctic mountain midges (Diptera: Deuterophlebiidae). *Journal of Natural History* 24: 81–188.

Courtney, G.W. (1994) Biosystematics of the Nymphomyiidae (Insecta: Diptera): Life History, Morphology, and Phylogenetic Relationships. *Smithsonian Contributions to Zoology* 550, 41 pp.

Cox, C.B. & Moore, P.D. (2005) *Biogeography: An ecological and evolutionary approach.* 7th ed. Blackwell Scientific Publication, London, xii + 428 pp.

Curran, C.H. (1934) *The families and genera of North American Diptera.* Ballou Press, New York, 512 pp.

Curran, C.H. (1965) [Reprint of the above.] Henry Tripp, New York, [ii] + 514 pp.

Danks, H.V. (ed.) (1979) Canada and its insect fauna. *Memoirs, Entomological Society of Canada* 108, [ii] + 573 pp.

Danks, H.V. (1981a) *Arctic arthropods. A review of systematics and ecology with particular reference to the North American fauna.* Entomological Society of Canada, Ottawa, v + 608 pp.

Danks, H.V. (1981b) *Bibliography of the Arctic arthropods of the Nearctic Region.* Entomological Society of Canada, Ottawa, [iv] + 125 pp.

Danks, H.V. (1993) Patterns of diversity in the Canadian insect fauna. Pages 51–74 *in*: Ball, G.E. & Danks, H.V. (eds), Systematics and Entomology: Diversity, distribution, adaptation, and application. *Memoirs, Entomological Society of Canada* 165, 272 pp.

Danks, H.V. & Downes, J.A. (eds) (1997) *Insects of the Yukon.* Biological Survey of Canada (Terrestrial Arthropods), Ottawa, x + 1034 pp.

Darlington, P.J., Jr. (1957) *Zoogeography: The geographic distribution of animals.* John Wiley & Sons, New York, xv + 675 pp.

Darsie, R.F., Jr. & Ward, R.A. (2005) *Identification and geographical distribution of the mosquitoes of North America, north of Mexico.* University Press of Florida, Gainesville, xiv + 384 pp.

Downes, J.A. (1981) Temporal and spatial changes in the Canadian insect fauna. *Canadian Entomologist* 112: [unnumbered page before 1089]

Downes, J.A. & Kavanaugh, D.H. (1988) Origins of the North American insect fauna. *Memoirs, Entomological Society of Canada* 144, [ii] + 168 pp.

Evenhuis, N.E. (2007) Helping solve the "other" taxonomic impediment: Completing the Eight Steps to Total Enlightenment and Taxonomic Nirvana. *Zootaxa* 1407: 3–12.

Evenhuis, N.E., Pape, T., Pont, A.C. & Thompson, F.C. (eds) (2008) *BioSystematic Database of World Diptera.* Version 10.5. Available at http://www.diptera.org/ names/, accessed 15 June 2008.

Fabricius, J.C. (1775) *Systema entomologiae, sistens insectorum classes, ordines, genera, species adiectis synonymis, locis, descriptionibus, observationibus.* Officina Libraria Kortii, Flensburgi et Lipsiae [=Flensburg and Leipzig], 832 pp.

Fabricius, J.C. (1805) *Systema antliatorum secundum ordines, genera, species.* Carolum Reichard, Brunsviage [=Brunswick], 373 + 30 pp.

Feijen, H.R. (1989) Diopsidae. *In*: Griffiths, G.C.D. (ed.), *Flies of the Nearctic Region* 9(12), 122 pp.

Ferrar, P. (1987) Guide to the breeding habits and immature stages of Diptera Cyclorrhapha. *Entomonograph* 8, 907 pp. [2 vols.]

Finnamore, A.T. & Marshall, S.A. (eds) (1994) Terrestrial arthropods of peatlands, with particular reference to Canada. *Memoirs, Entomological Society of Canada* 169, 289 pp.

Frank, J.H. & Curtis, G.A. (1979) Trend lines and the number of species of Staphylinidae. *Coleopterists Bulletin* 33: 133–149.

Gagné, R.J. (1983) Biology and taxonomy of the *Rhopalomyia* gall midges (Diptera: Cecidomyiidae) of *Artemisia tridentata* Nuttall (Compositae) in Idaho. *Contributions of the American Entomological Institute* 21, 90 pp.

Gagné, R.J. (1989) *The plant-feeding gall midges of North America.* Cornell University Press, Ithaca, xi + 356 pp.

George, J.N. & Mattoni, R. (2006) *Rhaphiomidas terminatus terminatus* Cazier, 1985 (Diptera, Mydidae): notes on the rediscovery and conservation biology of a presumed extinct species. *Pan-Pacific Entomologist* 82: 30–35.

Griffiths, G.C.D. (1972) The phylogenetic classification of Diptera Cyclorrhapha with special reference to the structure of the male postabdomen. *Series Entomologica* 8. Dr. W. Junk, The Hague, 340 pp.

Griffiths, G.C.D. (1980) Preface. *In*: Griffiths, G.C.D. (ed.), *Flies of the Nearctic Region* 1(1): v–viii.

Griffiths, G.C.D. (1981) Book review: Manual of Nearctic Diptera. *Bulletin of the Entomological Society of Canada* 13: 49–55.

Griffiths, G.C.D. (1982–2004) Anthomyiidae. *In*: Griffiths, G.C.D. (ed.), *Flies of the Nearctic Region* 8 (2): 1–169 [1982], 161–228 [1983], 289–408 [1984], 409–600 [1984], 601–728 [1986], 729–952 [1987], 953–1048 [1991], 1049–1240 [1991], 1241–1416 [1992], 1417–1632 [1993], 1873–2120 [1998], 2121–228 [2001], 2289–2484 [2003], 2485–2635 [2004].

Griffiths, G.C.D. (1984) Note on characterization of Eremoneura, Orthogenya and Cyclorrhapha. *In*: Griffiths, G.C.D. For discussion on "Male genitalia in the classification of Chloropidae". XVII International Congress of Entomology, Hamburg, August 1984. Hamburg. Privately printed, 2 pp.

Griffiths, G.C.D. (1987) Unpublished manuscripts from lecture given at the First International Congress of Dipterology, Budapest (August 1986) and lecture given at the Smithsonian Institution (May 1987).

Grimaldi, D.A. (2000) *Studies on fossils in amber, with particular reference to the Cretaceous of New Jersey.* Backhuys Publishers, Leiden, viii + 498 pp.

Grimaldi, D.A. & Cumming, J.M. (1999) Brachyceran Diptera in Cretaceous ambers and Mesozoic diversification of the Eremoneura. *Bulletin of the American Museum of Natural History* 239, 124 pp.

Grogan, W.L., Jr. & Philips, R.A. (2008) A new species of biting midge in the subgenus *Monoculicoides* of *Culicoides* from Utah (Diptera: Ceratopogonidae). *Proceedings of the Entomological Society of Washington* 110: 196–203.

Groombridge, B. (ed.) (1992) *Global biodiversity. Status of the Earth's living resources.* Chapman & Hall, London, xxx + 585 pp.

Hall, J.C. & Evenhuis, N.L. (1980–2004) Bombyliidae. *In*: Griffiths, G.C.D. (ed.), *Flies of the Nearctic Region* 5(13): 1–716.

Hardy, D.E. (1943) A revision of Nearctic Dorilaidae (Pipunculidae). *University of Kansas Science Bulletin* 29, 231 pp.

Hardy, D.E. (1945) Revision of Nearctic Bibionidae including neotropical *Plecia* and *Penthetria* (Diptera). *University of Kansas Science Bulletin* 30: 367–547.

Hardy, J.D. (ed.) (1993) *NODC Taxonomic Code.* Version 7.0. CD-ROM NODC 35.

Hennig, W. (1948) *Die Larvenformen der Diptera.* Pt. 1. Akademie Verlag, Berlin, 185 pp.

Hennig, W. (1950) *Die Larvenformen der Diptera.* Pt. 2. Akademie Verlag, Berlin, 458 pp.

Hennig, W. (1952) *Die Larvenformen der Diptera.* Pt. 3. Akademie Verlag, Berlin, , 628 pp.

Hennig, W. (1972) Insektenfossilien aus der unteren Kreide. IV. Psychodidae (Phlebotominae), mit einer kritischen Übersicht über das phylogenetische System der Familie und die bisher beschriebenen Fossilien (Diptera). *Stuttgarter Beiträge zur Naturkunde* 241, 69 pp.

Heywood, V.H. (ed.) (1995) *Global biodiversity assessment.* UNEP, Cambridge University Press, Cambridge, xi + 1140 pp.

Hogue, C.L. (1973) A taxonomic review of the genus *Maruina* (Diptera: Psychodidae). *Science Bulletin of the Natural History Museum of Los Angeles County* 17, 69 pp.

Hogue, C. L. (1987) Blephariceridae. *In* Griffiths, G.C.D. (ed.), *Flies of the Nearctic Region* 2(4): 1–172.

Hull, F.M. (1960) New species of Syrphidae and Asilidae. *Pan-Pacific Entomologist* 36: 69–74.

International Commission on Zoological Nomenclature [ICZN] (1989) General session of the Commission, Canberra, 15–19 October 1988. *Bulletin of Zoological Nomenclature* 46: 7–13.

International Commission on Zoological Nomenclature [ICZN] (1999) *International Code of Zoological Nomenclature.* Fourth Edition adopted by the XX General Assembly of the International Union of Biological Sciences. International Trust for Zoological Nomenclature, London, xx + 338 pp.

International Union of Biological Sciences [IUBS] (1989) Section of Zoological Nomenclature. Report of meeting, Canberra, 14–18 October 1988. *Bulletin of Zoological Nomenclature* 46: 14–18.

International Union for Conservation of Nature [IUCN] (2007) 2007 IUCN Red List of Threatened species. Available at http://www.iucnredlist.org/search/details.php/20867/all, accessed 2 September 2008.

Kits, J.H., Marshall, S.A. & Evenhuis, N.L. (2008) The bee flies (Diptera: Bombyliidae) of Ontario, with a key to species of eastern Canada. *Canadian Journal of Arthropod Identification* 6, 52 pp.

Knutson, L., Thompson, F.C. & Carlson, R.W. (1987) Biosystematics and biological control information systems in entomology. *Agricultural Zoology Review* 2: 361–412.

Kosztarab, M. & Schaefer, C.W. (1990) Systematics of the North American insects and arachnids: Status and needs. *Virginia Agricultural Experiment Station, Information Series* 90-1, xii + 247 pp.

Krivosheina, N.P. (1969) [*Ontogenesis and evolution of Diptera.*] Nauk, Moscow, 290 pp. [In Russian, see Krivosheina (1978) for translation.]

Krivosheina, N.P. (1978) Ontogensis and evolution of the two-winged insects. Franklin Book Program, Cairo, iii + 415 pp. [English translation of Krivosheina (1969).]

Krivosheina, N.P. (1986) Families Hesperinidae, Pleciidae and Bibionidae. Pages 314–316, 318–330 *in*: Soós, Á. & Papp, L. (eds), *Catalogue of Palaearctic Diptera*, Vol. 4. Akademiai Kiado, Budapest.

Krivosheina, N.P. (1988) Approaches to solutions of questions of classification of the Diptera. *Entomologicheskoe Obozrenie* 67: 378–390.

Krombein, K.V., Hurd, P.D., Jr., Smith, D.R. & Burks, B.D. (eds) (1979) *Catalog of Hymenoptera in America north of Mexico.* Vols 1–3. Smithsonian Institution Press, Washington, D.C., 2735 pp. [Vol 1: xvi + 1198 pp., Vol. 2: xvi + 1199–2209, Vol. 3: xxx + 2210–2735.]

Lane, R. [chair] (2008) Taxonomy in Europe in the 21st century. *Report of the Board of Directors, European Distributed Institute of Taxonomy,* 7 pp. [Available at http://ww2.bgbm.org/EditDocumentRepository/Taxonomy21report.pdf]

Lindner, E. (ed.) (1924–1993) *Die Fliegen der paläarktischen Region. Band I-XII.* Schweizerbart'sche Verlagsbuchhandlung, Stuttgart.

Linnaeus, C. (1758) *Systema naturae per regna tria naturae.* 10th ed., Vol. 1. Laurentii Salvii, Holmiae [=Stockholm], 824 pp.

Llorente-Bousquets, J.E., García Aldrete, A. & González Soriano, E. (eds) (1996) Biodiversidad, taxonomía y biogeografía de Arthrópodos de México: Hacia uni Síntesis de su Conocimiento. Universidad Nacional Autónoma de México, Mexico, xvi + 661 pp.

Loew, H. (1862) Monographs of the Diptera of North America. Part I. *Smithsonian Miscellaneous Collections* 6(1), 221 pp.

Loew, H. (1864) Monographs of the Diptera of North America. Part II. *Smithsonian Miscellaneous Collections* 6(2), 360 pp.

Loew, H. (1873) Monographs of the Diptera of North America. Part III. *Smithsonian Miscellaneous Collections* 11(2), 351 pp.

Lonsdale, O. & Marshall, S.A. (2007) Revision of the North American *Sobarocephala* (Diptera: Clusiidae, Sobarocephalinae). *Journal of the Entomological Society of Ontario* 138: 65–106.

Maa, T.C. & Peterson, R.V. (1987) Hippoboscidae. Pages 1271–1281 *in*: McAlpine, J.F. (ed.), Manual of Nearctic Diptera, vol. 2, *Research Branch, Agriculture Canada, Monograph* 28.

Mathis, W.N. (1975) A systematic study of *Coenia* and *Paracoenia* (Diptera: Ephydridae). *Great Basin Naturalist* 35: 65–85.

Mathis, W.N. (1982) Studies on the Ephydridae (Diptera: Ephydridae), VII; Revision of the genus *Steacera* Cresson. *Smithsonian Contributions to Zoology* 380, 57 pp.

Matile, L. (1990) Recherches sur la systématique et l'évolution des Keroplatidae (Diptera Mycetophiloidea). *Mémoires du Muséum national d'Histoire naturelle (Série A, Zoologie)* 148, 682 pp.

McAlpine, D.K. (1985) The Australian genera of Heleomyzidae (Diptera: Schizophora) and a reclassification of the family into tribes. *Records of the Australian Museum* 36: 203–251.

McAlpine, J.F. (1979) Diptera. Pages 389–424 *in*: Danks, H.V. (ed.), Canada and its insect fauna. *Memoirs, Entomological Society of Canada* 108, [ii] + 573 pp.

McAlpine, J.F. (ed.) (1981) *Manual of Nearctic Diptera.* Vol. 1. Research Branch, Agriculture Canada. Monograph 27, 1–674 pp.

McAlpine, J.F. (ed.) (1987) *Manual of Nearctic Diptera*. Vol. 2. Research Branch, Agriculture Canada, Monograph 28, 675–1332 pp.

McAlpine, J.F. (ed.) (1989) *Manual of Nearctic Diptera*. Vol. 3. Research Branch, Agriculture Canada, Monograph 32, 1333–1581 pp.

Meyer, H.W. (2003) *The fossils of Florissant*. Smithsonian Books, Washington, D.C., xiii + 258 pp.

Michelsen, V. (1988) A world revision of *Strobilomyia* gen. n.: the anthomyiid seed pests of conifers (Diptera: Anthomyiidae). *Systematic Entomology* 13: 271–314.

Middlekauff, W.W. & Lane, R.S. (1980) Adult and immature Tabanidae (Diptera) of California. *Bulletin of the California Insect Survey* 22, 99 pp.

Nagatomi, A. & Liu, N. (1994) Apystomyiidae, a new family of Asiloidea (Diptera). *Acta Zoologica Academiae Scientiarium Hungaricae* 40: 203–218.

O'Brien, C.W. & Wibmer, G.J. (1979) The use of trend curves of rate of species descriptions: Examples from the Curculionidae (Coleoptera). Coleoptera). *Colepterists Bulletin* 33: 151–166.

Oosterbroek, P. (1986) A phylogenetic classification of the tribes of the Tipuloidea, based on pre-imaginal characters. Page 178 *in*: Darvas, B. & Papp, L. (eds), *Abstracts of the First International Congress of Dipterology*. University of Veterinary Science Press, Budapest.

Osten Sacken, C.R.R. (1858) Catalogue of the described Diptera of North America. *Smithsonian Miscellaneous Collections* 3(1), 92 pp.

Osten Sacken, C.R. (1869) Monographs of the Diptera of North America. Part IV. *Smithsonian Miscellaneous Collections* 8(1), 345 pp.

Osten Sacken, C.R. (1878) Catalogue of the described Diptera of North America. [ed. 2]. *Smithsonian Miscellaneous Collections* 16(2), 276 pp.

Peterson, R.V. (1987) Phoridae. Pages 689–712 *in*: McAlpine, J.F. (ed.), Manual of Nearctic Diptera, *Research Branch, Agriculture Canada, Monograph* 28, Ottawa.

Peterson, R.V. & Wenzel, R.L. (1987) Nycteribiidae. Pages 1283–1292 *in*: McAlpine, J.F. (ed.), Manual of Nearctic Diptera, *Research Branch, Agriculture Canada, Monograph* 28, Ottawa.

Polaszek, A., Agosti, D., Alonso-Zarazaga, M.A., Beccaloni, G., Bjorn, P. de Place, Bouchet, P., Brothers, D.J., Earl of Cranbrook, Evenhuis, N.L., Godfray, H.C.J., Johnson, N.F., Krell, F.-T., Lipscomb, D., Lyal, C.H.C., Mace, G.M., Mawatari, S.F., Miller, S.E., Minelli, A., Morris, S. Ng, P.K.L., Patterson, D.J., Pyle, R.L., Robinson, N., Rogo, L., Taverne, J., Thompson, F.C., Tol, J. van, Wheeler, Q.D. & Wilson, E.O. (2005a) Commentary: A universal register for animal names. *Nature* 437: 477.

Polaszek, A., Alonso-Zarazaga, M.A. Bouchet, P., Brothers, D.J., Evenhuis, N.L., Krell, F.-T., Lyal, C.H.C., Minelli, A., Pyle, R.L., Robinson, N., Thompson, F.C. & Tol, J. van (2005b) ZooBank: The open-access register for zoological taxonomy: Technical discussion paper. *Bulletin of Zoological Nomenclature* 62: 210–220.

Polaszek, A., Pyle, R.L. & Yanega, D. (2008) Animal names for all: ZooBank and the New Taxonomy. *The Systematics Association Special Volume series* 76: 129–141.

Poole, R.W. (1996) Diptera. Pages 15–604 *in*: Poole, R.W. (ed.), *Nomina Insecta Nearc-tica. A checklist of the insects of North America*. Vol. 3. Entomological Information Service, Rockville, 1143 pp.

Pratt, G.K. & Pratt, H.D. (1980) Notes on Nearctic *Sylvicola* (Diptera: Anisopodidae). *Proceedings of the Entomological Society Washington* 82: 86–98.

Ride, W.D.L. & Yones, T. (eds) (1986) Biological Nomenclature Today. A review of the present state and current issues of biological nomenclature of animals, plants, bacteria and viruses. *International Union of Biological Sciences, Monograph 2*, 70 pp.

Robinson, H. & Knowles, M. (2008) The Robinson expeditions to Spout Run, Arling-ton County, Virginia: Notes on *Gymnopternus* (Diptera: Dolichopodidae). *Proceedings of the Entomological Society of Washington* 110: 562–576.

Rosenberg, D.M. & Danks, H.V. (1987) Aquatic insects of peatlands and marshes in Canada. *Memoirs, Entomological Society of Canada* 140, 174 pp.

Ross, H.H. (1953) On the origin and composition of the Nearctic insect fauna. *Evolu-tion* 7: 145–158.

Rotheray, G.E. & Gilbert, F. (2008) Phylogenetic relationships and the larval head of the lower Cyclorrhapha (Diptera). *Zoological Journal of the Linnean* Society 153: 287–323.

Sclater, P.L. (1858) On the general geographic distribution of the members of the Class Aves. *Journal and Proceedings of the Linnean Society, Zoology, London* 2: 130–145.

Scoble, M.J. (2008) Networks and their role in e-Taxonomy. *The Systematics Associa-tion Special Volume series* 76: 19–31.

Shor, E.N. (1971) *Fossils and flies. The life of a compleat scientist, Samuel Wendell Wil-liston (1851–1918)*. University of Oklahoma Press, Norman. xiv + 285 pp.

Smith, K.G.V. (1977) An appreciation and introductory preface. Pages 1-10 *in*: Osten-Sacken, C.R., *Record of my life-work in entomology. A facsimile Reprint*. Classey, Oxon, 253 pp., 4 pls.

Spencer, K.A. (1981) A revisionary study of the leaf-mining flies (Agromyzidae) of California. *University of California, Berkeley, Division of Agricultural Sciences, Special Publication* 3273, iv + 489 pp.

Spencer, K.A. & Stegmaier, C.E. (1973) Agromyzidae of Florida with a supplement on species from the Caribbean. *Arthropods of Florida and Neighboring Land Areas* 7, iv + 205 pp.

Spencer, K.A. & Steyskal, G.C. (1986) Manual of the Agromyzidae (Diptera) of the United States. *United States Department of Agriculture, Agricultural Handbook* 638, 478 pp.

Steyskal. G.C. (1965) Trend curves of the rate of species description in zoology. *Sci-ence* 149: 880–882.

Steyskal, G.C. (1988) How big the list, catalogue, database? *Antenna* 12: 38–39.

Stone, A. (1980) History of Nearctic dipterology. *In*: Griffiths, G.C.D. (ed.), *Flies of the Nearctic Region* 1(1): 1–62.

Stone, A., Sabrosky, C.W., Wirth, W.W., Foote, R.H. & Coulson, J.R. (1965) A catalog of the Diptera of America north of Mexico. *United States Department of Agriculture, Agricultural Handbook* 276, 1696 pp.

Stone, A., Sabrosky, C.W., Wirth, W.W., Foote, R.H. & Coulson, J.R. (1983) A catalog of the Diptera of America north of Mexico. Washington, D.C.: Smithsonian Institution Press. [Reprint of original from 1965.]

Telford, H.S. (1970) *Eristalis* (Diptera: Syrphidae) from America north of Mexico. *Annals of the Entomological Society of America* 63: 1201–1210.

Thompson, F.C. (1972) A contribution to a generic revision of the Neotropical Milesinae (Diptera: Syrphidae). *Arquivos de Zoologia* 23: 73–215.

Thompson, F.C. (1980) The North American species of *Callicera* Panzer (Diptera: Syrpidae). *Proceedings of the Entomological Society Washington* 82: 195–221.

Thompson, F.C. (1981) Revisionary notes on Nearctic *Microdon* flies (Diptera: Syrphidae). *Proceedings of the Entomological Society Washington* 83: 725–758.

Thompson, F.C. (ed.) (1988) *Nearctic Diptera Dataset.* Incorporated into Hardy (1993) *q. v.*

Thompson, F.C. (1990) Biosystematic information: Dipterists ride the third wave. Pages 179–201 *in*: Kosztarab, M. & Schaefer, C.W. (eds), Systematics of the North American insects and arachnids: Status and Needs. *Virginia Agricultural Experiment Station Information series* 90-1, xii + 247 pp. Blacksburg.

Thompson, F.C. (1999a) Data dictionary and standards. *Myia* 9: 49–63.

Thompson, F.C. (ed.) (1999b) Fruit fly expert identification system and systematic information database. A resource for identification and information on fruit flies and maggots, with information on their classification, distribution and documentation. *Myia* 9, 524 pp.

Thompson, F.C. & Knutson, L. (1987) Catalogues, checklists and lists: A need for some definitions, new words and ideas. *Antenna* 11: 131–134.

Vockeroth, J.R. (1986) Revision of the New World species of *Paragus* Latreille (Diptera: Syrphidae). *Canadian Entomologist* 118: 183–198.

Wallace, A.R. (1876) *The geographical distribution of animals animals; with a study of the relations of living and extinct faunas as elucidating the past changes of the Earth's surface.* 2 vols. Macmillan, London, xxiv + 503, viii + 602 pp.

Webb, D.W. (1984) A revision of the Nearctic species of the family Solvidae (Insects: Diptera). *Transactions of the American Entomological Society* 110: 245–293.

Wenzel, R.L. & Peterson, R.V. (1987) Streblidae. Pages 1293–1320 *in*: McAlpine, J.F. (ed.), Manual of Nearctic Diptera. *Research Branch, Agriculture Canada, Monograph* 28, Ottawa.

West, L.S. & Peters, O.B. (1973) *An annotated bibliography of* Musca domestica *Linnaeus.* Folkestone and London, England: Dawsons of Pall Mall, xiii + 743 pp.

White, R.E. (1975) Trend curves of the rate of species description for certain North American Coleoptera. *Coleopterists Bulletin* 29: 281–295.

White, R.E. (1979) Response to the use of trend curves by Erwin, Frank and Curtis, and O'Brien and Wibmer. *Coleopterists Bulletin* 33: 167–168.

Whiting, M.F. (2005) Phylogenetic position of Diptera: Review of the evidence. Pages 3–13 *in*: Yeates, D.K. & Wiegmann, B.M. (eds), *The evolutionary biology of flies*. Columbia University Press, New York. x + 430 pp.

Wilcox, J. (1964) The genus *Metapogon* (Diptera: Asilidae). *Pan-Pacific Entomologist* 40: 191–200.

Wilcox, J. (1966) *Efferia* Coquillett in America north of Mexico (Diptera: Asilidae). *Proceedings of the California Academy of Sciences.* 4th Series, 34: 85–234.

Wiley, E.O. (1981) *Phylogenetics. The theory and practice of phylogenetic systematics.* John Wiley & Sons, New York, 456 pp.

Willams, D.D. & Danks, H.V. (1991) Arthropods of springs, with particular reference to Canada. *Memoirs, Entomological Society of Canada* 155, 217 pp.

Williston, S.W. (1879) An anomalous bombylid. *Canadian Entomologist* 11: 215–216.

Williston, S.W. (1888) *Synopsis of the families and genera of North America Diptera, exclusive of the genera of the Nematocera and Muscidae, with bibliography and new species, 1878–88.* J. T. Hathaway, New Haven, 84 pp.

Williston, S.W. (1896) *Manual of the families and genera of North American Diptera.* 2nd edition, rewritten and enlarged. J. T. Hathaway, New Haven, iv + 167 pp.

Williston, S.W. (1908) *Manual of North American Diptera.* 3rd edition. J. T. Hathaway, New Haven, 405 pp.

Wilson, E.O. (1985a) The biological diversity crisis. A challenge to science. *Issues in Science and Technology* 1: 20–29.

Wilson, E.O. (1985b) Time to revive Systematics. *Science* 230: 1227. [Editorial.]

Wilson, E.O. (1986) The value of systematics. *Science* 231: 1057. [Response to a letter.]

Wilson, E.O. (1987a) An urgent need to map biodiversity. *The Scientist* 1(6): 11.

Wilson, E.O. (1978b) The little things that run the world. *Wings* 12(3): 4–8. [Also in *Conservation Biology* 1: 344–346.]

Wilson, E.O. (1988) The current state of biological diversity. Pages 3–18 *in*: Wilson, E.O. & Peter, F.M. (eds), *Biodiversity.* National Academy Press, Washington, D.C.

Winterton, S. (2009) Bioinformatics and Dipteran Diversity. Pages 381-407 in: Pape, T., Bickel, D. & Rudolf, M. (eds), *Diptera Diversity: Status, Challenges and Tools.* Brill, Leiden.

Wood, D.M. & Borkent, A. (1989) Phylogeny and classification of Nematocera. Pages 1333–1370 *in*: McAlpine, J.F. (ed.), *Manual of Nearctic Diptera*, Research Branch, Agriculture Canada, Monograph 32, Ottawa.

Wood, D.M., Dang, P.T. & Ellis, R.A. (1979) The mosquitoes of Canada (Diptera: Culicidae). *In*: The insects and arachnids of Canada. Part 6. *Agriculture Canada Publication* 1686, 390 pp.

Woodley, N.E. (1989) Phylogeny and classification of "Orthorrhaphous" Brachycera. Pages 1371–1395 *in*: McAlpine, J.F. (ed.), Manual of Nearctic Diptera, *Research Branch, Agriculture Canada, Monograph* 32, Ottawa.

Woodley, N.E. & Hilburn, D.J. (1994) The Diptera of Bermuda. *Contributions of the American Entomological Institute* 28(2), 64 pp.

Yeates, D.K. & Wiegmann, B.M. (2005) Phylogeny and evolution of Diptera: Recent insights and new perspectives. Pages 14–44 *in*: Yeates, D.K. & Wiegmann, B.M. (eds), *The evolutionary biology of flies*. Columbia University Press, New York, x + 430 pp.

Zloty, J., Sinclair, B.J. & Pritchard, G. (2005) Discovered in our backyard: A new genus and species of a new family from the Rocky Mountains of North America (Diptera, Tabanomorpha). *Systematic Entomology* 30: 248–266.

CHAPTER TWO

HAWAII'S DIPTERA BIODIVERSITY

NEAL L. EVENHUIS

Hawaii Biological Survey, Bishop Museum, Honolulu, Hawai'i, USA

INTRODUCTION

Hawai'i harbors the highest diversity of endemic Diptera per land unit of any area in the world. A number of factors have led to this phenomenon, which, in combination with other exemplary groups of organisms with similar diversity and adaptive radiation, has ultimately led to Hawai'i being given the moniker of 'a living laboratory of evolution'. The interplay of extreme geographical isolation, geologic youth, and an abundance and variety of ecosystems on high islands has resulted in some of the more unique and unusual adaptations seen anywhere, such as stinkless stink-bugs, carnivorous caterpillars, nettle-less nettles, no-eyed big-eyed spiders, and flightless flies.

PART 1. BIOGEOGRAPHY AND DIVERSITY OF THE HAWAIIAN DIPTERA FAUNA

1. Background

1.1 Physical environment

The Hawaiian Island chain, which extends from a subtropical latitude (Kure at about 28°N) to a tropical one (Hawai'i Island at 19°N) as a 2,400 km-long archipelago of volcanic islands, contains the most isolated high islands in the world. They lie some 3,800 km from the nearest continental land mass and about the same distance from the nearest high islands, the Marquesas in French Polynesia. The entire island chain, comprising just

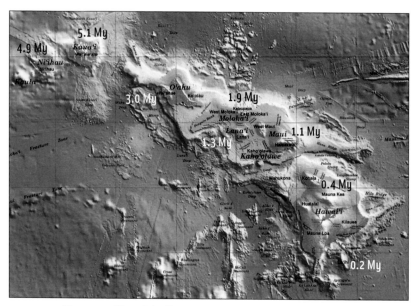

Figure 2.1. Map of the Hawaiian Islands showing geologic ages.
(Courtesy U.S. Geological Survey.)

over 16,000 square kilometers of emerged land, originated some 85 million years ago (Mya) (Clague 1996) and consists of three parts: the 8 main
southeastern islands (Kaua'i and Ni'ihau in the northwest to Hawai'i Island in the southeast), the Northwestern Hawaiian Islands (emerged and
submerged eroded remnants of previous high volcanic islands), and the
Emperor Seamounts (a line of submerged seamounts and guyots that extend northward to the Kamchatka Peninsula, under which they are subducted and/or accreted).

All the islands in the Hawaiian-Emperor chain are the summits of
submarine volcanoes originating from a hotspot beneath the Pacific Plate.
The hotspot has remained relatively stationary through time and the more
mobile Pacific Plate has produced islands in a conveyor-belt fashion as it
moved northwest (Fig. 2.1). Each island is thus progressively older in a
northwest direction with Kure being the oldest emerged island (29.8 Mya)
and Hawai'i Island being the youngest (0.6 Mya). Because of the 'Kure gap'
where no emerged islands existed north of Kure to act as source points,
the oldest the fauna of the existing Hawaiian Islands can be is about 32
million years (Clague 1996).

DIPTERA DIVERSITY: STATUS, CHALLENGES AND TOOLS
(EDS T. PAPE, D. BICKEL & R. MEIER). © 2009 KONINKLIJKE BRILL NV.

1.2 Natural Environment

The Hawaiian Islands support an incredible array of different habitats due to their diverse topography and climate. More than 150 distinct natural ecosystems have been recognized from the Hawaiian Islands representing virtually every kind except arctic alpine tundra. The major zones are littoral (both rocky and sandy shores), coastal strand, lowland dry scrub, desert, grassland, partly deciduous dry forest, mesic forest, rain forest, and alpine desert (Howarth & Mull 1992). Significantly, many of these ecosystems have evolved independently on each island in the chain. This ecological richness in combination with mild tropical climate and cool trade winds, have allowed a diverse array of organisms to become established, adapt, and evolve over millions of years.

2. Origin of the Hawaiian Fauna

The extreme isolation and geologic youth of the islands has limited the number of organisms that could have arrived to evolve into the more than 25,600 species of plants and animals that are currently known to exist in Hawai'i (Eldredge & Evenhuis 2003). Successful populations have all evolved from waif elements that arrived across the vast ocean distances from other areas throughout the geologic history of the islands. Potential colonizers of native populations could have initially arrived via air currents (storms, the jet stream, aerial plankton, phoresy on birds) or the ocean (rafted on floating debris, swam, or drifted with the currents). Zimmermann (1948) and Howarth (1990) both surmised that a large majority of colonizers of the native land biota currently in Hawai'i came via the air. Success in colonization would be rare and contingent upon the following critical factors: surviving the long distance of travel across the ocean, finding suitable habitat and sustainable resources upon arrival, and producing viable offspring. Despite these obstacles to success, at least 350–400 colonizations or founding events must have taken place to account for the current native insect fauna (Zimmerman 1948, Gagné 1988). Based on the figures in Table 2.1, it can be calculated that at least 75 founding events (assuming one founder per genus) have led to the current native Hawaiian Diptera fauna.

Interestingly, most of the genera with native species have not diversified. Of the 82 Diptera genera listed that have endemic species, only 10 have more than 10 species, and only 4 have over 100 species. The majority

(49) have only 1 or 2 endemic species. Thus, it appears from this data that most successful Diptera colonists to Hawai'i do not diversify.

3. Diversity of the Hawaiian Diptera Fauna

Hawai'i harbors the highest number of endemic species per unit of land area of any insular setting in the world (62.4/10 km²). Comparison of the data of selected Diptera groups from Hawai'i with other selected insular areas is given in Table 2.3 and shows that endemic Diptera from Hawai'i are 6.6/10 km². The Galápagos commonly receive a large share of the scientific publicity when it comes to biodiversity, but after scientists come

Table 2.1. Families and Genera of Endemic Species of Hawaiian Diptera.
An asterisk denotes a genus endemic to Hawai'i.

Family/Genus	Endemic spp.	Family/Genus	Endemic spp.	Family/Genus	Endemic spp.
Asteiidae		*Paralellodiplosis*	1	*Asyndetus*	1
Asteia	9	**Ceratopogonidae**		*Campsicnemus*	163
*Bryania**	1	*Dasyhelea*	4	*Diaphorus*	1
Loewimyia	1	*Forcipomyia*	5	*Elmoia**	8
Calliphoridae		**Chironomidae**		*Emperoptera**	5
*Dyscritomyia**	25	*Chironomus*	2	*Eurynogaster**	24
Lucilia	1	*Clunio*	3	*Hydrophorus*	2
Canacidae		*Metriocnemus*	1	*Major**	1
Canaceoides	1	*Micropsectra*	4	*Paraliancalus**	1
Procanace	9	*Orthocladius*	9	*Sigmatineurum**	11
Cecidomyiidae		*Pontomyia*	1	*Sweziella**	7
Arthrocnodax	1	*Pseudosmittia*	1	*Thambemyia*	1
Coccodiplosis	1	*Telmatogeton*	5	*Uropachys**	7
Giardomyia	2	*Thalassomyia*	1	**Drosophilidae**	
Heterocontarinia	1	**Chyromyidae**		*Celidosoma**	1
Lestodiplosis	2	*Aphaniosoma*	2	*Drosophila*	403
Lestremia	1	*Gymnochiromyia*	1	*Scaptomyza*	155
Mayetiola	1	**Dolichopodidae**		**Ephydridae**	
Monardia	1	*Adachia**	6	*Atissa*	2
Mycodiplosis	1	*Arciellia**	3	*Brachydeutera*	1

Table 2.1 (continued)

Family/Genus Endemic spp.		Family/Genus Endemic spp.		Family/Genus Endemic spp.	
Ephydridae (cont'd)		Muscidae		Plastosciara	3
Hydrellia	1	Lispe	3	Scaptosciara	1
Scatella	15	Lispocephala	102	Sciara	1
Hippoboscidae		Mycetophilidae		Sphaeroceridae	
Ornithoctona	1	Leia[1]	1	Opacifrons	1
Hybotidae		Phoridae		Pterogramma	1
Chersodromia	2	Megaselia	9	Trachyopella	3
Keroplatidae		Pipunculidae		Tephritidae	
Trigemma*	2	Cephalops	36	Neotephritis	2
Tylparua	4	Psychodidae		Phaeogramma*	3
Limoniidae		Psychoda	4	Trupanea	21
Geranomyia	1	Trichomyia	2	Xenasteiidae	
Gonomyia	2	Sciaridae		Xenasteia	2
Dicranomyia	13	Bradysia	3		
Milichiidae		Ctenosciara	1	Totals	1104
Leptometopa	1	Hyperlasion	1		
Milichiella	2	Lycoriella	1	The totals are 24 families,
		Phytosciara	1	82 genera and 1104 species[2]

[1] May not be an endemic (F. Howarth, pers. comm.).
[2] Total endemic species includes updated information since Nishida (2002).

to Hawai'i and conduct research on the incredible diversity that abounds throughout the islands, they inevitably come away with the same refrain: 'Had Darwin visited Hawai'i instead of the Galápagos, we would have had examples of speciation, diversity, and adaptive radiation from Hawai'i in our textbooks today'.

The examples of diverse groups of Diptera in Hawai'i span the entire order from Nematocera through muscoid brachycerans and are representative of various ecological niches as well. Table 2.2 shows examples of families with diverse genera of endemic Hawaiian Diptera and the ecological niches they occupy in the food web.

Diptera Diversity: Status, Challenges and Tools
(eds T. Pape, D. Bickel & R. Meier). © 2009 Koninklijke Brill NV.

Figure 2.2. The picture-winged *Drosophila conspicua* Grimshaw (right)
and the much smaller *D. melanogaster* Meigen (left)
to show size gigantism in the picture-winged group.
(Photo W.P. Mull.)

3.1 Drosophilidae

Much of the literature surrounding the diversity and speciation mechanisms of Hawaiian Diptera has focused on the Hawaiian *Drosophila*
Fallén. The genus is represented in Hawai'i by some of the most striking
examples of gigantism and host adaptation of any dipteran lineage. The
picture-winged group of *Drosophila* includes over 100 species (such as *D.
crucigera* Grimshaw) that are sometimes as much as 10 x the size of the
common *Drosophila melanogaster* Meigen used in genetic research (see
Fig. 2.2 for a comparison of sizes of these two species). Currently there are
424 species of *Drosophila* known from the Hawaiian Islands, with 403 of
these being endemic (N.L. Evenhuis, unpubl.). The total number of species
of *Drosophila* worldwide is about 1,200, thus Hawai'i comprises about 33%
of the world's known *Drosophila* fauna. Since estimates of the numbers of
Drosophila in Hawai'i show hundreds more endemic species awaiting descriptions, this percentage will go higher (K.Y. Kaneshiro, pers. comm.).

3.2 Dolichopodidae

Second to the Drosophilidae in ranking of diverse groups of Diptera are
the Dolichopodidae or long-legged flies. Currently, there are 244 species

in the family recorded from Hawaiʻi, the majority of which (163) are found in the genus *Campsicnemus* Haliday. Current taxonomic research on this genus by the author shows that there are probably an additional 100–150 species of *Campsicnemus* yet to be described. With the world fauna of this genus at around 225, Hawaiʻi currently harbors about 72% of the world's species of this genus and the percentage will no doubt increase as more Hawaiian species are discovered and described.

All the dolichopodids are predatory, so the fact that *Campsicnemus* has such a rich and speciose nature in Hawaiʻi makes for interesting study ecologically (e.g., an experimental pan trapping survey resulted in 11 species of *Campsicnemus* being collected that occurred sympatrically in the Olaʻa rainforest on Hawaiʻi Island). Of additional interest is that the genus is primarily a boreal one, found primarily in the northern temperate latitudes with only a few populations found in tropical areas (Hawaiʻi, French Polynesia as far south as Rapa, Sri Lanka, Nepal, Congo, and Cameroon).

Table 2.2. Diverse Hawaiian Diptera Genera (more than 20 species).

Family/Genus	Species	Est. Species[1]	Food Web Position
Calliphoridae			
Dyscritomyia	25	25	Scavenger
Dolichopodidae			
Campsicnemus	163	300	Predator
Eurynogaster	24	30	Predator
Drosophilidae			
Drosophila	403	800	Herbivore[2]
Scaptomyza	155	180	Herbivore[2]
Muscidae			
Lispocephala	102	200	Predator
Pipunculidae			
Cephalops	36	40	Parasitoid
Tephritidae			
Trupanea	21	25	Herbivore
Total estimated new species		**700**	

[1] Total species taking into account estimated undescribed native species.
[2] A few are scavengers and predators.

Diptera Diversity: Status, Challenges and Tools
(eds T. Pape, D. Bickel & R. Meier). © 2009 Koninklijke Brill NV.

Table 2.3. Comparison of Number of Species of Hawaiian Diptera with the
Galápagos and Canary Islands Faunas (using selected groups).

Family/Genus	Hawaii[1]		Galápagos[2]		Canary Is[3]	
	total	endemic	total	endemic	total	endemic
Asilidae	0	0	2	2	24	21
Bombyliidae	1	0	6	6	25	21
Calliphoridae	41	26	7	2?	10	1
Chironomidae	40	27	8	?	57	3
Orthocladius	9	9	0	0	4	0
Dolichopodidae	247	231	18	12?	41	17
Campsicnemus	163	163	0	0	2	0
Drosophilidae	595	541	17	?	25	2
Drosophila	424	403	11	?	12	1
Scaptomyza	146	142	1	1	4	0
Hybotidae	4	2	3	3	24	17
Chersodromia	2	2	1	1	3	1
Ephydridae	45	19	16	?	45	2
Scatella	18	15	3	2?	2	0
Limoniidae	23	16	5	3	24	9
Dicranomyia	13	13	1	1	5	1
Phoridae	22	9	4	0	52	14
Pipunculidae	36	36	2	2	12	2
Syrphidae	18	0	9	4?	33	11
Tephritidae	38	26	5	2	32	7
Trupanea	21	21	0	0	4	0
Total Diptera	**1518**	**1108**	**264**	**?**	**1024**	**297**

[1] Data from Nishida (2002) with updates. Total land area is 16,636 km²
   (= 62.4 endemic Diptera per 10 km²).
[2] Data from Sinclair (2009). Total land area is 45,000 km²
   (= < 0.5 endemic Diptera per 10 km²).
[3] Data from Báez & Garcia (2001). Total land area is 7,242 km²
   (= 4.1 endemic Diptera per 10 km²).

The continental species are chiefly black or gray colored, while the insular
and tropical species have brown and yellow patterns. DNA studies will
have to be conducted to ascertain the origin of the Pacific faunas of this

genus, but it appears from morphological examination that the Hawaiian and French Polynesian species may have originated from southwestern Polynesia or eastern Melanesia, possibly from the *Sympycnus* Loew lineage.

The genus *Eurynogaster* Parent and allies have 55 endemic species known from Hawai'i. Recent study of this genus by the author (Evenhuis 2005) indicated that a number of species previously placed in the heterogeneous *Eurynogaster* required separate genera to house them. Despite the removal of species to seven other genera, there still remain 24 species in *Eurynogaster* (*sensu stricto*).

Although speciation in *Eurynogaster* (*sensu lato*) appears to be a result of geographic isolation, the extensive speciation observed in Hawaiian *Campsicnemus* is an example of adaptive radiation (cf. Gillespie *et al.* 2001) in that a concurrent diversification into new ecological roles appears to be happening during the speciation process. Mainland *Campsicnemus* species appear to be found primarily in association with streams and ponds (normally around the edges). In Hawai'i, there are species that are water skaters, while others are found in leaf litter, on low-growing vegetation such as ferns, and others are only found in the high canopy. Biologies of the immatures are, for the most part, unknown, but some published rearings have shown them to be found under bark of *Cheirodendron* (Montgomery 1975). Further rearings may show that the immatures partition themselves into specific host habitats (i.e., various vegetation substrata) while the adults are free-living predators and can coexist with other species in the same area without competing for resources.

3.3 Muscidae

For the most part, muscids in Hawai'i are represented by nonindigenous synanthropic species. Some may have arrived with Polynesians, but the vast majority of species have been introduced more recently through further human intervention. An exception to this is the predaceous genus *Lispocephala* Pokorny where a striking amount of speciation has taken place. There are 102 species known in the genus from Hawai'i, all of them endemic. The amount of speciation in this otherwise cosmopolitan genus is remarkable. There are roughly 150 species of the genus worldwide, thus Hawai'i harbors 67% of the world's fauna of *Lispocephala*. Hardy (1981) estimated that the total Hawaiian endemic species of *Lispocephala* might number 150–200 after all species are described.

Diptera Diversity: Status, Challenges and Tools
(eds T. Pape, D. Bickel & R. Meier). © 2009 Koninklijke Brill NV.

Both the larval and adult stages of *Lispocephala* are predaceous, especially on other Diptera. In some cases, the degree of specificity of *Lispocephala* predation on species of *Drosophila* is remarkable. Ken Kaneshiro (pers. comm.) has indicated that there seems to be an almost one-to-one correlation with *Drosophila* species and their *Lispocephala* predators, even including an undescribed aquatic *Drosophila* that has been observed being preyed upon by an aquatic *Lispocephala* larva.

3.4 Calliphoridae

As with muscids, many of the calliphorids in Hawai'i are nonindigenous synanthropic flies that breed in carrion of various animals and arrived in tandem with humans. However, one genus, *Dyscritomyia* Grimshaw, is entirely endemic and is comprised of 25 species. Unfortunately, since their discovery at the turn of the 20th century, many species have not been seen and may have gone extinct, especially those on O'ahu. Calliphorids in general are not known to have extensive speciation within genera, so having one with this many species is unusual in itself, but one with such a large number of species and restricted to such a small area of land is even more fascinating.

Species of *Dyscritomyia* are scavengers in carrion. Upon finding a suitable host, the female larviposits directly onto the animal carcass where the larva feeds until it pupates. Discovering what their natural hosts were before the arrival of Polynesians is difficult. Because there were no mammals in Hawai'i aside from two native bats (one has since gone extinct), it has been surmised that the flies were at one time either host-specific on land snails or forest birds and that the subsequent demise of the land snail fauna and the forest birds in the 20th century, especially on O'ahu, has also led to a corresponding reduction in numbers of species of *Dyscritomyia*.

Another potential host has been speculated recently. Hawaiian entomologist, Dr. Stephen Montgomery, left a recently killed sphingid moth in the crotch of a tree to be picked up by him later. When he arrived the next day to pick up the moth, it had been larviposited upon by a *Dyscritomyia*.

3.5 Pipunculidae

The genus *Cephalops* Fallén has radiated extensively in Hawai'i. Currently there are 36 species known in the genus. All are parasitoids of auchenorrhynchous Homoptera, in Hawai'i primarily on genera in the families Cicadellidae and Delphacidae. Hardy (1964) surmised that these flies were parasitic on the native species of these families, which have themselves radiated extensively in Hawai'i. Collection records have shown that these flies were collected in association with the genera *Nesophrosyne* and *Nesophryne* (both Cicadellidae) and *Nesodyne* (Delphacidae). Unfortunately, little biological work or rearings have been done, so the association of species and possible native hosts has yet to be ascertained.

Collecting records in the early 1900s showed an abundance of pipunculids in sugarcane fields, and young nymphs of the invasive alien sugarcane planthopper, *Perkinsiella saccharicida*, were recorded as hosts. However, this was most likely a host-shift to a new resource soon after the planthopper was introduced into Hawai'i (ca. 1900) since pipunculids have not been observed in cane fields much after 1920.

3.6 Tephritidae

Tephritids are notorious worldwide as agricultural pests causing millions of dollars worth of damage to crops each year. In Hawai'i, four agricultural pests have been introduced since the turn of the 20th century.

However, one genus of tephritids in Hawai'i that has radiated extensively does not contain any species that are considered pests. The endemic species have co-evolved with a group of endemic plants, a few of which are on the brink of extinction, thereby sealing the fate of the flies unless conservation efforts can help resurrect populations to sustainable levels once again. *Trupanea* Schrank, a cosmopolitan genus, is represented in Hawai'i by 21 endemic species whose larvae breed in the various parts of composites, including many found in species of the silversword alliance. The silversword alliance is yet another example of adaptive radiation in Hawai'i (Baldwin & Robichaux 1995), this one comprising 28 endemic species in genera *Argyroxiphum*, *Dubautia*, and *Wilkesia*. Work is currently being done on the association of *Trupanea* with their plant hosts to research possible host specificity, adaptations, and molecular evolution (Brown 2003).

3.7 Flightless flies

Flightlessness in insects is not restricted to oceanic islands or wind-swept harsh environments; examples can be found in virtually every ecosystem (Howarth 1990). However, island ecosystems (whether islands in oceans

Figure 2.3. *Dicranomyia gloria* (Byers) from Mt. Ka'ala on O'ahu.
(Photo W.P. Mull.)

or high mountains whose ecosystems act as 'island' ecosystems) are the
most common areas in which flightlessness is known to occur.

Hawai'i is home to more species of flightless flies than any other place
in the world. It is interesting that such a rare phenomenon in other areas
of the globe is so much more common in Hawai'i. Conspicuous examples
of flightlessness in Hawaiian Diptera include those in the Limoniidae (3
spp.) and Dolichopodidae (8 spp.). What is fascinating about flightlessness
in the Hawaiian Islands is that both alate and flightless forms of certain
groups can be found living side-by-side.

Among the limoniids, species of the genus *Dicranomyia* Stephens have
lost wings and forage on the ground and leaf litter atop the highest peaks
on O'ahu and Kaua'i (Fig. 2.3). On O'ahu the tallest peaks in the Ko'olau
Range have one species (*Dicranomyia hardyana* (Byers)), and the tallest
peak in the Waia'nae Range, Mt. Ka'ala, has another species (*Dicranomyia
gloria* (Byers)).

In the dolichopodids, two genera have species that are flightless (Evenhuis 1997). One genus, *Emperoptera* Grimshaw, is endemic to the Hawaiian Islands, and possesses five species, four of which have not been collected again since their first observation. Only *Emperoptera montgomeryi*

Figure 2.4. *Emperoptera montgomeryi* Evenhuis from Mt. Ka'ala on O'ahu.
(Photo Y. Imomori.)

Evenhuis on Mt. Ka'ala (Fig. 2.4) seems to be surviving; its larvae have
been found foraging in the mesophyll layer of leaves of *Melicope* in the
leaf litter (N.L. Evenhuis, unpubl.). *Campsicnemus* has three species that
are flightless, two of which have been commonly collected since their first
discovery.

4. Biogeographic Affinities of the Hawaiian Diptera Fauna

There are currently 1,518 species of Diptera recorded from Hawai'i with
1,108 of these endemic — a fairly high rate of endemicity (73%) when compared to an overall rate of endemicity of 58% for the entire insect fauna
(Eldredge & Evenhuis 2003). Of these 1,518 species, 396 introduced species
of Diptera in Hawai'i are recorded in the literature. Many of these 'aliens'
have arrived in Hawai'i through human intervention (some purposefully introduced as biological control agents; others inadvertently through
commerce), although some may have arrived naturally from storms.

Hawaii's freshwater aquatic Diptera have been extensively studied
and some conspicuous groups, such as Ephydridae, Chironomidae, and
Canacidae, have been hypothesized as originating from marine species

(Howarth & Polhemus 1991). Indeed, some species in the genera *Procanace* Hendel (Canacidae), *Telmatogeton* Schiner and *Clunio* Haliday (Chironomidae), and *Scatella* Robineau-Desvoidy (Ephydridae) inhabit coastline strand habitats while others in those genera are found in cool mountain streams at much higher elevations.

No broad sweeping statements can be made as to the origin of the entire Hawaiian Diptera fauna. Different groups arrived under different circumstances from different areas. The preponderance of the genera seem to have been founded by elements from the Oriental Region, although there have been some sister-group relationships identified with North American elements as well. True affinities are best ascertained through molecular analysis in comparison with sister groups from other areas. Until this is done, we can only guess at the source areas from which the various Hawaiian groups have evolved.

In a few groups, molecular studies have been conducted to determine origin. In the Calliphoridae, the lone native species of *Lucilia* Robineau-Desvoidy, *L. graphita* Shannon, is only found in the low islands and atolls of the Northwestern Hawaiian Islands. The larvae have been collected from the skulls of the Federally Endangered Hawaiian Monk Seal (*Monachus shauinslandi*), which is also restricted primarily to these islands. Preliminary molecular studies have shown *L. graphita* to be possibly derived from the Palaearctic and Oriental *L. ampullacea* Villeneuve (J.R. Stevens, pers. comm.).

In his morphological cladistic analysis of Drosophilidae, Grimaldi (1990) concluded that the sister group to the Hawaiian *Drosophila* lineage is probably a mycophagous group from North America, and that *Zaprionus* Coquillett is the sister group to the *Scaptomyza* Hardy lineage. However, DeSalle (1995) stated that comparisons of these hypotheses as to the origin with those based on molecular analysis were equivocal. Instead, molecular information combined with morphological data showed that the Hawaiian Drosophilidae (*Scaptomyza* and *Drosophila* lineages) are in a clade of the subgenus *Drosophila* and originated from a member of the *Drosophila virilis-repleta-robusta-melanica* group (DeSalle *et al.* 1997). All four of these species are known from North America, with three of them found in Mexico.

5. Threats to the Fauna

Despite the high diversity of some groups of Diptera in Hawai'i, there are still pervasive elements that threaten their survival and also threaten the very diversity we seek to discover and research.

The continual introduction of aliens into Hawai'i is a serious problem that is being addressed by various groups. As pointed out by Allison & Evenhuis (2001):

> 'Ants, termites, mosquitoes, and other insect pests, together with garden weeds, are among the better-known alien species. In addition to these, there is an enormous number of others that are bringing about the decline and extinction of the unique native plants and animals of the Islands and are causing millions of dollars of damage to agricultural crops.'

Hawai'i is a 'living laboratory of evolution', but it also unfortunately carries the dubious distinction of 'extinction capital of the world' (Eyre 2000). Approximately 20–30 new alien insect introductions reach Hawaii's shores each year. Their presence causes pressure on depleting resources, displacement of populations of native species, and possible extinctions of vulnerable fauna and flora that cannot adapt or shift hosts quickly enough to survive.

5.1 Predators

Among the Diptera, flightless species seem to have been the most conspicuous victims of introduced insects. *Emperoptera mirabilis* Grimshaw was originally collected by R.C.L. Perkins in 1900 on Mt. Tantalus at roughly 280 meters. This locale is very near to and overlooks today's urban Honolulu, but when Perkins first collected there, the environment held high hopes for a robust fauna of insects, including this flightless dolichopodid. However, revisiting the locale just a few years later, Perkins was dismayed that no specimens of *Emperoptera* ('locally abundant' when he first collected there in December 1900) could be found (Perkins 1907).

A notorious suspect with regard to the demise of the Tantalus insect fauna is the alien big-headed ant, *Pheidole megacephala*. This ant has been responsible for the extreme reductions and even extinctions of many populations of ground dwelling insects throughout the Hawaiian Islands. In this case, it has been implicated in the extinction of *Emperoptera mirabilis* (Hardy 1961, Zimmerman 1970). Trips to the area by biologists for the next 100 years have been unsuccessful in finding the species again. Over-all, there are 41 species of ants known from Hawai'i. All of them are alien and virtually all have had a deleterious effect on the fauna, especially on ground dwelling and flightless forms, as well as on the remaining environment.

Aside from ants, other introduced predators and parasitoids are serious threats to the survival of the native Diptera fauna. Of these, the western yellowjacket, *Vespula pensylvanica* is probably the worst culprit. Its predatory habits are more general than those of the ants since it can prey on flying insects as well as ground dwelling ones. Additionally, Jackson's chameleon (*Chameleo jacksonii*) and the coqui frog (*Eleutherodactylus coqui*), two vertebrate predators that have been recently introduced to the Islands, have become well established and are currently eating their way through the lowland forest invertebrates with no natural controls on their populations (cf. Tummons 2003).

5.2 Habitat and resource displacement

Incipient threats to the health of the native Hawaiian ecosystems are introductions that cause disturbance in the environment through reduction of habitat and resources. Two of the most serious of these are the plant *Miconia calvescens* and the aquatic insect *Cheumatopsyche analis*.

The uncontrolled introduction of *Miconia* into certain lowland portions of Hawai'i as an ornamental has resulted in serious environmental change where it has reproduced and proliferated to create monocultures that crowd out native species and destroy native understory vegetation that are host and home to many native Diptera and other invertebrates.

The introduction in 1965 of the sister sedge caddisfly, *Cheumatopsyche analis,* has led to it becoming the dominant invertebrate in lowland streams in Hawai'i today. Its presence has depleted resources of other herbivorous, native, aquatic immatures [e.g., the native *Telmatogeton torrenticola* Terry was once plentiful in certain streams and even irrigation ditches on O'ahu but since the invasion of caddisflies into those streams the flies have been extirpated (J.W. Beardsley, pers. comm.)], but just as serious is that its abundance as a food source has possibly led to a consequential increase in the populations of introduced predatory fish in these streams (Flint *et al.* 2003), thereby increasing the likelihood of the potential reduction or extirpation of any native immatures that still may be clinging to survival.

PART 2. HISTORY AND FUTURE OF HAWAIIAN DIPTEROLOGY

6. History of Collections

Our knowledge of the diversity of the Diptera of Hawai'i could not be told to the degree that we can today without the collections that have taken place throughout the young history of these islands.

The first recorded mention in the literature of flies in Hawai'i is found in the journals of William Ellis (1783). Ellis was the assistant surgeon on Captain James Cook's voyage that took him to the shores of Kaua'i and Hawai'i Island in 1778 and 1779. Ellis writes:

> They have also a kind of fly-flap, made of a bunch of feathers fixed to the end of a thin piece of smooth and polished wood; they are generally made of the tail feathers of the cock but the better sort of people have them made of the tropick bird's feathers, or those belonging to a black and yellow bird called Mo-ho.

These 'fly-flaps' are the Hawaiian *kahili* that evolved from an original use as fly swatters to become ceremonial icons of royalty (the highest royalty having the ones with the rarest feathers).

Cook's voyages did not describe any Diptera from Hawai'i and it is not known if any were collected (no record exists of such collections). The first descriptions of Diptera from Hawai'i were by Thomson (1869), who described the flies collected on the world voyage of the Swedish frigate *Eugenie*. The *Eugenie* stopped in Hawai'i in August 1852 and collected in and around Honolulu. Thomson described a paltry 6 species from that expedition, only two of which were endemic.

The first comprehensive survey of the Diptera of the Hawaiian Islands was conducted by R.C.L. Perkins between 1892 and 1901 and the results of these collections were subsequently published in the *Fauna Hawaiiensis* [see Manning (1986) for full details on the history and publication of this singularly significant research in Hawaiian entomology]. Perkins spent months at a time alone in the rainforests of Hawai'i, and after seven years of field collecting amassed probably the best collection of native insects ever assembled in Hawai'i (Evenhuis 2007).

The efforts of Perkins sparked the interests of many other entomologists and they came from many parts of the world to collect the unique endemic fauna found in the islands. But it wasn't until the arrival in 1948 of D. Elmo Hardy that Diptera research in Hawai'i began in earnest. Hardy's *raison d'être* was to revise the current knowledge of the Diptera fauna of Hawai'i for the series *Insects of Hawai'i*, begun that same year by E.C. Zimmerman (Zimmerman 1948).

Hardy was instrumental in garnering financial and scientific support for the survey of the diverse Diptera fauna that hitherto had only been little publicized in *Fauna Hawaiiensis* or the checklist of Bryan (1934). He started the Hawaiian *Drosophila* Project in 1963 (still continuing today) and embarked on taxonomic research of Pipunculidae, Dolichopodidae, and Tephritidae of the islands and increased the numbers of species in those families manifold.

Hardy's teaching attracted a cadre of graduate students over the years, who also contributed significantly to the collection and knowledge of Hawaiian Diptera. Prime among these who are still active and prominent in the Hawaiian scientific community are Kenneth Kaneshiro, Stephen Montgomery, and Francis Howarth.

Recent studies on aquatic ecosystems have added yet another complement of specimens and increased knowledge of an environment in which Diptera are dominant and an essential component in water quality assessments. Dan Polhemus and Ronald Englund have contributed significant new information on aquatic Diptera and many specimens of new species from the freshwater stream habitats throughout the islands, especially those that had been inaccessible to previous collectors.

7. Collecting Methods

Significant advances have been made over the decades in collecting methods for flies in Hawai'i. Each advance in methodology has resulted in the discovery of an incredibly rich fauna of flies that had not been known previously or known only from a few examples that had been otherwise collected by accident along with other insects.

Aerial sweep nets have always been the mainstay of collectors and this is what had been used for the most part with most Diptera collecting until the middle of the 1900s. With the advent of the Malaise traps in 1937, collecting Diptera burgeoned from collecting dozens of specimens at a time to hundreds and thousands of specimens at a time. Malaise trapping also exposed new species that had not been seen previously because of their otherwise uncommonness, cryptic nature, or elusiveness. However, the

Figures 2.5–2.6. Hawaiian habitats.
Left: Sacred Falls canyon, O'ahu.
Right: Kaipapa'u Falls, O'ahu.

coarse mesh of the early nettings for the Malaise traps and aerial nets missed the very small insects. In the late 1960s, fine mesh netting in both aerial sweep nets and Malaise traps resulted in even more undescribed species, especially in Diptera and Hymenoptera.

Other collecting methods that have been employed in recent decades include yellow pan traps (water traps), small-canister pyrethrin fogging, window pane flight intercept traps (used in association with water traps placed below them), and vacuuming vegetation.

In Hawai'i, the extreme topography of many locales (e.g., Figs 2.5–2.6) makes them virtually impossible to be accessed on foot. Helicopter access to such localities has allowed for an increase in the number of areas that can be explored and the results of collections from these rarely reached habitats have increased the number of new endemic species that other-wise would not have been discovered.

Diptera Diversity: Status, Challenges and Tools
(eds T. Pape, D. Bickel & R. Meier). © 2009 Koninklijke Brill NV.

8. Inventorying the Data

Early voyages of discovery and subsequent published descriptions of Hawaiian Diptera appeared in books, then later in scientific journal articles. After over 200 years of scientific exploration of the Hawaiian Diptera, the results are to be found in hundreds of scattered references. In 1992 the Hawaii Biological Survey (HBS) was established by State law to collect, inventory, and disseminate the results of research on the plants and animals of the Hawaiian Islands. The HBS had a mandate to survey the fauna and flora of the islands, but it also realized at its conception that the information provided from the surveys and research should reach the widest audience possible, thereby validating its scientific merit at the same time as publicizing the results of its work to the public.

The first task of the HBS was to database all of the literature that made mention of any plant or animal found in Hawai'i. As that bibliography was being researched, the taxa noted in those papers and books as being found in Hawai'i were also databased. After completing the literature database and the database of species found in the literature, the first resulting checklist of terrestrial arthropods from these efforts was published (Nishida 1992). In that checklist, 1,426 species of Diptera are recorded including 1,045 endemic and 353 nonindigenous species. Further editions were published in subsequent years (Nishida 1994, 1997) and simultaneous access to electronic versions of this checklist was made available on the World Wide Web. The most recent published edition (Nishida 2002) records 1,545 species of Diptera known from Hawai'i, of which 1,061 are endemic and 391 are nonindigenous.

To augment the online databases and published checklists, HBS staff go out to schools to present results of surveys and engage schoolchildren in field surveys and the wonders of scientific discovery. By educating children about the Hawaiian environment and helping them embrace their natural heritage, future generations will hopefully continue to conserve the native Hawaiian landscape.

9. Future of Hawaiian Diptera Biodiversity

When Elmo Hardy first arrived in Hawai'i in 1948 and began his work on collecting and describing the Hawaiian Diptera, 160 years of exploration in Hawai'i had passed, yet there were only 197 species of flies known (roughly equal to discovery of 1 species per year). After the last Diptera volume of the *Insects of Hawai'i* was published by Hardy in 1981, that number had risen to 1,209 (roughly equal to an increase of 30 species per year from 1948 to 1981). Today, there are 1,518 species of Diptera and it is expected that an additional 700 new species of Diptera await description or discovery (Table 2.2).

The numbers of Diptera in Hawai'i will steadily increase as long as taxonomists show an interest in the Diptera fauna of these islands. There are still many areas in hard-to-reach places in the islands that have not been explored for flies, especially rainforest canopies, high elevation streams and bogs, and slot canyons and waterfalls; and additional collecting techniques exist or will be invented that are still to be employed that will expose new species that were missed by previous collecting methods.

As long as conservation efforts protect the environment from unwanted threats of urban change and introductions of invasive pests, the fertile grounds for Hawaiian Diptera will yield many more species to the checklist of Hawaii's biodiverse Diptera fauna.

ACKNOWLEDGEMENTS

I thank Ken Kaneshiro, Patrick O'Grady, Dan Polhemus, Jamie Stevens, Stephen Montgomery, Francis Howarth, Dan Bickel, and the late D. Elmo Hardy for helpful discussions over the years that led to this paper. Ronald Englund, Dan Polhemus, and Francis Howarth kindly reviewed early drafts of this paper. This paper constitutes Contribution No. 2008-009 to the Hawaii Biological Survey.

References

Allison, A. & Evenhuis, N.L. (2001) Foreword. Page vii *in*: Staples, G.W. & Cowie, R.H. (eds), *Hawaii's invasive aliens. A guide to invasive plants and animals in the Hawaiian Islands*. Mutual Publishing & Bishop Museum Press, Honolulu.

Báez, M. & Garcia, A. (2001) Order Diptera. Pages 249–267 *in*: Izquierdo, I., Martin, J.L., Zurita, N. & Arechavaleta, M. (eds), *Lista de especies silvestres de Canarias (hongos, plants y animales terrestres)*. Consejería de Política Territorial y Medio Ambiente Gobierno de Canarias. La Laguna, Tenerife.

Baldwin, B.G. & Robichaux, R.H. (1995) Historical biogeography and ecology of the Hawaiian silversword alliance (Asteraceae): new molecular phylogenetic perspectives. Pages 259–285 *in*: Wagner, W.L. & Funk, V.A. (eds), *Hawaiian biogeography: evolution on a hot spot archipelago*. Smithsonian Institution Press, Washington, D.C.

Brown, J.M. (2003) Sexual selection and radiation in Hawaiian tephritid flies. Available at http://web.grinnell.edu/individuals/brownj/research.html, accessed 8 September 2003.

Bryan, E.H., Jr. (1934) A review of the Hawaiian Diptera, with descriptions of new species. *Proceedings of the Hawaiian Entomological Society* 8: 399–468.

Clague, D.A. (1996) The growth and subsidence of the Hawaiian-Emperor volcanic chain. Pages 35–50 *in*: Keast, A. & Miller, S.E. (eds), *The origin and evolution of Pacific Island biotas, New Guinea to Eastern Polynesia: patterns and processes*. SPB Academic Publishing, Amsterdam.

DeSalle, R. (1995) Molecular approaches to biogeographic analysis of Hawaiian Drosophilidae. Pages 72–89 *in*: Wagner, W.L. & Funk, V.A. (eds), *Hawaiian biogeography: evolution on a hot spot archipelago*. Smithsonian Institution Press, Washington, D.C.

DeSalle, R., Brower, A.V.Z., Baker, R. & Remsen, J. (1997) A hierarchical view of the Hawaiian Drosophilidae (Diptera). *Pacific Science* 51: 462–474.

Eldredge, L.G. & Evenhuis, N.L. (2003) Hawaii's biodiversity: a detailed assessment of the numbers of species in the Hawaiian Islands. *Bishop Museum Occasional Paper* 76: 1–30.

Ellis, W. (1783) *An authentic narrative of a voyage performed by Captain Cook and Captain Clerke, in his majesty's ships* Resolution *and* Discovery, *during the years 1776, 1777, 1778, 1779 and 1780; in search of a North-West passage between the continents of Asia and America. Including a faithful account of all their discoveries, and the unfortunate death of Captain Cook*. 2 vols. G. Robinson, J. Sewell & J. Debrett, London.

Evenhuis, N.L. (1997) Review of flightless Dolichopodidae (Diptera) in the Hawaiian Islands. *Bishop Museum Occasional Papers* 53, 29 pp.

Evenhuis, N.L. (2005) A review of the genera comprising species of the genus *Eurynogaster sensu* Hardy & Kohn, 1964 in Hawai'i (Diptera: Dolichopodidae). *Zootaxa* 1017: 39–60.

Evenhuis, N.L. (2007) Barefoot on lava. The journals and correspondence of naturalist R.C.L. Perkins in Hawai'i, 1892–1901. *Bishop Museum Bulletin in Zoology* 7: 1–412.

Eyre, D.E. (2000) *By wind, by wave: an introduction to Hawaii's natural history*. Bess Press, Honolulu, 178 pp.

Flint, O.S., Englund, R.A. & Kumashiro, B. (2003) A reassessment and new state records of Trichoptera occurring in Hawai'i with discussion on origins and potential ecological impacts. *Bishop Museum Occasional Papers* 73: 31–40.

Gagne, W.C. (1988) Conservation priorities in Hawaiian natural systems. *BioScience* 38: 264–271.

Gillespie, R.G., Howarth, F.G. & Roderick, G.K. (2001) Adaptive radiation. Pages 25–44 *in*: Levin, S. (ed.), *Encyclopedia of biodiversity*. Vol. 1. Academic Press, New York.

Grimaldi, D.A. (1990) A phylogenetic revised classification of genera in the Drosophilidae. *Bulletin of the American Museum of Natural History* 197: 1–137.

Hardy, D.E. (1961) The Diptera of Hawaii. *Verhandlungen der XI Internationaler Kongres der Entomologie* (Wien) 1: 167–168.

Hardy, D.E. (1964) Pipunculidae. *Insects of Hawaii* 11: 302–379.

Hardy, D.E. (1981) Diptera: Cyclorrhapha IV. *Insects of Hawaii* 14: 1–491.

Howarth, F.G. (1990) Hawaiian terrestrial arthropods: an overview. *Bishop Museum Occasional Paper* 30: 4–26.

Howarth, F.G. & Mull, W.P. (1992) *Hawaiian insects and their kin*. University of Hawai'i Press, Honolulu, 160 pp.

Howarth, F.G. & Polhemus, D.A. (1991) A review of the Hawaiian stream insect fauna. Pages 40–50 *in*: Devick, W. (ed.), *Proceedings of the 1990 Symposium on Freshwater Stream Biology and Fisheries Management*. State of Hawai'i, Department of Land and Natural Resources, division of Aquatic Resources, Honolulu.

Manning, A. (1986) The Sandwich Islands Committee, Bishop Museum, and R.C.L. Perkins: cooperative zoological exploration and publication. *Bishop Museum Occasional Papers* 26: 1–46.

Montgomery, S.L. (1975) Comparative breeding site ecology and adaptive radiation of picture-winged *Drosophila* (Diptera: Drosophilidae) in Hawai'i. *Proceedings of the Hawaiian Entomological Society* 22: 65–103.

Nishida, G.M. (1992) Hawaiian terrestrial arthropod checklist. *Bishop Museum Technical Report* 1, viii + 262 pp.

Nishida, G.M. (1994) Hawaiian terrestrial arthropod checklist. Second edition. *Bishop Museum Technical Report* 4, iv + 287 pp.

Nishida, G.M. (1997) Hawaiian terrestrial arthropod checklist. Third edition. *Bishop Museum Technical Report* 12, iv + 263 pp.

Nishida, G.M. (2002) Hawaiian terrestrial arthropod checklist. Fourth edition. *Bishop Museum Technical Report* 22, iv + 313 pp.

Perkins, R.C.L. [1907]. The insects of Tantalus. *Proceedings of the Hawaiian Entomological Society* 1[1906]: 38–51.

Sinclair, B.J. (2009) Dipteran Biodiversity of the Galápagos. Pages 98-120 *in*: Pape, T., Bickel, D. & Meier, R. (eds), *Diptera Diversity: Status, Challenges and Tools*. Brill, Leiden.

Thomson, C.G. [1869]. Diptera. Species nova descripsit. Pages 443–614 *in: Kongliga svenska fregatten Eugenies resa omkring jorden under befäl af C.A. Virgin, åren 1851-1853*. Vol. 2 (Zoologi), [section] 1, (Insecta). P.A. Norstedt & Söner, Stockholm.

Tummons, P. (2003) More than a noisy nuisance, coqui are eyed as threat to island fauna. *Environment Hawai'i* 13(12): 1, 7–8.

Zimmerman, E.C. (1948) Introduction. *Insects of Hawaii* 1, xx + 206 pp.

Zimmerman, E.C. (1970) Adaptive radiation in Hawaii with special reference to insects. *Biotropica* 2: 32–38.

CHAPTER THREE

NEOTROPICAL DIPTERA DIVERSITY: RICHNESS, PATTERNS, AND PERSPECTIVES

DALTON DE SOUZA AMORIM

Universidade de São Paulo, São Paulo, Brazil

INTRODUCTION

It seems reasonable to assume that there were Diptera species on the South American continent just after the origin of the dipterans, presumably in the Upper Permian or in the Lower Triassic. The dipteran radiation, hence, must have been well underway in South America long before the beginning of the breakup of Gondwana in the Jurassic. In the Lower Cretaceous there are already some unique forms in South America, as the extinct family Cratomyiidae. Throughout most of the Cretaceous and the Tertiary, South America was an island continent, disjunct from other tropical areas, and extraordinary life forms evolved during this 'splendid isolation', not only among mammals (Simpson 1980), but certainly also within Diptera, and the Neotropical Region now contains clades with such awesome species as the largest and most bulky dipterans: the Pantophthalmidae. The Neotropical Region is renowned for global highs of species richness in groups like canopy trees, birds and butterflies (Gentry 1988, Robbins & Opler 1997). And Diptera are no exception (Brown 2005).

1. Neotropical Region: Delimitation and Complexity

The term Neotropical Region was apparently used for the first time by Sclater (1858), even though Prichard (1826) already pointed to equatorial America being faunistically and floristically connected to Africa and In-

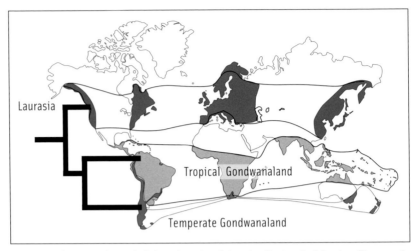

Figure 3.1. Intercontinental disjunctions involving the Neotropical Region. Three main components are present, even though a circum-Pacific component may also be present. There is overlap of circum-tropical elements and temperate circum-Antarctic elements in southeastern Brazil and in some areas along the Andes.

dia. It was up to Wallace (1876), however, to consolidate an evolutionary perspective of the Neotropical Region as one of the major biogeographic divisions of the Earth's terrestrial biome. In this sense, the Neotropical Region extends from the extreme south of South America, including the Falkland Islands, north to tropical parts of Mexico, also enclosing the Galápagos Islands, the Caribbean and the Antilles.

The Neotropical Region is actually a composite both in a geological and a biogeographic sense. Gill (1885), Allen (1892) and Lydekker (1896) already in the 19th century kept the southern portion of South America as a separate biogeographic region, indicating the similarities with the New Zealand and Australian biotas. Jeannel (1942) coined the term *Archiplata* for the South American temperate subregion, which in connection with the rest of the austral temperate areas collectively were referred to as *Paléantarctide*. Also, mountains and deserts in northern Mexico are largely Nearctic in their faunal and floral composition (Thompson 2009), while the Neotropical forests seem more related to tropical areas in Africa, Southeast Asia and northwest Australia (Fig. 3.1).

The Neotropical Region has a complex vegetation structure (Fig. 3.2). The forest biomes are the most well known and comprise, in South America, the Atlantic Forest and Amazonia Forest, to which should be added

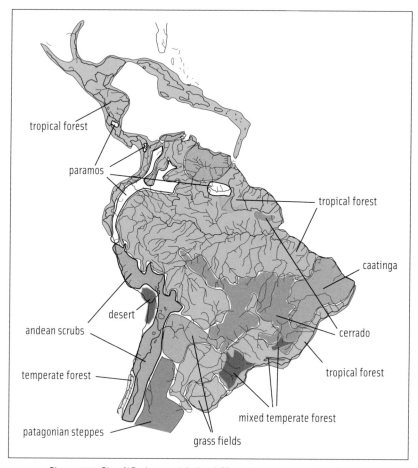

Figure 3.2. Simplified map with the different main vegetational types
in the Neotropical region.

the Central American tropical forests and the Caribbean tropical forests.
Forest elements in Amazonia actually extend outside the Amazon Basin
into other independent basins in northern South America (e.g., Orinoco),
as posed by Haffer (1978) and Nores (1999), but also in Central America
and the Caribbean, implying the artificial nature of the 'Amazonia Forest'
(Amorim 2001). There are also some forest areas to the south, extending
into the Paraná Basin, that are basically Amazonian in their fauna and
flora.

The expression 'open vegetation' refers to a spectrum of biomes mostly
running northeast to southwest in South America. The *cerrado*, a savanna-

like environment, occupies large expanses mostly in central Brazil, with
some spots in the south and in the north, within the Amazonian forest.
The *caatinga*, a dry or semi-arid environment. occupies large portions of
northeastern Brazil. More to the southwest, the *pampas*, a kind of grassland, occupy part of Argentina, Uruguay, Paraguay and southern Brazil.
Southeastern Brazil has grassland biomes over rocky soil called *rupestre*
fields, usually above 1,000 m elevation but also present in some rocky
lowlands. Higher terranes along the Andes are also quite specialized in
their fauna and flora, with the *paramos* in the northern part of the Andes
(including some spots in Central America, northern Brazil, and southern Venezuela), and the *Andean scrubs* in Peru, Bolivia, and northern
Argentina and Chile. Finally, the Patagonian steppes and the well characterized Atacama Desert in northern Chile and southern Peru should
be mentioned to recapitulate the open vegetation types of the neotropics.
Most of these open environments are rich in strong dipteran flyers, such
as Tabanidae, Nemestrinidae, Acroceridae, Asilidae, Bombyliidae, and
Therevidae, the diversity of which remains grossly understudied.

Some of the floral elements in the *cerrado* have their sister clades in
forest environments, not in other open vegetation formations. This supports the opinion of some authors, who consider the *cerrado* a modified
forest biome (e.g., de Vivo 1997). Even though the *rupestre* fields share
some faunal and floral elements with the *cerrado*, some of their endemic
elements, e.g., the plant species of the family Velloziaceae, have their sister group in the Afrotropical Region in similar environments of South
Africa and Madagascar (Mello-Silva 2005). Floral and faunal elements on
beach sands of high-precipitation areas along the Atlantic coast of South
America are connected to the *caatinga* and other dry environments rather
than to the neighboring forest biomes.

Southern temperate Neotropical taxa with circum-Antarctic distribution are well known in Diptera. Many of these groups are present in temperate forests in Chile and southern Argentina, but they are often found
to extend their distribution to the north along the Andes, reaching as
far as Colombia (e.g., in different mycetophilid genera), and into areas
of higher altitude in southern and southeastern Brazil, as in *Nervijuncta*
Marshall and *Australosymmerus* Freeman in the Ditomyiidae (Munroe
1974) and *Chiletricha* Chandler in the Rangomaramidae (Fig. 3.3). Elements of this fauna shared with New Zealand and southeast Australia are
widespread across the Diptera phylogeny, from the Tipulomorpha to at

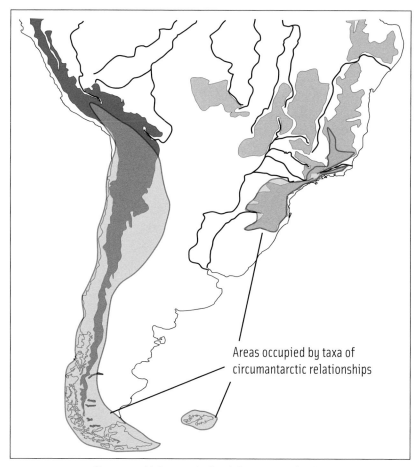

Figure 3.3. Main areas in South America involved in
circum-Antarctic relationships.

least the Empidoidea (Amorim & Silva 2002). Some groups, as *Diamphidi-cus* Cook (Scatopsidae) (Amorim 1989), *Perissomma* Colless (Perissom-matidae) (Colless 1962) and *Austroleptis* Hardy (Austroleptidae) have a Chile-Australia disjunction but are not known in southern Brazil. Know-ledge about the Triassic and Jurassic diversity of Diptera (Evenhuis 1994, Krzemiński 1992, Krzemińsky & Krzemińska 2004) gives strong support to the idea that many of these cases are truly Gondwanan elements, even though higher dipterans may have secondarily occupied Gondwanan ter-ranes (Grimaldi & Cumming 1999). The overlap of tropical elements with these temperate, circum-Antarctic elements at the southern extreme of the Atlantic forest generates a biogeographic node, called by Morrone (2004) the 'South American transition zone', which may be considered evolutionarily as relevant as the transition zone between the neotropics and the Nearctic in Mexico.

The forest canopy corresponds to an environment with dipteran elements associated with open areas, in strong contrast to the dipteran fauna in shaded parts of the forests. For example, Malaise traps at the ground level collect a wide variety of dipterans, such as sciarids, mycetophilids, gall midges, tipulids, limoniids, and many brachycerans associated with the soil or with detritivorous microhabitats. Malaise traps placed at the canopy will collect strong flyers, such as tabanids, asilids, bombyliids and many calyptrates, but also species associated with flowers, such as some Ulidiidae and Tephritidae, that cannot be collected otherwise. This is relevant in this discussion to demonstrate the complexity of the biogeographic and ecological structure of the fauna in the region.

A last component of the Neotropical dipteran diversity includes the species associated with aquatic environments. The diversity in freshwater environments is particularly high in Psychodidae, most families of the Culicomorpha, and Ephydridae. Marine dipterans are less common, but there are some cases like the chironomid genera *Telmatogeton* Schiner and *Clunio* Haliday (e.g., Oliveira 1950a,b). Some acalyptrates, such as *Tethina brasiliensis* Prado & Tavares, breed in decaying algae on sandy beaches (Artigas *et al.* 1992).

2. Neotropical Diptera Richness

Knowledge on dipteran diversity in the Neotropical Region has gradually increased since the 18th century. Early taxonomic efforts came from authors such as C. Linnaeus, J. C. Fabricius, P. A. Latreille, C. R. W. Wiedemann, J. B. Robineau-Desvoidy, P. J. M. Macquart, C. H. Blanchard, J. M. F. Bigot, J. O. Westwood, F. Walker, R. A. Philippi, H. Loew, V. von Roeder, E. Rübsaamen, I. R. Schiner, F. and E. Lynch Arribálzaga, C. Rondani, L. Bellardi, E. Giglio-Tos, F. M. van der Wulp, F. Müller, E. A. Goeldi, C. R. von Osten Sacken, S. W. Williston, and J. M. Aldrich (cf. Papavero & Guimarães 2000). This history of Neotropical dipterology in the 18th and 19th centuries has been described by Papavero (1971, 1973), including the field trips of early collectors, which may help solving a number of problems on the exact provenance for the types of some species described in the older literature.

The organization of the dipterological information took the first steps by the publication of the catalogue of Enrique Lynch Arribálzaga (1883) covering the Diptera of parts of Argentina and Uruguay. W. D. Hunter (1900–1901) initiated an ambitious new catalogue of all South American Diptera, but his work remained incomplete — only the parts relative to the Nematocera and part of the Brachycera ('Homeodactyla' and Mydidae) came to print. From 1966 to 1984, Nelson Papavero edited the *Catalogue of the Diptera of the Americas south of the United States* (Papavero 1966–1984), with a total of 102 published fascicles and 2877 printed pages — still with some families missing.

The *BioSystematic Database of World Diptera* (Evenhuis *et al.* 2007) presently records slightly more than 31,000 species for the Neotropical Region. This is only one fifth of the world diversity of the entire order, of about 153,000 described species of flies (Evenhuis *et al.* 2007). The families with the highest number of described species in the region are Tachinidae and Limoniidae (with more than 2,600 species), followed by Syrphidae, Phoridae, Dolichopodidae, Tabanidae, and Mycetophilidae, all with 1,000 or more species.

The proportion between the known and the actual diversity of Neotropical dipterans certainly varies according to the family. The ratio of undescribed species known after intensive collection for some groups of Mycetophilidae in some areas in Brazil reaches more than 10:1 (D.S. Amorim, unpubl.). This agrees fully with Brown's (2005) estimate for the Phoridae of Costa Rica, which is strong evidence that the Neotropical Diptera fauna is very superficially known. Diversity estimations are usually produced from accumulation curves from given sites or more vaguely as 'informed guesses' based on the literature. The estimation of overall diversity of an entire biogeographic region, however, needs to consider, apart from the in-site species richness, the question of geographical diversity. The large number of areas of high local endemism in the Neotropical Region (probably higher than in any other region) in combination with a less extreme effect of late Cenozoic glacial fluctuations may explain the high overall diversity in the region. Considering the number of areas of endemism, Amorim & Pires (1996) projected the actual number of species of Bibionomorpha to be about 30 times larger than the presently known diversity. Hammond (1992), however, estimated the actual diversity of Diptera

DIPTERA DIVERSITY: STATUS, CHALLENGES AND TOOLS
(EDS T. PAPE, D. BICKEL & R. MEIER). © 2009 KONINKLIJKE BRILL NV.

species in the world to be 1,600,000. If we consider that the ratio of the present number of described Neotropical species in relation to the world dipteran fauna is about one fifth, Hammond's figure would result in a projection of 320,000 Neotropical species. This, however, would be strongly affected by the fact that the Neotropical Diptera fauna is proportionally much more poorly known, particularly when compared to the Nearctic and Palaearctic faunas. However, with a more conservative projection, it does not seem to be an overestimation to consider the Neotropical fauna of Diptera as containing somewhere above 150,000 species, or at least five times the number known today.

Simuliidae (Coscarón & Coscarón-Arias 2007), Culicidae, and possibly the Tephritidae are seemingly the best known families in the Neotropical Region, obviously due to the fact that they are groups of applied interest. At the other end of the spectrum, and considering the diversity of species and genera in unsorted material, the Cecidomyiidae seem to be the family with the least developed taxonomic knowledge in the neotropics, as also mentioned by Brown (2005). Despite the efforts made to date, there seems to be a huge hidden diversity of Limoniidae and Tachinidae. This probably also applies to the Chironomidae, Ceratopogonidae, Sciaridae, Mycetophilidae, Lauxaniidae, and Sphaeroceridae. The Museu de Zoologia da Universidade de São Paulo, for example, holds about 600 undescribed species of Sciaridae, with specimens basically from south and southeastern Brazil, some few parts of Amazonia, and from Chile. Other families that are still much underexplored in the region, in terms of number of species, are the Ulidiidae, Stratiomyidae, Chloropidae, and Syrphidae.

3. Diptera Families Absent from or Poorly Represented in the Neotropical Region

A number of Diptera families are absent in the neotropics, belonging to different levels in the phylogeny of the Diptera and as such to clades of different age. In the Tipulomorpha, all families are present, but the Cylindrotomidae are represented by a single species only. In the Bibionomorpha, the families Pachyneuridae and Bolitophilidae are entirely absent. Rangomaramidae, originally described including only the genus *Rangomarama* Jaschhof & Didham from New Zealand (Jaschhof & Didham 2002), was expanded by Amorim & Rindal (2007) to also include the Heterotrichinae — only with *Heterotricha* Loew — the Ohakuneinae (with *Ohakunea*

Tonnoir & Edwards and related genera), and Chiletrichinae (including *Chiletricha* and related genera). The Ohakuneinae and the Chiletrichinae have representatives in the Neotropical Region, so the Rangomaramidae certainly are present. *Hesperinus* Walker, ranked by some authors as a family, is absent in the neotropics, but it is here considered as the sister clade of the remaining bibionids and belonging to this family.

In the Blephariceromorpha, both families Nymphomyiidae and Deuterophlebiidae are absent in the neotropics, as well as the Axymyiidae, the only family of the Axymyiomorpha. In the Ptychopteromorpha, Ptychopteridae are represented by a single species. In the Psychodomorpha, Psychodidae, Scatopsidae, and Canthyloscelidae are known from the region, while the Perissommatidae are known from only a single species. Valeseguyidae, recently transferred to the Psychodomorpha (Amorim & Grimaldi 2006), are known in the region from a single Miocene Dominican amber fossil (Grimaldi 1991). None of the Culicomorpha families are absent in the neotropics.

The classification of the 'lower' Brachycera faces some instability because of the uncertain monophyly of taxa such as the Rhagionidae and Xylophagidae, divided by some authors into smaller taxa. The Spaniinae, with a single neotropic representative, are here considered as a subfamily of Rhagionidae. The Vermileonidae have three Neotropical species in Mexico and the Caribbean, but basically as a southern extension of a Nearctic clade. The Coenomyiidae and the Oreoleptidae are both absent in the neotropics. In the Xylophagidae, *Exeretonevra* Macquart (sometimes ranked as family) does not have Neotropical representatives, while *Xylophagus* Meigen has marginal species in Mexico belonging to a Nearctic clade.

In the Asiloidea, Apsilocephalidae have a single extant species in the southern Nearctic Region, that extends into southern areas of Nearctic affinities in Mexico. Similarly, the Apystomyiidae and Hilarimorphidae are known from some species in California and may extend their distribution slightly to the south into Nearctic Mexico. Evocoidae have been recently erected for a single species from Chile (Yeates *et al.* 2003, 2006).

There is still no consensus on the higher classification of the Empidoidea except that Empididae *s.l.* are paraphyletic in relation to the Dolichopodidae. Taxa referred to as families include Empididae *s.s.*, Hybotidae, Atelestidae, Brachystomatidae, Microphoridae, Oreogetonidae. Of these, only the Microphoridae (ranked by some authors as a subfamily of Dolichopodidae) are absent in the neotropics, while Atelestidae has two known Neotropical species.

In the Aschiza (which is possibly paraphyletic), Ironomyiidae, known in three species from Australia, and Opetiidae, endemic to the Palaearctic region, are absent in the neotropics. The Calyptratae families missing in the Neotropical Region are the Glossinidae, Mystacinobiidae, and Rhiniidae (the latter often as a subfamily of a non-monophyletic Calliphoridae). Within the Acalyptratae, the number of absent families is higher, including Nothybidae, Diopsidae, Acarthophthalmidae, Fergusoninidae, Opomyzidae, Xenasteidae, Australimyzidae, Mormotomyiidae, and Chyromyidae. *Braula schmitzi* Örösi Pal is possibly an introduced species in the neotropics, and the Braulidae are otherwise absent in the region. Helosciomyzidae, Homalocnemiidae, and Nannodastiidae are known from two species each in the neotropics, while Camillidae, Cryptochetidae, Dryomyzidae, Lonchopteridae, Megamerinidae, Strongylophthalmyiidae, and Tachiniscidae are known from a single species each.

4. Endemic or Near-Endemic Families in the Neotropics

Four extant dipteran families are endemic (or near-endemic) to the Neotropical Region. Pantophthalmidae, with the genera *Pantophthalmus* Thunberg, *Opetiops* Enderlein, and *Rhaphiorhynchus* Wiedemann, are widespread in the neotropics but absent in other regions. Evocoidae, as commented above, is known from a single species from Chile. Syringogastridae are known from nine species from southern Brazil to Mexico, and a single species reaching the southern parts of the Nearctic Region. The family Somatiidae has a single genus restricted to the Neotropical Region, with eight species extending from southern Brazil to Central America. In this context it seems worth noting the extinct family of Lower Brachycera, Cratomyiidae, known from a single species — *Cratomyia macrorrhyncha* Mazzarolo & Amorim — from the Lower Cretaceous of northeastern Brazil. The family belongs with the Stratiomyidae, Panthophthalmidae, and Xylomyiidae in the Stratiomyomorpha (Mazzarolo & Amorim 2000).

5. Fossil Dipterans in the Neotropics

Fossil dipterans in the Neotropical Region are known from three main deposits. The most important in terms of number of described species is the Oligo-Miocene amber of the Dominican Republic (see, e.g., Lambert *et al.* 1985, Grimaldi 1995a, Iturralde-Vincent & MacPhee 1996).

The Dominican amber fauna basically belongs to extant genera, with some notable exceptions. Grimaldi (1991), for example, described some extinct species of *Mycetobia* Meigen and *Mesochria* Enderlein (both Anisopodidae), extant species of which are found in northern Neotropical and southern Nearctic terranes, but also *Valeseguya disjuncta* Grimaldi (Valeseguyidae), which has a single Australian extant relative. Species have also been described from the families Psychodidae (Schlüter 1978), Limoniidae (Krzemiński 1992), Corethrellidae (Borkent & Szadziewski 1992), Ceratopogonidae (Szadziewski & Grogan 1994, 1997, 1998), Scatopsidae (Amorim 1998), Keroplatidae (Schmalfuss 1979), Tabanidae (Lane *et al.* 1988), Acroceridae (Grimaldi 1995b), Scenopinidae (Yeates & Grimaldi 1993), Asilidae (Scarbrough & Poinar 1992), Mythicomyiidae (Schlüter 1976, Evenhuis 2002), Empidoidea (Cumming & Cooper 1992), Phoridae (Brown 1999), Anthomyiidae (Michelsen 1996), Periscelididae (Grimaldi & Mathis 1993), and Carnidae (Grimaldi 1997). Fossil material of species from a number of different families, Bibionidae, Sciaridae, Cecidomyiidae, Rhagionidae, Acroceridae, Dolichopodidae, Phoridae, Muscidae, Tachinidae, Micropezidae, Ulidiidae, Lauxaniidae, Chloropidae, Odiniidae, Milichiidae, etc., have been referred to in the literature but are still undescribed (see, e.g., Poinar 1992, Brown 1992).

From Mexican amber, there are species described from the families Ceratopogonidae (Szadziewski & Grogan 1996), Psychodidae (Quate 1961), Bibionidae (Hardy 1971), Scatopsidae (Cook 1971), Stratiomyidae (James 1971), and Periscelididae (Sturtevant 1963), and undescribed species of Mycetophilidae, Sciaridae, and Cecidomyiidae (Gagné 1973, 1980), as well as Asilidae (Poinar 1992). Cockerell (1923) refers to fossil dipterans from Colombia.

Dipterans from the Lower Cretaceous in the Neotropical Santana Formation have been revised by Grimaldi (1990). Species of Chironomidae (Borkent 1993), Asilidae (Grimaldi 1990), and of the endemic family Cratomyiidae (Mazzarolo & Amorim 2002) have been formally described, but there are additional undescribed species of Limoniidae, Chironomoidea, Keroplatidae, Bibionidae, and Sciaroidea (D.S. Amorim, unpubl.).

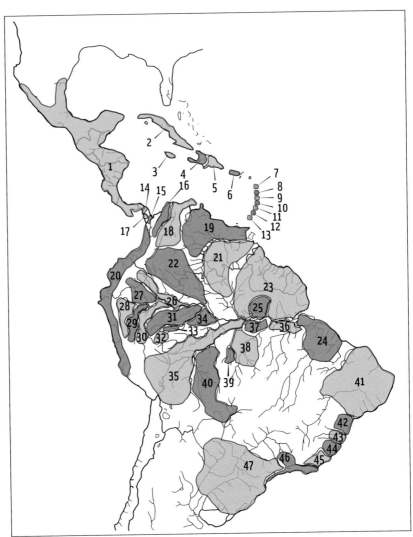

Figure 3.4. Simplified picture of main areas of endemism for groups of tropical forest environment in the Neotropical Region based on vertebrates, insects, and other groups (Amorim & Pires 1996). The mere existence and the limits of areas of endemism are always hypotheses that may be improved with additional studies. There are possibly additional areas, but there is insufficient data to attain a minimally reliable hypothesis.

6. Biogeographic Patterns in Neotropical Tropical Forests

Some species in the Neotropical Region are widely distributed, occupying large expanses of South and Central America. Most species, however, have a more restricted distribution, and these distributions are often matching or broadly overlapping with other such distributions. These 'areas of endemism' proposed by dispersionists, refuge theory biogeographers, and vicariant biogeographers and based on studies of as different groups as insects, arachnids, mammals, vascular plants, etc., are to a large extent congruent in their distribution and limits (Fig. 3.4).

The study of the relationship between these areas of endemism began to clarify the biogeographic history of the region, using mainly information on dipterans. Amorim (1987), Amorim & Pires (1996), and Amorim (2001, Figs 3–21) considered the relationships between species of the genera *Rhynchosciara* Rübsaamen (Sciaridae), *Rhipidita* Edwards and *Calliceratomyia* Lane (Ditomyiidae) as evidence of a general pattern of fragmentation of the biota in the region (Fig. 3.5). The generality of this pattern has been corroborated more recently by data for heteropterans, hymenopterans, lepidopterans, birds, primates, and cycads (Silveira 2003, Fernandes 1998, Campos 1999, Camargo & Pedro 1996, Racheli & Racheli 2004, Cracraft & Prum 1998). Even some freshwater fish groups, such as the genus *Roeboides* (Lucena 2003), in a certain sense mirror the patterns of terrestrial groups. This strongly suggests common causes for the origin of these patterns.

The congruence between individual biogeographic patterns points to a sequence of events with a first division between Caribbean and continental elements of the Neotropical Region. Later, the continental component was divided along a northeast/southwest axis through Amazonia. The northwestern component had a first separation between Central America and northwest South America, which later divided further into a southwestern area in the Amazonian Basin separated from another area including northern South America. Northern South America was then divided along the Orinoco Basin and by the more recent change in the opening of the Amazon River to the west (Petri & Fúlfaro 1983). The southeast component, on the other hand, separated southeast Amazonia from the Atlantic Forest (Fig. 3.6) before each of these areas underwent a number of more recent divisions. There is considerable overlap of the main South American geological barriers and these biogeographic components in the

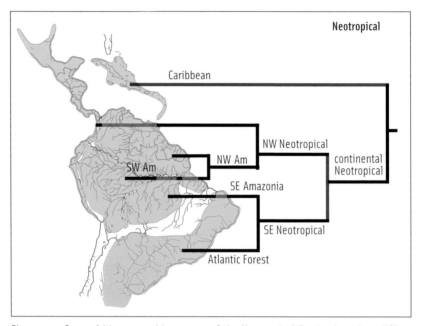

Figure 3.5. General biogeographic pattern of the Neotropical Region based on different groups of vertebrates, insects and plants. First vicariance event corresponds to the separation of the Caribbean arc from the continental Neotropical Region. Second event divides northwest South America, Central America and coastal Mexico (NW) from southeast South America (SE). Third event separates Central America and the Chocó regions from the Amazonian forest in the NW Neotropical component, and southeast Amazonia from the Atlantic Forest in the SE Neotropical component.

Neotropical Region. This reconstruction is entirely consistent with the proposal by Carvalho *et al.* (2004) that sea introgressions in the Amazonia led to the isolation of the freshwater stingrays Potamotrigonidae at least 50 million years ago. Summing up, ancient Gondwanan groups in the region have survived together with groups arriving later through Proto-Central America, and have been exposed to the same sequence of events occurring all through the late Cretaceous-Cenozoic.

Even though there is geological evidence that this sequence of divisions in the biogeographic pattern could have occurred in the Cretaceous (Amorim & Pires 1996), it seems more probable that these events have been generated during the Cenozoic. Rossetti & Góes (2001) point to important sea transgressions in South America in the Eocene, mid Miocene, late Miocene and in the Plio-Pleistocene (see also Räsänen *et al.* 1995).

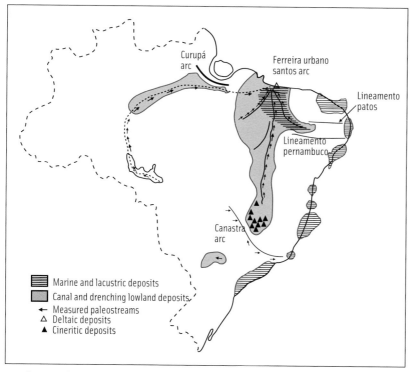

Figure 3.6. Major geological Cretaceous events in Brazil (Petri & Fúlfaro 1983).

Gondwanan and post-Gondwanan dipteran elements compose the present diversity of the Neotropical Region. A similar perspective also applies to the Afrotropical region (Jacob 2001) and the Australian region (Kitching *et al.* 2004). Congruent distributions in different groups of insects, vascular plants, and vertebrates, and the Cretaceous-Tertiary geological history of the Neotropical Region, show that a long history of vicariance-induced disjunctions due mostly to tectonic movement and sea transgressions may be assumed to have caused the distribution patterns of extant dipterans of the Neotropical Region. This is congruent with the demonstration that predictions of the refuge theory are not observable in the present diversity of the region (Amorim 1991).

A relevant question is which of the dipteran groups are actually Gondwanan and which arrived more recently due to dispersion, joining the groups in the subsequent vicariance events (Amorim & Silva 2002). A number of dipteran groups known from the Triassic and Lower Jurassic — including Tipulomorpha, Bibionomorpha, Culicomorpha, Psychodo-

morpha, and early Lower Brachycera families and genera — give palaeon-
tological support for a hypothesis of Pangaeic or Gondwanan origin for
the clades to which they belong, but higher dipteran lineages, especially in
the Cyclorrhapha (Grimaldi & Cumming 1999), may only secondarily occupy Gondwanan terranes in the Neotropical Region, having originated
elsewhere (Amorim *et al.* in press).

A more robust understanding of the evolution of dipteran diversity in
the Neotropical Region is dependent on progress in a number of activities: (1) sampling in areas never systematically collected; (2) sampling in
different kinds of environments; (3) establishment of sound phylogenies
at the species and genus level; (4) gathering detailed information about
relevant tectonics, orogenies, and sea transgressions; (5) the development
of analytical methods considering the complexity of the biogeographic
processes, including the effects of reiterative barriers.

7. Highly Diverse Areas in the Neotropical Region

There is still not enough comparative data that will allow for robust statements about differences in species diversity between areas within the Neotropical Region. The composition of the fauna varies between different
habitats, tropical forests probably being the richest. The traditional study
of Hammond (1992) points to an enormous diversity in Central America,
but this should be considered in comparison with other areas of the world.
The generic composition of the dipteran fauna in different tropical forests
is considerably similar, from Mexico to northern Argentina. However, biotic overlap may put focus on areas of particular richness. The overlap of
elements of the Nearctic and Neotropical faunas may be responsible for
the particular richness at higher levels in some parts of Mexico. Southern
Brazil (e.g., Nova Teutônia, intensely collected by Fritz Plaumann) also
has an overlap between the usual tropical fauna and circum-Antarctic elements, usually present in Chile. Colombia contains both tropical elements
and circum-Antarctic elements that dispersed to the north with the uplift
of the Andes (Oliveira *et al.* 2007). An area of particularly high species diversity worth mentioning is Serra do Mar in southeastern Brazil at higher
altitudes (e.g., at the Reserva Biológica de Boracéia, Salesópolis in the State
of São Paulo). This area has a particularly rich faunal composition because
of at least marginal overlap between different areas of endemism.

Diptera Diversity: Status, Challenges and Tools
(eds T. Pape, D. Bickel & R. Meier). © 2009 Koninklijke Brill NV.

8. Perspectives and Needs

Brown (2005) considered the problem of low activity and sparse taxonomic knowledge of the Diptera, with a special focus on the Neotropical Region. His comments fit into a larger discussion in the literature about the nature of systematics and of taxonomic information and its role in conservation (e.g., Cotterill 1995; Thiele & Yeates 2002; Knapp *et al.* 2002; McNeely 2002; Godfray 2002, 2007; Godfray & Knapp 2004; Janzen 2004; Blaxter 2004; M.R. Carvalho *et al.* 2005, 2007; Wheeler 2005; DeSalle *et al.* 2005; Ebach & Holdrege 2005; Evenhuis 2007; Santos & Amorim 2007; Miller 2007). On one hand, there is an urgent need for funding for taxonomy and for using available systematic information for the purposes of conservation of the habitats that are being rapidly and irreversibly destroyed (Carvalho *et al.* 2008). On the other hand, there is an urgent need for quality taxonomic information, since for whatever purpose taxa are strictly scientific hypotheses, and poor taxonomy will by no means help neither ecology nor conservation. In this context, some recent efforts made in dipterology are worth mentioning. The *BioSystematic Database of World Diptera* (Thompson 1990, Evenhuis *et al.* 2007) is certainly the most comprehensive, authoritative, taxonomic electronic database available on a megadiverse group on a worldwide scale. This database has in the last years been a source of information for both systematists and non-systematists. The extraordinary effort of general taxonomic research made in Costa Rica through INBio has resulted in exceptionally important collections of Diptera and has been catalytic, if not instrumental, for the organization of a *Manual of Central American Diptera* (Brown *et al.* 2008). The Colombia Arthropod Project (see Brown 2005) generated a major collection of Diptera from different altitudes and environments. Identification of this material extended the known distribution of a number of typically circum-Antarctic mycetophilid genera — such as *Neoaphelomera* Miller, *Allocotocera* Mik, *Eudicrana* Loew, *Neuratelia* Rondani, *Parvicellula* Marshall, *Stenophragma* Skuse, *Phthinia* Winnertz — to the extreme north of the Andes (Oliveira *et al.* 2007). In Brazil, a large collection made with Malaise traps along the Atlantic Forest by Carlos Roberto Ferreira Brandão and Eliana Cancello, from the Museu de Zoologia da USP, resulted in about 200,000 dipterans (unpublished data). At least half of the extension of the Atlantic Forest had never been sampled before for most dipteran groups. The study of this material, funded by the Fundação de

Amparo à Pesquisa do Estado de São Paulo, has resulted in a project of an electronic *Manual of Neotropical Diptera*, and a new journal, named *Neotropical Diptera* (Amorim & Papavero 2008). Finally, it is necessary to mention the role of the journal *Zootaxa*, which in the last few years has facilitated taxonomic output for zoology worldwide. Examples of large contributions to Neotropical dipterology published in *Zootaxa* are the catalogues of the Neotropical Muscidae and Fanniidae (C.J.B. Carvalho *et al.* 2003, 2005) and works on Ceratopogonidae (Borkent & Picado 2004) and Corethrellidae (Borkent 2008).

High quality systematic research (including both alpha-taxonomy and phylogenetically supported hypotheses), fulfilling the demands of taxonomic information from other parts of science and of the society, obviously depends on many issues. This begins with field work planned with geographical criteria and well preserved collections, but also the formation of a new generation of systematists, better availability of information, and peer-reviewed, fast publishing journals. Despite the crisis due to the pressure on the natural environments (e.g., Brown 2005) and the loss of positions for taxonomists in Europe and North America, dipterology has now available some new tools that allows high quality and fast dissemination of taxonomic information (e.g., Winterton 2009).

It is difficult to indicate priorities in terms of collection areas. Some parts of the Neotropical Region have been intensely collected: Costa Rica, through the recent efforts of INBio, some parts of Chile, especially from of the collections made by Luis Peña, and southern Brazil, due to the collection effort by Fritz Plaumann spanning several decades. The huge collection with Malaise traps recently made in Colombia has already been mentioned. When the material resulting from these collections is processed, identified, and described as needed, much more will be known about the diversity of Neotropical Diptera. But even these areas have species known from single specimens, and certainly many species even in these areas still remain to be collected. Areas of tropical forests in southeastern Brazil have been more intensely collected in the last twenty years, as well as some areas around Manaus in Amazonia. Most other areas have been poorly collected (including, e.g., Mexico, the Caribbean and Antillean islands, Ecuador, Venezuela, Guiana, the Brazilian state of Roraima, etc.), and in some areas practically nothing is known, which holds particularly for large parts of Amazonia, Peru, and Bolivia.

Acknowledgements

This manuscript benefited from suggestions, comments, and help from different colleagues to whom I am deeply indebted. Charles Morphy D. Santos and Adolfo R. Calor, of the Departamento de Biologia, Faculdade de Filosofia, Ciências e Letras de Ribeirão Preto, Universidade de São Paulo, carefully read earlier versions and made a number of useful comments. Guilherme C. Ribeiro, of the same institution, helped with discussion and useful literature. Brian Brown, Los Angeles County Museum, made useful suggestions, and Dan Bickel, Australian Museum, and Thomas Pape, Natural History Museum of Denmark, thoroughly revised the text, with very useful corrections and suggestions.

References

Allen, J.A. (1892) The geographical distribution of North American Mammals. *Bulletin of the American Museum of Natural History* 4: 199–243.

Amorim, D.S. (1987) *Refúgios quaternários e mares epicontinentais: Uma análise dos modelos, métodos e reconstruções biogeográficas da região Neotropical, incluindo o estudo de grupos de Mycetophiliformia (Diptera: Bibionomorpha).* Tese de Doutoramento, Universidade de São Paulo, São Paulo.

Amorim, D.S. (1989) A new species of *Diamphidicus* Cook (Diptera, Bibionomorpha, Scatopsidae) from Chile, with comments on the phylogenetic relationships of the genus. *Revista Brasileira de Entomologia* 33: 477–482.

Amorim, D.S. (1991) Refuge model simulations: Testing the theory. *Revista Brasileira de Entomologia* 35: 803–812.

Amorim, D.S. (1998) Amber fossil Scatopsidae (Diptera: Scatopsidae). I. Considerations on Scatopsidae described fossils, *Procolobostema roseni* sp.n. from Dominican amber and the position of *Procolobostema* in the family. *American Museum Novitates* 3227: 1–17.

Amorim, D.S. (2001) Dos Amazonias. Pages 245–255 *in*: Llorente-Bousquets, J. & Morrone, J.J. (eds), *Introducción a la biogeografia en Latinoamérica: teorías, conceptos, métodos y aplicaciones.* Facultad de Ciencias, UNAM, México, D.F.

Amorim, D.S. & Grimaldi, D. (2006) Valeseguyidae, a new family of Diptera in the Scatopsoidea, with a new genus in Cretaceous amber from Myanmar. *Systematic Entomology* 31: 508–516.

Amorim, D.S. & Papavero, N. (2008) A journal for the systematics and biogeography of Neotropical Diptera, 250 years after the publication of the *Systema Naturae.* *Neotropical Diptera* 1: 1–5.

Amorim, D.S. & Pires, M.R.S. (1996) Neotropical biogeography and a method for maximum biodiversity estimation. Pages 183–219 *in*: Bicudo, C.E.M. & Menezes, N.A. (eds), *Biodiversity in Brazil: A First approach.* CNPq, São Paulo.

Amorim, D.S. & Rindal, E. (2007) A phylogenetic study of the Mycetophiliformia, with creation of the subfamilies Heterotrichinae, Ohakuneinae, and Chiletrichinae for the Rangomaramidae (Diptera, Bibionomorpha). *Zootaxa* 1535: 1–92.

Amorim, D.S., Santos, C.M.D. & Oliveira, S.S. In press. Circumantarctic disjunctions, Gondwana, transoceanic dispersal. *Systematic Entomology.*

Amorim, D.S. & Silva, V.C. (2002) How far advanced was Diptera evolution in Pangaea. *Annales de la Société Entomologique de France* 38: 177–200.

Artigas, J.N., Papavero, N. & Amorim, D.S. (1992) On the puparium of *Tethina brasiliensis* Prado & Tavares (Diptera, Tethinidae). *Gayana Zoologia* 56(3-4): 127–129.

Blaxter, M.L. (2004) The promise of a DNA taxonomy. *Philosophical Transactions of the Royal Society of London B,* 359: 669–679.

Borkent, A. (1993) A world catalogue of fossil and extant Corethrellidae and Chaoboridae (Diptera), with a listing of references to keys, bionomic information and descriptions of each known life stage. *Entomologica Scandinavica* 24: 1–24.

Borkent, A. (2008) The Frog-Biting Midges of the World (Corethrellidae: Diptera). *Zootaxa* 1804: 1–456

Borkent, A. & Picado, A. (2004) Distinctive new species of *Atrichopogon* Kieffer (Diptera: Ceratopogonidae) from Costa Rica. *Zootaxa* 637: 1–68.

Borkent, A. & Szadziewski, R. (1992) The first records of fossil Corethrellidae. *Entomologica Scandinavica* 22: 457–63.

Brown, B.V. (1992) Generic revision of Phoridae of the Nearctic Region and phylogenetic classification of Phoridae, Sciadoceridae and Ironomyiidae (Diptera: Phoridea). *Memoirs of the Entomological Society of Canada* 164: 1–144.

Brown, B.V. (1999) Re-evaluation of the fossil Phoridae. *Journal of Natural History* 33: 1561–1573.

Brown, B.V. (2005) Malaise trap catches and the crisis in Neotropical Dipterology *American Entomologist* 51: 180–183.

Brown, B.V., Borkent, A., Cumming, J.M., Wood, D.M., Woodley, N.E. & Zumbado, M. (eds) (2008) *Manual of Central American Diptera*. Vol. 1. NRC Press, Ottawa, 752 pp.

Camargo, J.M. & Pedro, S. (1996) Meliponini neotropicais (Apinae, Apidae, Hymenoptera): Biogeografia histórica. *Anais do Encontro sobre Abelhas* 2: 107–121.

Campos, L.A. (1999) *Análise cladística de Ochlerini Rolston, 1981 e descrição de dois novos gêneros (Heteroptera: Pentatomidae)*. Tese de Doutoramento, Universidade de São Paulo, São Paulo.

Carvalho, C.J.B. de, Pont, A.C., Couri, M.S. & Pamplona, D. (2003) A catalogue of the Fanniidae (Diptera) of the Neotropical Region. *Zootaxa* 219: 1–32.

Carvalho, C.J.B. de, Couri, M.S., Pont, A.C., Pamplona, D. & Lopes, S.M. (2005) A catalogue of the Muscidae (Diptera) of the Neotropical Region. *Zootaxa* 860: 1–282.

Carvalho, M.R. de, Bockmann, F.A., Amorim, D.S., de Vivo, M., de Toledo-Piza, M., Menezes, N.A., de Figueiredo, J.L., Castro, R.M.C., Gill, A.C., McEachran, J.D., Compagno, L.J.V., Schelly, R.C., Britz, R., Lundberg, J.G., Vari, R.P. & Nelson, G. (2005) Revisiting the taxonomic impediment. *Science* 307: 353.

Carvalho, M.R. de, Bockmann, F.A., Amorim, D.S., Brandão, C.R.F, de Vivo, M., de Figueiredo, J.L., Britski, H.A., de Pinna, M.C.C., Menezes, N.A., Marques, F.P.L., Papavero, N., Cancello, E.M., Crisci, J.V., McEachran, J.D., Schelly, R.C., Lundberg, J.G., Gill, A.C., Britz, R., Wheeler, Q.D., Stiassny, M.L.J., Parenti, L.R., Page, L.M., Wheeler, W.C., Faivovich, J., Vari, R.P., Grande, L., Humphries, C.J., DeSalle, R., Ebach, M.C. & Nelson, G.J. (2007) Taxonomic impediment or impediment to taxonomy? A commentary on systematics and the cybertaxonomic-automation paradigm. *Evolutionary Biology* 34: 140–143.

Carvalho, M.R. de, Bockmann, F.A., Amorim, D.S. & Brandão, C.R.F. (2008) Systematics must embrace comparative biology and evolution, not speed and automation. *Evolutionary Biology* 35: 97–104.

Carvalho, M.R., Maisey, J.G. & Grande, L. (2004) Freshwater stringrays of the Green River Formation of Wyoming (Early Eocene), with the description of a new genus and species and an analysis of its phylogenetic relationships (Chondrichthyes: Myliobatiformes). *Bulletin of the American Museum of Natural History* 284: 1–136.

Cockerell, T.D.A. (1923) Insects in amber from South America. *American Journal of Science* 5: 331–333.

Colless, D.H. (1962) A new Australian genus and family of Diptera (Nematocera: Perissommatidae). *Australian Journal of Zoology* 10: 519–535.

Cook, E.F. (1971) Studies of the fossiliferous amber arthropods of Chiapas, Mexico. 2. Fossil Scatopsidae in Mexican amber (Diptera: Insecta). *University of California Publications on Entomology* 63: 57–61.

Coscaron, S. & Coscaron Arias, C.L. (2007) *Neotropical Simuliidae (Diptera: Insecta)*. Aquatic Biodiversity of Latin America. Vol. 3. Pensoft Publishers, Sofia-Moscow, 700 pp.

Cotterill, F.P.D. (1995. Systematics, biological knowledge and environmental conservation. *Biodiversity & Conservation* 4: 183–205.

Cracraft, J. & Prum, R.O. (1988) Patterns and processes of diversification: speciation and historical congruence in some neotropical birds. *Evolution* 42: 603–620.

Cumming, J.M. & Cooper, B.E. (1992) A revision of the Nearctic species of the tachydromiine fly genus *Stilpon* Loew (Diptera: Empidoidea). *Canadian Entomologist* 124: 951–998.

DeSalle, R., Egan, M.G. & Siddall, M. (2005) The unholy trinity: taxonomy, species delimitation and DNA barcoding. *Philosophical Transactions of the Royal Society of London B* 360: 1905–1916.

de Vivo, M. (1997) Mammalian evidence of historical ecological change in the Caatinga semi-arid vegetation of northeastern Brazil. *Journal of Comparative Biology* 2: 63–74.

Ebach, M.C. & Holdrege, C. (2005) DNA barcoding is no substitute for taxonomy. *Nature* 434: 697.

Evenhuis, N.L. (1994) *Catalogue of the fossil flies of the world (Insecta: Diptera)*. Backhuys Publishers, Leiden, 600 pp.

Evenhuis, N.L. (2002) Catalog of the Mythicomyiidae of the world (Insecta: Diptera). *Bishop Museum Bulletin in Entomology* 10: 1–85.

Evenhuis, N.L. (2007) Helping solve the "other" taxonomic impediment: completing the eight steps to total enlightenment and taxonomic Nirvana. *Zootaxa* 1407: 3–12.

Evenhuis, N.L., Pape, T., Pont, A.C. & Thompson, F.C. (eds.) 2007) *BioSystematic Database of World Diptera, Version 10*. Available at http://www.diptera.org/biosys.htm, accessed 20 January 2008.

Fernandes, J.A.M. (1998) *Análise cladística e revisão do gênero* Antiteuchus *Dallas 1851 (Heteroptera, Pentatomidae, Discocephalinae).* Tese de Doutoramento, Universidade de São Paulo, São Paulo.

Gagné, R.J. (1973) Cecidomyiidae from Mexican Tertiary amber (Diptera). *Proceedings of the Entomological Society of Washington* 75: 169–171.

Gagné, R.J. (1980) Mycetophilidae and Sciaridae (Diptera) in Mexican amber. *Proceedings of the Entomological Society of Washington* 82: 152.

Gentry, A.H. (1988. Tree species richness of upper Amazonian forests. *Proceedings of the National Academy of Sciences* 85: 156–159.

Gill, T. (1885) The principles of zoogeography. *Proceedings of the Biological Society of Washington* 2: 1–23.

Godfray, H.C.J. (2002) Challenges for taxonomy. *Nature* 417: 17–19.

Godfray, H.C.J. (2007) Linnaeus in the information age. *Nature* 446: 259–260.

Godfray, H.C.J. & Knapp, S. (2004) Introduction. *Philosophical Transactions of the Royal Society of London B* 359: 559–569.

Grimaldi, D. (1990) Diptera. *In*: Grimaldi, D. Insects from the. Santana Formation, Lower Cretaceous of Brazil. *Bulletin of the American Museum of Natural History* 195: 164–183.

Grimaldi, D. (1991) Mycetobiine Woodgnats (Diptera: Anisopodidae) from the Oligo-Miocene Amber of the Dominican Republic, and Old World Affinities. *American Museum Novitates* 3014, 24 pp.

Grimaldi, D.A. (1995a) The age of Dominican amber. Pages 203–217 *in*: Anderson, K.B. & Crelling, J.C. (eds), *Amber, resinites, and fossil resins.* ACS Symposium Series 617, American Chemical Society, Washington, DC.

Grimaldi, D.A. (1995b) A remarkable new species of *Ogcodes* (Diptera: Acroceridae) in Dominican amber. *American Museum Novitates* 3127, 8 pp.

Grimaldi, D. (1997) The bird flies, genus *Carnus*: Species revision, generic relationships, and a fossil *Meoneura* in amber (Diptera: Carnidae). *American Museum Novitates* 3190, 30 pp.

Grimaldi, D.A. & Cumming, J.M. (1999) Brachyceran Diptera in Cretaceous ambers and Mesozoic diversification of the Eremoneura. *Bulletin of the American Museum of Natural History* 239: 1–124.

Grimaldi, D.A. & Mathis, W.N. (1993) Fossil Periscelididae (Diptera). *Proceedings of the Entomological Society of Washington* 95: 383–403.

Haffer, J. (1978) Distribution of Amazon forest birds. *Bonner Zoologische Beiträge* 29: 38–78.

Hammond, P.M. (1992) Uncharted realms of species richness. Pages 26–39 *in*: Groombridge, B. (ed.), *Global biodiversity: status of the Earth's living resources.* World Conservation Monitoring Centre, Chapman and Hall, London.

Hardy, D.E. (1971) A new *Plecia* (Diptera: Bibionidae) from Mexican amber. *University of California Publications in Entomology* 63: 65–67.

Hunter, W.D. (1900–1901) A catalogue of the Diptera of South America. Part I. Bibliography and Nematocera. Part II. Homodactyla and Mydiadae [*sic*]. *Transactions of the American Entomological Society* 26: 260–298 (1900); 27: 121–155 (1901).

Iturralde-Vincent, M. & MacPhee, R.D.E. (1996) Age and paleogeographic origin of Dominican amber. *Science* 273: 1850–1852.

Jacob, U. (2001. Africa and its Ephemeroptera: Remarks from a biogeographic view. Pages 317–325 *in*: Gaino, E. (ed.), *Research update on Ephemeroptera & Plecoptera*. Perugia, Università di Perugia.

James, M.T. (1971) A stratiomyid fly (Diptera) from the amber of Chiapas, Mexico. *University of California Publications in Entomology* 63: 71–73.

Janzen, D.H. (2004) Now is the time. *Philosophical Transactions of the Royal Society of London B* 359: 731–732.

Jaschhof, M. & Didham, R.K. (2002) Rangomaramidae fam. nov. from New Zealand and implications for the phylogeny of the Sciaroidea (Diptera: Bibionomorpha). *Studia dipterologica Supplement* 11: 1–60.

Jeannel, R. (1942. *La genèse des faunes terrestres. Éléments de biogéographie.* Prèsses Universitaires de France, Paris, 514 pp.

Kitching, R.L., Bickel, D., Creagh, A.C., Hurley, K. & Symonds, C. (2004) The biodiversity of Diptera in Old World rain forest surveys: a comparative faunistic analysis. *Journal of Biogeography* 31: 1185–1200.

Knapp, S., Bateman, R.M., Chalmers, N.R., Humphries, C.J., Rainbow, P.S., Smith, A.B., Taylor, P.D., Vane-Wright, R.I. & Wilkinson, M. (2002) Taxonomy needs evolution, not revolution. *Nature* 419: 559.

Krzemiński, W. (1992) Triassic and Lower Jurassic stage of Diptera evolution. *Mitteilungen der Schweizerischen Entomologischen Gesellschaft* 65: 39–59.

Krzemiński, W. & Krzemiński, E. (2004) Triassic Diptera: descriptions, revisions and phylogenetic relations. *Acta Zoologica Cracoviensia* 46: 153–184.

Lambert, J.B., Frye, J.S. & Poinar, G.O., Jr. (1985) Amber from the Dominican Republic: an analysis by nuclear magnetic resonance spectroscopy. *Archaeometry* 27: 43–51.

Lane, R.S., Poinar, G.O., Jr. & Fairchild, G.B. (1988) A fossil horsefly (Diptera: Tabanidae) in Dominican amber. *Florida Entomologist* 71: 593–96.

Lucena, C.A.S. (2003) Revisão taxonômica e relações filogenéticas das espécies de *Roeboides* grupo-*microlepis* (Ostariophysi, Characiformes, Characidae). *Iheringia, Série Zoologia* 93: 283–308.

Lydekker, B.A. (1896) *A geographical history of mammals.* Cambridge University Press, Cambridge, 400 pp.

Lynch Arribálzaga, E. (1883) Catálogo de los dípteros hasta ahora descritos que se encuentran en las repúblicas del Río de La Plata. *Boletin de la Academia Ciencias Naturales, Córdoba* 4: 109–152.

Mazzarolo, L.A. & Amorim, D.S. (2000) *Cratomyia macrorrhyncha*, a Lower Cretaceous brachyceran fossil from the Santana Formation, Brazil, representing a new species, genus and family of the Stratiomyomorpha (Diptera). *Insect Systematics and Evolution* 31: 91–102.

McNeely, J.A. (2002) The role of taxonomy in conserving biodiversity. *Journal of Nature Conservation* 10: 145–153.

Mello-Silva, R. de. (2005) Morphological analysis, phylogenies and classification in Velloziaceae. *Botanical Journal of the Linnean Society* 148: 157–173.

Michelsen, V. (1996) First reliable record of a fossil species of Anthomyiidae (Diptera), with comments on the definition of recent and fossil clades in phylogenetic classification *Biological Journal of the Linnean Society* 58: 441–451.

Miller, S.E. (2007) DNA barcoding and the renaissance of taxonomy. *Proceedings of the National Academy of Sciences* 104: 4775–4776.

Morrone, J. (2004) La zona de transición sudamericana: caracterización y relevancia evolutiva. *Acta Entomologica Chilena* 28: 41–50.

Munroe, D.D. (1974) The systematics, phylogeny, and zoogeography of *Symmerus* Walker and *Australosymmerus* Freeman (Diptera: Mycetophilidae: Ditomyiinae). *Mémoirs of the Entomological Society of Canada* 92: 1–183.

Nores, M. (1999) An alternative hypothesis for the origin of Amazonian bird diversity. *Journal of Biogeography* 26: 475–485.

Oliveira, S.J. (1950a) Sobre duas novas espécies neotrópicas do gênero *Telmatogeton* Schiner, 1866 (Diptera, Chironomidae). *Memórias do Instituto Oswaldo Cruz* 48: 470–485.

Oliveira, S.J. (1950b) Sobre uma nova espécie neotrópica do gênero *Clunio* Haliday, 1855 (Diptera, Chironomidae). *Revista Brasileira de Biologia* 10(4): 493–500.

Oliveira, S.S., Silva, P.C.A. & Amorim, D.S. (2007) Neotropical, circum-antarctic and neartic overlap? Mycetophilidae (Diptera) of Colombia and its biogeographic implications. *Darwiniana* 45(Supplement): 106–107.

Papavero, N. (1966–1984) *A catalogue of the Diptera of the Americas south of the United States.* Museu de Zoologia, Universidade de São Paulo, São Paulo, 2877 pp. [102 separate fascicles.]

Papavero, N. (1971) *Essays on the history of Neotropical Dipterology, with special reference to collectors (1750-1905).* Vol. 1. Museu de Zoologia, Universidade de São Paulo, São Paulo, vii + 216 pp.

Papavero, N. (1973) *Essays on the history of Neotropical Dipterology, with special reference to collectors (1750-1905).* Vol. 2. Museu de Zoologia, Universidade de São Paulo, São Paulo, iii + 217–446.

Papavero, N. & Guimarães, J.H. (2000) The taxonomy of Brazilian insect vectors of transmissible diseases (1900–2000) — then and now. *Memorias do Instituto Oswaldo Cruz* 95 (Supplement 1): 109–118.

Petri, S. & Fúlfaro, V.J. (1983) *Geologia do Brasil (Fanerozóico).* T.A. Queiroz, Editor & EDUSP, São Paulo, 631 pp.

Poinar, G.O., Jr. (1992) *Life in amber.* Stanford University Press, Stanford, xiii + 350 pp.

Prichard, J.C. (1826) *Researches into the physical history of the mankind.* 2nd ed, John and Arthur Arch, London, 2 vols.

Quate, L.W. (1961) Fossil Psychodidae (Diptera: Insecta) in Mexican amber; part I. *Journal of Paleontology* 35: 949–951.

Racheli, L. & Racheli, T. (2004) Patterns of Amazonian área relationships based on raw distributions of papilionid butterflies (Lepidóptera: Papilionidae). *Biological Journal of the Linnean Society* 82: 345–357.

Räsänen, M.E., Linna, A.M., Santos, J.C.R. & Negri, F.R. (1995) Late Miocene tidal deposits in the Amazonian foreland basin. *Science* 269: 386–390.

Robbins, R.K. & Opler, P.A. (1997) Butterfly diversity and a preliminary comparison with bird and mammal diversity. Pages 69–82 *in*: Reaka-Kudla, M.L., Wilson, D.E. & Wilson, E.O. (eds), *Biodiversity II: Understanding and Protecting our Biological Resources.* Joseph Henry Press, Washington, DC.

Rossetti, D.F. & Góes, A.M. (2001) Imaging Upper Tertiary to Quaternary deposits from northern Brazil applying ground penetrating radar. *Revista Brasileira de Geociências* 31: 195–202.

Santos, C.M.D. & Amorim, D.S. (2007) Why biogeographic hypotheses need a well supported phylogenetic framework: a conceptual evaluation. *Papéis Avulsos do Museu de Zoologia* 47: 63–73.

Scarbrough, A.G. & Poinar, G.O., Jr. (1992) Upper Eocene robber flies of the genus Ommatius (Diptera: Asilidae) in Dominican amber. *Insecta Mundi* 6: 13–18.

Schlüter, T. (1976) Die Wollschweber-Gattung *Glabellula* (Diptera: Bombyliidae) aus dem oligozänen Harz der Dominikanischen Republik. *Entomologica Germanica* 2: 355–363.

Schlüter, T. (1978) Die Schmetterlingsmücken-Gattung *Nemopalpus* (Diptera: Psychodidae) aus dem oligozänen Harz der Dominikanischen Republik. *Entomologica Germanica* 4: 242–249.

Schmalfuss, H. (1979) *Proceroplatus hennigi* n. sp., die erste Pilzmücke aus dem Dominikanischen Bernstein (Stuttgarter Bernsteinsammlung: Diptera, Mycetophiloidea, Keroplatidae). *Stuttgarter Beiträge zur Naturkunde (B)* 49: 1–9.

Sclater, P.L. (1858) On the general geographic distribution of the members of the Class Aves. *Journal and Proceedings of the Linnean Society, Zoology, London* 2: 130–145.

Silveira, F.G. (2003) *Revisão sistemática e análise cladística do gênero Arniticus Pascoe, 1881 (Coleoptera, Curculionidae).* Tese de Doutoramento, Universidade de São Paulo, São Paulo.

Simpson, G.G. (1980) *Splendid isolation: the curious history of South American mammals.* Yale University Press, New Haven, ix + 266 pp.

Sturtevant, A.H. (1963) A fossil periscelid (Diptera) from the amber of Chiapas, Mexico. *Journal of Paleontology* 37: 121–122.

Szadziewski, R. & Grogan, W.L., Jr. (1994) Biting midges from Dominican amber. I. A new fossil species of *Baeodasymyia* (Diptera: Ceratopogonidae). *Proceedings of the Entomological Society of Washington* 96: 219–229.

Szadziewski, R. & Grogan, W.L., Jr. (1996) Biting midges (Diptera: Ceratopogonidae) from Mexican amber. *Polskie Pismo Entomologiczne* 65: 291–295.

Szadziewski, R. & Grogan, W.L., Jr. (1997) Biting midges from Dominican amber. II. Species of the tribes Heteromyiini and Palpomyiini (Diptera: Ceratopogonidae). *Memoirs of the Entomological Society of Washington* 18: 254–260.

Szadziewski, R. & Grogan, W.L., Jr. (1998) Biting midges from Dominican amber. III. Species of the tribes Culicoidini and Ceratopogonini (Diptera: Ceratopogonidae). *Insecta Mundi* 12: 39–52.

Thiele, K. & Yeates, D. (2002) Tension arises from duality at the heart of taxonomy. *Nature 419*: 337.

Thompson, F.C. (1990) Biosystematic Information — Dipterists ride the Third Wave. Pages 179–201 *in*: Kosztarab, M. & Schaefer, C.W. (eds), *Systematics of the North American Insects and Arachnids: Staus and Needs*. Virginia Agricultural Experiment Station Information Series 90-1.

Thompson, F.C. (2009) Nearctic Diptera: Twenty years later. Pages 3-46 *in*: Pape, T., Bickel, D. & Rudolf, M. (eds), *Diptera Diversity: Status, Challenges and Tools*. Brill, Leiden.

Wallace, A.R. (1876) *The geographical distribution of animals; with a study of the relations of living and extinct faunas as elucidating the past changes of the Earth's surface*. Vol. 1. MacMillan & Co., London; 503 pp.

Wheeler, Q.D. (2005) Losing the plot: DNA "barcodes" and taxonomy. *Cladistics* 21: 405–407.

Winterton, S. (2009) Bioinformatics and Dipteran Diversity. Pages 381-407 in: Pape, T., Bickel, D. & Rudolf, M. (eds), *Diptera Diversity: Status, Challenges and Tools*. Brill, Leiden.

Yeates, D.K. & Grimaldi, D.A. (1993) A new *Metatrichia* window fly (Diptera, Scenopinidae) in Dominican amber: with a review of the systematics and biogeography of the genus. *American Museum Novitates* 3078, 8 pp.

Yeates, D.K., Irwin, M.E. & Wiegmann, B.M. (2003) Ocoidae, a new family of asiloid flies (Diptera: Barachycera: Asiloidea), based on *Ocoa chilensis* gen. and sp. n. from Chile, South America. *Systematic Entomology* 28: 417–431.

Yeates, D.K., Irwin, M.E. & Wiegmann, B.M. (2006) Evocoidae (Diptera: Asiloidea), a new family name for Ocoidae, based on *Evocoa*, a replacement name for the Chilean genus *Ocoa* Yeates, Irwin, and Wiegmann 2003. *Systematic Entomology* 31: 373.

CHAPTER FOUR

DIPTERAN BIODIVERSITY OF THE GALÁPAGOS

Bradley J. Sinclair

Canadian Food Inspection Agency, Ottawa, Canada

Introduction

The Galápagos Archipelago (Ecuador) is famous worldwide for its unique flora and fauna and its contribution to the understanding of evolution. The islands are now one of the most intensively studied regions in the Neotropics, especially the marine biota, seed plants and vertebrate animals.

Part 1. Biogeography and Diversity of the Galápagos Diptera Fauna

1. Physical Environment

The Galápagos Islands are located 900–1,000 km west of the Ecuadorian coast, astride the equator. This tropical oceanic archipelago is large and complex, with 127 islands, including 19 large islands and some 108 smaller islets and many unnamed rocks (Fig. 4.1). Individual island areas and elevations are summarised by Peck (2001, table 1.1). The islands are entirely volcanic in origin, with the present emerged islands considered to have existed for 3–4 My (million years), where the southeastern islands are the oldest (Española and San Cristóbal) and the central islands of Santa Cruz, Floreana and others were available for colonisation from about 0.7–1.5 Mya (million years ago), and Isabela, Santiago and Fernandina appeared less than 0.7 Mya.

Volcanic activity remains on some islands (e.g., Isabela and Fernandina), where vegetated regions are separated by large expanses of relative-

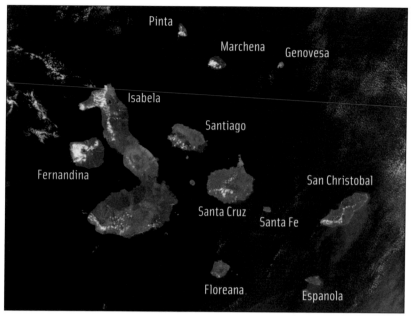

Figure 4.1. The principal islands of the Galápagos Archipelago,
with island contours (at 500 m intervals)
(NASA).

ly young barren larva (Fig. 4.2). Several of the volcanoes extend beyond
1,000 m in altitude; such as, Sierra Negra on Isabela, with its enormous
crater (9 x 7.2 km and 110 m deep). The islands are considered to be truly
oceanic and have never been connected directly to the mainland.

Drowned seamounts have been discovered with beach-worn cobbles
on their summits east of the Galápagos on the Carnegie Ridge (Christie
et al. 1992). This provides evidence that earlier volcanic islands existed
5–9 Mya or even earlier. The hotspot volcanism in the Galápagos region
has occurred for the past 15–20 My and Christie *et al.* (1992) considers it
likely that islands existed throughout the entire 80–90 My history of the
hotspot.

1.1 *Natural environment*

The islands are not at all what one may picture of tropical oceanic islands.
Upon arrival on the shores of the islands, one is struck by the harsh dry
lands, dominated by tall cacti and sparsely foliated trees (Figs 4.3, 4.4).

Figure 4.2. View north from Sierra Negra, Isla Isabela.

This harsh climate is strongly reflected in the insect fauna, which is generally viewed as depauperate, drab and usually quite small in size. The islands possess a strong or compressed floristic zonation, passing through six major vegetational zones in an elevational rise of only about 700 m (Jackson 1993). The Dipteran diversity also demonstrates zonation, although somewhat broader and less specific (Table 4.1).

Figure 4.3. Drought conditions, Isla Santa Fé.

Figure 4.4. Cactus forest, Charles Darwin Research Station, Isla Santa Cruz.
(Photo B. Landry.)

The littoral zone is characterised by high saline lagoons, sand beaches and intertidal rocks and extends 10–100 m inland. Species adapted to the littoral zone would be expected to have a high salt tolerance. Consequently, these species could be expected to have higher survival capability during colonisation events via rafting on vegetation or other flotsam (Wheeler

Figure 4.5. *Scalesia* forest, Isla Santa Cruz.

DIPTERA DIVERSITY: STATUS, CHALLENGES AND TOOLS
(EDS T. PAPE, D. BICKEL & R. MEIER). © 2009 KONINKLIJKE BRILL NV.

& Sinclair 1994). There is high species diversity in the littoral zone (Table 4.1; see also Bickel & Sinclair 1997), with at least 59 species restricted to the littoral zone, including 38 endemics. There is distinctly much greater littoral endemism compared to other vegetational zones.

The arid zone dominates all islands and in many cases comprises the remaining area of small or low islands. This zone is extremely hot and dry, with extended periods of drought and includes expanses of cactus forests (Figs 4.3, 4.4). The species diversity here is lowest, compared to the other two zones (Table 4.1). In contrast, the greatest number of species of Coleoptera, including endemics, occurs in the arid zone (Peck & Kukalová-Peck 1990; Peck 2006).

Table 4.1. Number of Diptera species by distributional categories and their occurrence in the different vegetation life zones of the Galápagos. The category 'Restricted' refers to species confined to a single life zone (note that many species occur in more than one life zone).

	Littoral	Arid	Humid
Single-island Endemics	11	11	10
Archipelago Endemics	38	28	24
Native (Indigenous)	33	28	51
Introduced (Adventive)	7	37	33
Totals	89	76	118
Restricted	59 (66%)	33 (43%)	51 (43%)

The humid zone (above 300 m) includes areas thickly covered with evergreen shrubs and trees such as *Miconia* and *Scalesia* (Fig. 4.5) and a pampa region dominated by tree ferns, sedges and bracken ferns. Treeline for the native and endemic plants is around 700 m, in contrast to a treeline at about 3,500 m in the Andes at the same latitude (Peck 2001). Unfortunately introduced trees, such as cinchona and guava, are spreading throughout the pampa regions and pose a serious threat to the native biota of these highlands. In addition, the humid zone of Floreana, southeastern Isabela, Santa Cruz and San Cristóbal have been extensively altered for agriculture and house a great variety of non-native plants. These alterations have also greatly modified the fly fauna. The greatest number of species is found in the humid zone; however, this number includes a large number of human introductions associated with agriculture.

Figures 4.6–4.9.
Examples of Galápagos Diptera.
4.6. *Condylostylus longicornis* (Fab.) (Dolichopodidae).
4.7. *Odinia williamsi* Johnson (Odiniidae).
4.8. *Euxesta galapagensis* Curran (Ulidiidae).
4.9. *Pareuxesta academica* Steyskal (Ulidiidae)
Scale bars = 1 mm.

There are two main seasons, a hot January to June phase with periodic heavy rains and warm ocean currents. From June to December it is cooler, with overcast skies and virtually no precipitation in the lowlands, while the highlands are almost continually wet (garua) and the oceans are cooler.

2. Origin of the Galápagos Fauna

It is estimated that there are 3,500 species of plants and animals present on the Galápagos (Peck 2001). Given the islands' volcanic origins, these species have evolved from successful terrestrial colonists that have had to cross a broad water barrier. The primary modes of dispersal to the islands include via the oceans (rafting, floating or drifting with the currents) and air currents (passive, active, or phoresy on birds) (Zimmerman 1948; Peck & Kukalová-Peck 1990; Peck 1996, 2006). It has been estimated that 60% of the Galápagos beetle colonists arrived via the oceans (Peck & Kuka-lová-Peck 1990). Peck (1996, 2001) estimates some 200 colonisations or founding events for the Galápagos Diptera fauna.

Aerial and sea-surface dispersal between the Galápagos Islands was investigated by Peck (1994a,b). This study documented both active and passive movement of insects and nematocerous Diptera dominated the catches. Several samples included exceptionally high numbers (several thousand specimens) of an undetermined species of Sciaridae.

3. Diversity of the Galápagos Diptera Fauna

As in most island faunas, the Diptera fauna of the Galápagos is unbalanced or disharmonic compared to the mainland fauna (Peck 1996). The diversity, ecology and vagility of the Diptera in the Neotropical source area determine which of them could and could not arrive and survive on the Galápagos. For example, the absence or poor representation of many aquatic groups is because most of the islands lack permanent water. An extensive review of the fauna of inland waters, including Diptera, was presented by Gerecke *et al.* (1995). Additional notable absences among terrestrial families include the Anthomyiidae, Lauxaniidae and Rhagionidae. The Mycetophiloidea are poorly represented, rarely occurring in samples.

The diversity of the Galápagos Diptera fauna is summarised in Table 4.2. A total of 49 families, 187 genera and 294 species are presently recorded. Of this total there are five endemic genera (families: Cecidomyiidae, Ulidiidae, Sarcophagidae and Tachinidae), 77 native (indigenous, also recorded from mainland Americas) and 95 endemic species (Figs 4.7–4.9). Some 32% of the species are considered to be endemic. The endemic beetle fauna is much greater (54%), with most endemism occurring in flightless forms (Peck 2006). The reduced dispersal ability of these groups between islands is considered to have promoted their differentiation (Peck & Kukalová-Peck 1990).

Table 4.2. Summary of the Diptera of the Galápagos Archipelago
(*recorded from Galápagos and Cocos Islands)

Family	Genera		Species				
	Total	Endemic	Total	Introduced	Native	Endemic	Estimated undetermined
Limoniidae	3		3		2	1	
Mycetophilidae	2		2				2
Sciaridae	4		5	4			1
Cecidomyiidae	6+	1	6+		1	2	3+
Psychodidae	4		6	3			3
Scatopsidae	5		5	1			4
Culicidae	3		3	2	1		
Simuliidae	1		1	1			
Chironomidae	7		8		3	1	4
Ceratopogonidae	3		11	2	4	4	1
Tabanidae	1		1			1	
Stratiomyidae	5		5	1		4	
Mythicomyiidae	1		1	1			
Bombyliidae	4		6		2	3	1
Scenopinidae	1		2			1	1
Asilidae	2		2			1	1
Hybotidae	3		5			1	4
Dolichopodidae	8		18		6	11	1
Phoridae	5		12	9		2	1
Syrphidae	7		9	1	4	3	1
Pipunculidae	2		2		1		1
Neriidae	1		1	1			
Lonchaeidae	2		2	1			1
Ulidiidae	4	1*	19	4		13	2
Tephritidae	4		7	1		4	2
Piophilidae	1		1	1			
Odinidae	1		1			1	
Agromyzidae	5		5	1	4		

Diptera Diversity: Status, Challenges and Tools
(eds T. Pape, D. Bickel & R. Meier). © 2009 Koninklijke Brill NV.

Family	Genera		Species				
	Total	Endemic	Total	Introduced	Native	Endemic	Estimated undetermined
Anthomyzidae	1		1				1
Periscelididae	1		1				1
Asteiidae	3		5			4	1
Milichiidae	3		4	2			2
Carnidae	1		1			1	
Sepsidae	1		1	1			
Trixoscelididae	1		1			1	
Chyromyidae	1		3		1	2	
Sphaeroceridae	14		19	3	10	2	4
Drosophilidae	3		17	9	5	2	1
Ephydridae	14		16	3	11	2	
Chloropidae	9		13	2		10	1
Tethinidae	1		3		2	1	
Canacidae	4		12		2	10	
Fanniidae	1		1	1			
Muscidae	12		12	8	2	1	1
Calliphoridae	2		5	2	1	2	
Sarcophagidae	8	2	14	6	2	6	
Tachinidae	7	1	8		2	1	5
Hippoboscidae	4		7		6	1	
Nycteribiidae	1		1		1		
Total 49	187	5	294	71	77	95	51

Overall, it is estimated that 53% of the terrestrial invertebrate species are endemic; consequently the Diptera fauna is significantly below this level. The lower Dipteran diversity could be due to the arid conditions, the large number of introduced species included in the total number of species, or also due to poor knowledge of the plant-feeding Diptera. Of the flies, 41 genera have been introduced; of the remaining 147 genera, 116 (79%) contain only one native or endemic species and 31 (21%) contain more than one. This is very similar to that estimated in the Heteroptera (23%) and Coleoptera (24%) (Peck & Kukalová-Peck 1990). These values have been interpreted to mean little speciation has occurred and represent relatively recent colonisation.

Only a few examples in Diptera display adaptive radiation or evolution of several species from a single colonist along the scale of the Galápagos tortoises and Darwin's finches. The best-documented case is the genus *Asyndetus* Loew (Dolichopodidae) (Bickel & Sinclair 1997). Other examples will certainly be documented in the endemic genus *Pareuxesta* Coquillett (Ulidiidae) (Fig. 4.9) and the Canacidae, but it remains to be shown whether the endemic species have evolved from single or multiple colonisation events. There are certainly no examples of the huge species radiations that have occurred on the Hawaiian Islands and no endemic flightless Diptera are known. Only wingless females of the endemic *Clunio*

Table 4.3. Comparison of the diversity of some island faunas of Diptera (n.a. = data not available)

	Present total area[6] (km²)	Number of families	Total number of genera	Number of endemic genera	Total number of species	Number of endemic species	% endemism
Continental Shelf Islands							
British Isles[1]	315,134	102	n.a.	0	7002	127	1.8
Oceanic Atlantic Archipelagos							
Bermuda[2]	54	44	174	0	258	17	6.6
Madeira[3]	790	63	n.a.	1	546	80	15
Oceanic Pacific Archipelagos							
Galápagos	7,856	49	187	5	294	95	32
Hawaii[4]	16,615	58	338	9	1518	1108	73
Juan Fernandez[5]	85	n.a.	n.a.	2	193	102	53

[1] Chandler (1998, 2008)
[2] Woodley & Hilburn (1994)
[3] Baez (1993) and Borges *et al.* (2008)
[4] http://www2.bishopmuseum.org/HBS/checklist/query.asp?grp=Arthropod, Eldredge & Evenhuis (2003) and Evenhuis (2008)
[5] Kuschel (1963)
[6] from Peck (2001).

Haliday (Chironomidae) and the introduced genera *Chonocephalus* Wandolleck and *Puliciphora* Dahl (Phoridae) occur.

Compared to other island faunas of the Pacific, the number of Galápagos Diptera is significantly lower. The Hawaiian Island chain has a tremendous biodiversity with more than 1,518 species (Table 4.3) (Evenhuis 2009). The Juan Fernandez Islands (650 km east of mainland) have a similar number of endemic species to the Galápagos, but an updated list of species is required for the latter group of islands before more detailed comparisons can be made. However, these islands have a land area of only 85 km², compared to 7,856 km² for the Galápagos. The Juan Fernandez Islands are more humid and did not experience a past arid period. Consequently the low diversity on the Galápagos is probably due to its overall aridity, distance from the mainland and difficulty for successful establishment of naturally dispersed taxa, and to the loss or diminished upper humid zone that was experienced during the late Pleistocene (Colinvaux 1972). Peck (2001, 2006) viewed the beetle fauna to be much lower than expected for islands of their size, and also considered that the harsh aridity was a major factor. Several groups of interest are discussed further below, organized by family.

3.1 Calliphoridae

There are two endemic species, which appear not affected by introduced species, but their phenology is unknown and assessments have not been made.

Known as the 'secondary screw-worm fly', *Cochliomyia macellaria* (Fabricius) was first recorded from the islands in 1849 (as *Musca phauda* Walker), but the last specimens currently known were collected in February and March 1970 and none were collected during the Peck expeditions starting in 1985. Probably the native Galápagos populations of *C. macellaria* were suppressed with the arrival of two other introduced species (*Chrysomya albiceps* (Wiedemann) and *C. megacephala* (Fabricius)), but detailed surveys of the Calliphoridae fauna are required to verify these observations. The latter two species were first collected from the Galápagos Islands during the 1985 and 1989 Peck expeditions, respectively (Causton *et al.* 2006). On the South American mainland, populations of *C. macellaria* were suppressed as populations of *Chrysomya albiceps* and *Ch. megacephala* increased (Baumgartner & Greenberg 1984, Wells & Kurahashi 1997). Throughout the New World, studies have shown that as populations of introduced species of *Chrysomya* increase, sympatric populations of the native calliphorid *C. macellaria* decrease (Wells & Kurahashi 1997).

3.2 Dolichopodidae

This family is one of the most diverse in the Galápagos, with eight genera, 18 species and eleven endemic species. The species are distributed in five subfamilies and ten species are confined to the littoral and/or arid zones.

Two endemic species (*Asyndetus versicolor* Johnson and *Paraclius desenderi* Bickel & Sinclair) are also recorded from Cocos Island (Costa Rica). Eight of 11 endemic species (including those species also found on Cocos Is.) are believed to have evolved from single colonisation events. In other words, their closest relative is living elsewhere than in the Galápagos and the ancestor of each arrived independently. The remaining three endemic species appear to be derived from the broadly distributed *Asyndetus tibialis* (Thomson), comprising *A. maelfaiti* Bickel & Sinclair from the outer island of Genovesa, *A. bursericola* Bickel & Sinclair, which is widespread in the archipelago, and *A. mystacinus* Bickel & Sinclair from the isolated island of Española (Bickel & Sinclair 1997). This small group of Diptera represents a species swarm, all evolving *in situ*, hypothesized to have evolved from a single colonisation event. The pattern of distribution where there is one widespread species and the others are isolated on single islands is similar to that noted for the lava lizards (Jackson 1993).

The species *Condylostylus longicornis* (Fabricius) (Fig. 4.6) is thought to have dispersed naturally from South America and is now known as far west as French Polynesia (Bickel 1996). Accidental introduction is unlikely since it occurs on several isolated archipelagos (Bickel 1996). Interestingly, a specimen of this species was collected on the sea surface in pleuston nets between islands of the Galápagos (Peck 1994b).

3.3 Drosophilidae

Compared to the Hawaiian Islands, the Drosophilidae diversity is extremely poor. There are 17 species, but 50% have been introduced (Carson *et al.* 1983). Only two endemic species are described, with a likely third endemic recently reared from leaf mines on *Scalesia* trees (Asteraceae). No key to species is currently available and much new material has been collected since the review by Carson *et al.* (1983).

3.4 Sarcophagidae

This family includes eight genera and 14 species, of which six are considered endemic and six introduced. In addition, two genera remain classified as endemic to the Galápagos Islands, *Galopagomyia* Bischof and *Sarothromyiops* Townsend.

Immature stages of *Sarcodexia lambens* (Wiedemann) were collected from nestlings (in nostrils, among the feathers), dead birds, or from nest material of Darwin's finches (Fessl *et al.* 2001). Lopes (1978) revised the Galápagos sarcophagids, but did not find this species in the material. This species occurs in most of the warmer parts of the New World, where it is a very common scavenger, and it has been introduced to Cook Is and French Polynesia (T. Pape, pers. comm.).

Galopagomyia inoa (Walker) was reared from puparia collected from exposed east Pacific green sea turtle (*Chelonia mydas agassizi*) eggs. Some eggs were completely filled with empty puparia. Flesh fly parasitism on sea turtle eggs is also known from Central America, where 30% parasitism by *Eumacronychia sternalis* Allen has been reported (Baumgartner 1988, Lopes 1982). Larvae are deposited on the soil surface, rapidly burrow to eggs and perforate previously undamaged shells (Baumgartner 1988).

3.5 Stratiomyidae

The soldier flies include five genera, four endemic species, plus the recently introduced *Hermetia illucens* (Linnaeus) (first recorded in 1998). Two species (*Brachycara digitata* James and *Nemotelus albiventris* Thomson) breed in leaf litter lying below salt-tolerant coastal herbaceous plants (e.g., *Sesuvium* spp.). Larvae of *Chrysochlorina fasciata* (Thomson) breed in wet, rotting cactus branches and trunks of *Jasminocereus* spp. and *Opuntia* spp.

3.6 Ulidiidae

This family is by far one of the most interesting Dipteran groups on the Galápagos, both in terms of diversity and evolutionary patterns and relationships. It includes some 17 described species and four genera, including the endemic genus *Pareuxesta*.

Pareuxesta comprises six described (Fig. 4.9) and one undescribed species reared from a fallen branch of a *Jasminocereus* cactus. In addition, two undescribed species are known from Cocos Island (Costa Rica) (V. Korneyev & E. Kameneva, pers. comm.). The widely distributed and poorly defined genus *Euxesta* Loew (Fig. 4.8) includes seven endemic species and an undescribed species recently reared from the stems of the rare endemic plant, *Calandrinia galapagosa* (Portulacaceae).

4. Biogeographic Affinities of the Galápagos Diptera Fauna

The Galápagos Diptera fauna has a predominant affinity or relationship with the Neotropical Region. In almost all modern revisions, the sister taxon was believed to occur on the New World mainland. The only exceptions are the following species:

> *Thinophilus hardyi* Grootaert & Evenhuis (Dolichopodidae) — known only from Hawaii and Galápagos, where it has been collected on sandy littoral habitats (Bickel & Sinclair 1997). However, the Hawaiian and Galápagos populations may actually not be conspecific (N.L. Evenhuis, pers. comm.).

> *Aphaniosoma galamarillum* Wheeler (Chyromyidae) — possible sister species to the Hawaiian species, *A. minutum* Hardy (Wheeler & Sinclair 1994).

An alternative interpretation of the biogeographic history of the Galápagos proposes that the islands were colonised by an ancestral biota inhabiting an eastern Pacific island arc sometime between the late Cretaceous and mid-Tertiary (Grehan 2001). The distribution of *Cymatopus* Kertész (Dolichopodidae) was used as an example of a Pacific basin track (Grehan 2001, fig. 4G), extending from Malaysia–Christmas Is–New Guinea–Queensland–Samoa–Galápagos–Cocos–Panama–Dominica. However, this broadly defined genus is common throughout the western tropical Pacific, and Mesoamerica/Caribbean as well (D. Bickel, pers. comm.). In addition, Grehan totally ignored the evidence from *Asyndetus*, which shows very strong New World affinities, to the level of identical species found on both the Galápagos and the American mainland (Bickel & Sinclair 1997).

5. Threats to the Fauna

There were no permanent human settlements on the Galápagos until after 1832 (Peck 2001). Today, five islands are inhabited with a population of some 28,000, of which 18,000 live on Santa Cruz. In addition, about

100,000 tourists visit the islands each year. Despite these pressures, they remain one of the world's least human-altered groups of tropical oceanic islands (Peck 2001).

The effects of alien species on the Diptera fauna are unknown. Introduction of non-native vertebrates has led to the loss of native plants and other animal hosts. But there are no actual studies on the effects on the insect fauna. There has also been extensive habitat alteration and destruction for agriculture and pastures. The replacement of large areas of native plants with introduced grasses and weeds must also have had an impact on the insects, but again this has not been investigated (Peck 2001).

5.1 Introduced flies

Peck *et al.* (1998) estimated that there were 39 species of Diptera introduced to the archipelago (16.5% of all species). They also extensively discussed the various possible modes of introduction and source regions. As of 2008 (Table 4.2), 71 introduced species (some 24% of the total number of species) have been identified and new introductions are predicted to occur each year (Causton *et al.* 2006).

An example of the speed of spread once established on the islands is the introduced sepsid, *Microsepsis armillata* (Melander & Spuler). Specimens collected in 1996 at the base of Sierra Negra represent the earliest collection record of this species, but the species was not found during extensive collecting in this region during the Peck expedition of 1989. In 2001, this species was very abundant around inhabited and agricultural regions of Santa Cruz, especially obvious on dung (from which it was absent in 1989). The impact of such species on the native insect populations is unknown, and possibly of no significance since it is associated with dung of non-native mammals.

One of the most significant recent introductions is *Aedes aegypti* (Linnaeus) and the associated disease Dengue Fever. The first record of this species was the interception of a male specimen by the quarantine service between June and August 2001. Established populations and Dengue Fever were confirmed in August 2002. The species is presently restricted to Santa Cruz and an eradication program is being investigated. Another introduced mosquito, *Culex quinquefasciatus* Say, is presently only known from urban zones, but it is a known vector of avian malaria and West Nile Virus (Whiteman *et al.* 2005). Although these diseases have not been recorded from the Galápagos, the probability of them arriving is considered great (Whiteman *et al.* 2005, Causton *et al.* 2006).

The black fly, *Simulium bipunctatum* Malloch (Simuliidae), is one of five introduced arthropods that Peck *et al.* (1998) considered to have been clearly harmful, and which has recently been classified as highly invasive (Causton *et al.* 2006). It has become a nuisance to humans in the moist uplands of San Cristobal since its introduction in 1989 (Abedrabbo *et al.* 1993). In addition, it may change the characteristics of the island's only stream ecosystem (Gerecke *et al.* 1995, Peck *et al.* 1998).

Another significant and highly invasive introduction is *Philornis downsi* Dodge & Aitken (Muscidae) (Causton *et al.* 2006). Mature larvae are free-living obligate external parasites upon nestlings in bird nests (Dodge & Aitken 1968; Fessl *et al.* 2001, 2006), including Darwin's finches. The lack of records from Daphne Major and high infestations on Santa Cruz suggest that this species is a recent introduction (Fessl & Tebbich 1998). The earliest known record of *P. downsi* is from 1964, collected from both arid and humid zones of Santa Cruz (Causton *et al.* 2006). The first cases of parasitism were not detected until March 1997 (Fessl & Tebbich 2002). This parasite was found in 97% of investigated nests, causing rather high nestling mortality (27%) and showed no host specificity (Fessl & Tebbich 2002). The life cycle and the fly's impact upon the finches are outlined in Fessl *et al.* (2006). There are fears, however, that currently threatened finch species (e.g., the Mangrove Finch) with low populations are at greatest risk.

The potential for more introductions to the islands is great. Present quarantine efforts must be continued and improved at seaports and airports to curb the flow of new and potential disastrous introductions (Peck *et al.* 1998). If these quarantine efforts are not effective, the numbers of introductions will continue to dilute the natural species compositions of the Galápagos, as is occurring on all other tropical oceanic islands worldwide.

Part 2. History and Future of Galápagos Dipterology

6. History of Collections

The study of the insects of the Galápagos Islands commenced with the first collections of Charles Darwin in 1835. He arrived during the dry season months of September and October. Darwin collected specimens of at least eight Diptera, including a bombyliid, syrphid, tephritid, piophilid, muscid, two calliphorids, two sarcophagids, and a hippoboscid (Smith 1987; Walker 1849).

Darwin (1845) stated in reference to his collections:

> 'I took great pains in collecting the insects, but, excepting Tierra del Fuego, I never saw in this respect so poor a country. Even in the upper damp region I produced very few, excepting some minute Diptera and Hymenoptera mostly of common mundane forms.'

Since Darwin, there have been more than 15 individuals and expeditions visiting the islands in search of insects (Peck & Kukalová-Peck 1990, fig. 3). The first comprehensive reports on Galápagos Diptera may be found in Coquillett (1901), Johnson (1924) and Curran (1932, 1934). These records were updated and expanded in the faunistic lists of Linsley & Usinger (1966) and a later supplement by Linsley (1977). The invertebrates were generally very poorly surveyed until the 1960s as demonstrated by Kuschel (1963) who was hard-pressed to compile data on the invertebrates, except for the Coleoptera.

In 1985, Stewart Peck started the first of five collecting expeditions to the Galápagos. There is little doubt that these surveys have had a tremendous impact on the knowledge of the terrestrial invertebrates. Peck (2001) estimated that the mass trapping techniques carried out during these expeditions, combined with published data, has resulted in sampling 90–95% of the arthropods. However, a vast number remain un-named.

I participated on Peck's second trip and focussed primarily on surveying the flies. This was likely the first time that Diptera was the primary focus, and since this time, I have strongly encouraged taxonomic studies of the Galápagos Diptera fauna. Sinclair (1993) provided a brief first analysis or summary of the Diptera fauna. An unpublished checklist of the Galápagos Diptera has been maintained by this author since the early 1990s and is updated on a regular basis. Major recent family revisions include Steyskal (1966), Borkent (1991), Wheeler & Sinclair (1994), Mathis (1995), Bickel & Sinclair (1997), Wheeler (2000), Forrest & Wheeler (2002), Wheeler & Forrest (2002, 2003), Jaschhof (2004), Sasakawa (2007), Disney & Sinclair (2008) and Foster & Mathis (2008).

7. Collecting Methods

The majority of Peck expedition material is presently housed in the Canadian National Insect Collection (Ottawa). The Diptera were surveyed mostly by sweep net and extensive use of wet Malaise and flight intercept traps. These collecting efforts were supplemented by the use of baited pitfall traps (dung, carrion), yellow pan traps and rearing of immature stages. Diptera were also commonly collected on white sheets illuminated by mercury-vapour lights.

Future efforts should focus on rearing flies from their breeding sites and host plants.

8. Inventorying the Data

The Diptera were never the primary focus of surveys until 1989. This is reflected in the first published checklist of the Galápagos insects (Table 4.4). Linsley & Usinger (1966) and Linsley (1977) reported 157 species and 31 families. Primarily due to the efforts of the Peck expeditions and increased priority on Diptera, more than 100 species have been added in the past 30 years. This is not only a reflection of an increased focus on Diptera, but also due to efficiency in collection and sampling techniques.

Table 4.4. Statistics on the diversity of Diptera of the Galápagos Archipelago.

Year	Families	Genera	Species	Genera/ family	Species/ genus	Species/ family	% endemism
1977	31	101	157	3.3	1.6	5.1	52.2
1993	46	149	237	3.2	1.6	5.2	47.0
2008	49	187	294	3.8	1.6	6.0	32.5

Scientists at the Charles Darwin Research Station on Santa Cruz conduct research leading to a better understanding of invertebrate species in the Galápagos Islands. The research is focused on three primary programs: (1) Research on the Galápagos invertebrate communities and amplification of the Invertebrate Reference Collection; (2) Invasive species eradication and control; (3) Technical assessment for the Inspection and Quarantine System for the Galápagos Islands (SICGAL). Material is frequently sent overseas for identification. Recently, a number of small surveys have been conducted on the inhabited islands in order to complete an inventory of agricultural pests and determine the presence of new invasive species. The material is used to help build a reference collection on site at the Darwin Station.

9. Future of Dipteran Biodiversity of the Galápagos

New species remain to be discovered, but probably at a much-reduced rate than in the past 30 years. Through increased focus on rearings, plant-feeding Diptera and unique micro-habitats, new species will be found. In addition, nematocerous Diptera are poorly known, specifically the families Sciaridae and Cecidomyiidae. In general, Diptera is a neglected group in comparison with other arthropods and much work remains, especially studies of immature stages, and the fauna of the smaller outer islands.

Acknowledgements

My sincere thanks and gratitude go to Stewart Peck (Carleton University) who invited me to join his expedition and has continued to support my studies. Bernard Landry (Museum d'Histoire Naturelle, Genève), Charlotte Causton (Darwin Station, Galápagos) and Mike Wilson (Cardiff University) encouraged my continued research. Charlotte Causton and Neal Evenhuis (Bishop Museum) kindly reviewed earlier drafts and the latter provided structural inspiration in compiling this review. Yde de Jong (Zoological Museum, Amsterdam) kindly provided an estimate of the endemic species of the British Isles based on the *Fauna Europaea* database. And finally, sincere thanks to all the Diptera systematists who were asked and cajoled to provide identifications or agreed to extensive studies in their particular taxon of interest.

REFERENCES

Abedrabbo, S., Le Pont, F., Shelley, A.J. & Mouchet, J. (1993) Introduction et acclimatation d'une simulie anthropophile dans l'île San Cristóbal, archipel des Galapagos (Diptera, Simulidae). *Bulletin de la Société entomologique de France* 98(2): 108.

Baez, M. (1993) Origins and affinities of the fauna of Madeira. *Boletim do Museu Municipal do Funchal, Supplement* No. 2: 9–40.

Baumgartner, D.L. (1988) Review of myiasis (Insecta: Diptera: Calliphoridae, Sarcophagidae) of Nearctic wildlife. Pages 3–46 *in*: Mackay, D. (ed.), *Selected papers presented at the Seventh Annual Symposium of the National Wildlife Rehabilitators Association, Denver, Colorado, March 9–13, 1988.*

Baumgartner, D.L. & Greenberg, B. (1984) The genus *Chrysomya* (Diptera: Calliphoridae) in the New World. *Journal of Medical Entomology* 21: 105–113.

Bickel, D.J. (1996) Restricted and widespread taxa in the Pacific: biogeographic processes in the fly family Dolichopodidae (Diptera). Pages 331–346 *in*: Keast, A. & Miller, S.E. (eds), *The origin and evolution of Pacific Island biotas, New Guinea to Eastern Polynesia: patterns and processes.* SPB Academic Publishing, Amsterdam, 531 pp.

Bickel, D.J. & Sinclair, B.J. (1997) The Dolichopodidae (Diptera) of the Galápagos Islands, with notes on the New World fauna. *Entomologica scandinavica* 28: 241–270.

Borges, P.A.V., Aguiar, A.M.F., Boieiro, M., Carles-Tolrá, M. & Serrano, A.R.M. (2008) The arthropods (Arthropoda) of the Madeira and Selvagens archipelagos. Pages 245–356 *in* Borges, P.A.V., Abreu, C., Aguiar, A.M.F., Carvalho, P., Jardim, R., Melo, I., Oliveira, P., Sérgio, C., Serrano, A.R.M. & Vieira, P. (eds), *A list of the terrestrial fungi, flora and fauna of Madeira and Selvagens archipelagos.* Direcção Regional do Ambiente da Madeira and Universidade dos Açores, Funchal and Angra do Heroísmo, 440 pp.

Borkent, A. (1991) The Ceratopogonidae (Diptera) of the Galápagos Islands, Ecuador with a discussion of their phylogenetic relationships and zoogeographic origins. *Entomologica scandinavica* 22: 97–122.

Carson, H.L., Val, F.C. & Wheeler, M.R. (1983) Drosophilidae of the Galápagos Islands, with descriptions of two new species. *International Journal of Entomology* 25: 239–248.

Causton, C.E., Peck, S.B., Sinclair, B.J., Roque-Albelo, L., Hodgson, C.J. & Landry, B. (2006) Alien insects: threats and implications for conservation of Galápagos Islands. *Annals of the Entomological Society of America* 99: 121–143.

Chandler, P.J. (ed.) (1998) Checklists of insects of the British Isles (New Series). Part 1: Diptera. *Handbooks for the Identification of British Insects* 12(1): 1–234. Royal Entomological Society of London.

Chandler, P.J. (2008) Corrections and changes to the Diptera Checklist (19). *Dipterists Digest* 15: 16–19.

DIPTERA DIVERSITY: STATUS, CHALLENGES AND TOOLS
(EDS T. PAPE, D. BICKEL & R. MEIER). © 2009 KONINKLIJKE BRILL NV.

Christie, D.M., Duncan, R.A., McBirney, A.R., Richards, M.A., White, W.M., Harpp, K.S. & Fox, C.G. (1992) Drowned islands downstream from the Galapagos hotspot imply extended speciation times. *Nature* 355: 246–248.

Colinvaux, P.A. (1972) Climate and the Galápagos Islands. *Nature* 240: 17–20.

Coquillett, D.W. (1901) Papers from the Hopkins Stanford Galapagos Expedition, 1898–99. II. Entomological results (2): Diptera. *Proceedings of the Washington Academy of Science* 3: 371–379.

Curran, C.H. (1932) [The Norwegian Zoological Expedition to the Galapagos Islands 1925, conducted by Alf Wollebaek. IV.] Diptera (excl. of Tipulidae and Culicidae). *Nyt Magazin for Naturvidenskaberne* 71: 347–366.

Curran, C.H. (1934) The Templeton Crocker Expedition of the California Academy of Sciences, 1932. no.13. Diptera. *Proceedings of the California Academy of Sciences,* ser. 4, 21: 147–172.

Darwin, C. (1845) *The Voyage of the 'Beagle'*. Heron Books, London, 551 pp.

Disney, R.H.L. & Sinclair, B.J. (2008) Some scuttle flies (Diptera: Phoridae) of the Galápagos Islands. *Tijdschrift voor Entomologie* 151: 115–132.

Dodge, H.R. & Aitken, T.H.G. (1968) *Philornis* flies from Trinidad (Diptera: Muscidae). *Journal of the Kansas Entomological Society* 41: 134–154.

Eldredge, L.G. & Evenhuis, N.L. (2003) Hawaii's biodiversity: a detailed assessment of the numbers of species in the Hawaiian Islands. *Bishop Museum Occasional Papers* 76: 1–28.

Evenhuis, N.L. (2009) Hawaii's Diptera Biodiversity. Pages 47-70 *in*: Pape, T., Bickel, D. & Meier, R. (eds), *Diptera Diversity: Status, Challenges and Tools*. Brill, Leiden.

Fessl, B., Couri, M.S. & Tebbich, S. (2001) *Philornis downsi* Dodge & Aitken, new to the Galapagos Islands (Diptera, Muscidae). *Studia dipterologica* 8: 317–322.

Fessl, B., Sinclair, B.J. & Kleindorfer, S. (2006) The life-cycle of *Philornis downsi* (Diptera: Muscidae) parasitizing Darwin's finches and its impacts on nestling survival. *Parasitology* 133: 739–747.

Fessl, B. & Tebbich, S. (1998) Larvae of *Philornis* Meinert (Muscidae) parasiting Galapagos Darwin Finches. Page 55 *in*: Ismay, J.W. (ed.), *Abstract Volume, 4th International Congress of Dipterology*. Oxford, UK.

Fessl, B. & Tebbich, S. (2002) *Philornis downsi* – a recently discovered parasite on the Galápagos archipelago – a threat for Darwin's finches? *Ibis* 144: 445–451.

Forrest, J. & Wheeler, T.A. (2002) Asteiidae (Diptera) of the Galápagos Islands, Ecuador. *Studia dipterologica* 9: 307–317.

Foster, G.A. & Mathis, W.N. (2008) A review of the Tethininae (Diptera: Canacidae) from the Galápagos Islands. *Proceedings of the Entomological Society of Washington* 110(3): 743–752.

Gerecke, R., Peck, S.B. & Pehofer, H.E. (1995) The invertebrate fauna of the inland waters of the Galápagos Archipeligo (Ecuador) — a limnological and zoogeographical summary. *Archiv für Hydrobiologie, Supplement* 107(2): 113–147.

Grehan, J. (2001) Biogeography and evolution of the Galapagos: integration of the biological and geological evidence. *Biological Journal of the Linnean Society* 74: 267–287.

Jackson, M.H. (1993) *Galápagos. A natural history.* Calgary University Press, 316 pp.

Jaschhof, M. (2004) Wood midges (Diptera: Cecidomyiidae: Lestremiinae) from the Galápagos Islands, Ecuador. *Faunistische Abhandlungen* 25: 99–106.

Johnson, C.W. (1924) Diptera of the Williams Galapagos Expedition. *Zoologica* 5: 85–92.

Kuschel, G. (1963) Composition and relationship of the terrestrial faunas of Easter, Juan Fernandez, Desventuradas, and Galapágos islands. *Occasional Papers of the California Academy of Sciences* 44: 79–95.

Linsley, E.G. (1977) Insects of the Galápagos (Supplement). *Occasional Papers of the California Academy of Sciences* 125: 1–55.

Linsley, E.G. & Usinger, R.L. (1966) Insects of the Galápagos Islands. *Proceedings of the California Academy of Sciences*, ser. 4, 33: 113–196.

Lopes, H. de Souza (1978) Sarcophagidae (Diptera) of Galapagos Islands. *Revista Brasileira de Biologia* 38: 595–611.

Lopes, H. de Souza (1982) On *Eumacronychia sternalis* Allen (Diptera: Sarcophagidae) with larvae living on eggs and hatchlings of the east Pacific green turtle. *Revista Brasileira de Biologia* 42: 425–429.

Mathis, W.N. (1995) Shore flies of the Galápagos Islands (Diptera: Ephydridae). *Annals of Entomological Society of America* 88(5): 627–640.

Peck, S.B. (1994a) Aerial dispersal of insects between and to islands in the Galápagos archipelago, Ecuador. *Annals of the Entomological Society of America* 87: 218–224.

Peck, S.B. (1994b) Sea-surface (pleuston) transport of insects between islands in the Galápagos archipelago, Ecuador. *Annals of the Entomological Society of America* 87: 576–582.

Peck, S.B. (1996) Origin and development of an insect fauna on a remote archipelago: The Galápagos Islands, Ecuador. Pages 91–122 *in*: Keast, A. & Miller, S.E. (eds), *The origin and evolution of Pacific Island biotas, New Guinea to Eastern Polynesia: patterns and processes.* SPB Academic Publishing, Amsterdam, 531 pp.

Peck, S.B. (2001) *Smaller orders of insects of the Galápagos Islands, Ecuador: evolution, ecology and diversity.* NRC Research Press, Ottawa, Ontario, Canada, 278 pp.

Peck, S.B. (2006) *The beetles of the Galápagos Islands, Ecuador: evolution, ecology and diversity (Insecta: Coleoptera).* NRC Research Press, Ottawa, 313 pp.

Peck, S.B., Heraty, J., Landry, B. & Sinclair, B.J. (1998) Introduced insect fauna of an oceanic archipelago: the Galápagos Islands, Ecuador. *American Entomologist* 44(4): 218–237.

Peck, S.B. & Kukalová-Peck, J. (1990. Origin and biogeography of the beetles (Coleoptera) of the Galápagos Archipelago, Ecuador. *Canadian Journal of Zoology* 68: 1617–1638.

Sasakawa, M. (2007) The Neotropical Agromyzidae (Insecta: Diptera) Part 7. Leaf-miners from the Galápagos Islands. *Species Diversity* 12: 193–198.

Sinclair, B.J. (1993) Diptera of the Galápagos archipelago (Ecuador). *Fly Times* 11: 4–6.

Smith, K.G.V. (1987) Darwin's Insects: Charles Darwin's entomological notes. *Bulletin of the British Museum (Natural History), Historical Series* 14(1): 1–143.

Steyskal, G.C. (1966) Otitidae from the Galápagos Islands (Diptera, Acalyptratae). *Proceedings of the California Academy of Sciences*, ser. 4, 34: 483–498.

Walker, F. (1848–49) *List of the specimens of Dipterous Insects in the collection of the British Museum*. London. Parts I–IV, 1172 pp.

Wells, J.D. & Kurahashi, H. (1997) *Chrysomya megacephala* (Fabr.) is more resistant to attack by *Ch. rufifacies* (Macquart) in a laboratory arena than is *Cochliomyia macellaria* (Fabr.) (Diptera: Calliphoridae). *Pan-Pacific Entomologist* 73(1): 16–20.

Wheeler, T.A. (2000) Carnidae of the Galapagos Islands, Ecuador: description and phylogenetic relationships for a new species of Neotropical *Meonura* Rondani, 1856 (Diptera: Carnidae). *Studia dipterologica* 7: 115–120.

Wheeler, T.A. & Forrest, J. (2002) A new species of *Elachiptera* Macquart from the Galápagos Islands, Ecuador, and the taxonomic status of *Ceratobarys* Coquillett (Diptera: Chloropidae). *Zootaxa* 98: 1–9.

Wheeler, T.A. & Forrest, J. (2003) The Chloropidae (Diptera) of the Galápagos Islands, Ecuador. *Insect Systematics and Evolution* 34: 265–280.

Wheeler, T.A. & Sinclair, B.J. (1994) Chyromyidae (Diptera) from the Galápagos Islands, Ecuador: Three new species of *Aphaniosoma* Becker. *Proceedings of the Entomological Society of Washington* 96: 440–453.

Whiteman, N.K., Goodman, S.J., Sinclair, B.J., Walsh, T., Cunningham, A.A., Kramer, L.D. & Palmer, P.G. (2005) Establishment of the avian disease vector *Culex quinquefasciatus* Say 1823 (Diptera: Culicidae) on the Galápagos Islands, Ecuador. *Ibis* 147: 844–847.

Woodley, N.E. & Hilburn, D.J. (1994) The Diptera of Bermuda. *Contributions of the American Entomological Institute* 28(2): 1–64.

Zimmerman, E.C. (1948) *Insects of Hawaii*. Vol. 1. Introduction. University of Hawai`i Press, Honolulu, 206 pp.

CHAPTER FIVE

PALAEARCTIC DIPTERA – FROM TUNDRA TO DESERT

Thomas Pape

Natural History Museum of Denmark, Copenhagen, Denmark

Introduction

Human life is a life in coexistence with Diptera, as these are ubiquitous and may be present in immeasurable myriads of individuals. While Diptera to most people probably are known mainly as a nuisance, not to mention the immense suffering brought about by the disease-carrying capacity of blood-sucking species, Diptera are also key players in the recycling of organic material in ecosystems, from the sewage of our urban communities to the leaf litter of the forest floor. And Diptera provide other general ecosystem services like pollination and pest control. Diptera have a particular relevance for the Palaearctic Region, as there is an increasing relative dominance of Diptera with latitude, until Diptera are reigning supreme among the insects in the high arctic.

1. Geology and Biogeography

The continental basis for the Palaearctic can be traced back to the late Jurassic, when the break-up of Pangea into Laurasia and Gondwana was well underway, and the Atlantic Ocean was born by the formation of the Mid-Atlantic Ridge. By mid to late Cretaceous (100–80 Mya), Laurasia was divided by epicontinental seaways into the two palaeocontinents Euramerica (Europe and eastern North America) and Asiamerica (Asia and western North America). With a growing Atlantic Ocean and the final closing of the epicontinental Turgai Strait about 30 Mya, the geological conditions for the Palaearctic Region as we know it today were largely set (Sanmartín *et al.* 2001), although events like the Alpine (Oligocene-

Diptera Diversity: Status, Challenges and Tools
(eds T. Pape, D. Bickel & R. Meier). © 2009 Koninklijke Brill NV.

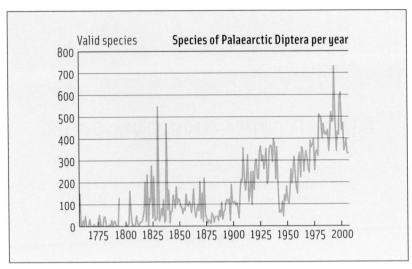

Figure 5.1. Number of valid species of Diptera described per year for the Palaearctic Region. Data from the *BioSystematic Database of World Diptera* (Evenhuis *et al.* 2007).

Miocene) and the Himalayan (Eocene-Miocene) orogenies and the opening of the Japanese Sea (Miocene) certainly have been shaping Diptera diversity at a more local geographical scale.

The Palaearctic Region includes all of Europe, Africa north of the Sahara, Asia north of the Himalayas, and Japan exclusive of the Ruykuy Islands. This is roughly the area from longitudes 10°–170°W and latitudes 30°–80°N, and with a total area of some 46 million km², the Palaearctic is the largest of the biogeographic regions, more than twice as large as each of the Nearctic, Neotropical and Afrotropical regions, and almost three times as large as the Oriental and Australasian regions taken together.

The Palaearctic Region has its northern border along the Arctic Ocean, where large stretches of marshes, bogs and lakes form the treeless tundra, and where the nutrient-poor topsoil is frozen for most of the year and the subsoil permafrost means very poor drainage. Approximately south of the 10°C July isotherm, the treeless tundra gives way to the taiga or boreal coniferous forest, which stretches across the entire region. South of the taiga is a belt of temperate broadleaf and mixed forests, the nemoral zone, and further south is the Mediterranean Basin and the Arabian deserts in the west, the steppe grasslands and desert basins of Central Asia, and the rich temperate broadleaf and mixed forests of China and Japan in the east (Olson *et al.* 2001).

The Palaearctic Region is separated from the Afrotropical Region by the barren Sahara desert, and the boundary is for practical reasons taken to follow the politically defined northern borders of Mauritania, Mali, Niger, Chad and Sudan (Crosskey 1980). The Arabian Peninsula is Palaearctic except for the extreme southern part, largely encompassed by Yemen, which is usually considered as Afrotropical due to a high number of plants and animals shared with the Ethiopian highlands (e.g., Puff & Nemomissa 2001, Delany 1989). Eastwards, the Himalayas provide a largely well-defined boundary between the Palaearctic and Oriental regions, although the politically defined borderlines interfere with the circumstances that the higher slopes of the Himalayas, e.g., in Kashmir and Nepal, show a largely Palaearctic Diptera fauna. Further east, there is no particular geophysical delimitation, but the southernmost provinces of continental China (Yunnan, Guangxi, Guangdong, Fujian and Zhejiang) are usually considered as Oriental.

2. Taxonomic History

During the Renaissance (14th–17th centuries), Europe was re-gaining scientific momentum. The natural world once again became a legitimate focus for study, old folklore was tested by direct observations, and new discoveries were made by new cadres of naturalists. Spurred by innovative technological advances like Gutenberg's movable type in the 15th century and the microscope at the turn of the 16th century, extraction of new knowledge and its dissemination through society was possible to an extent never seen before. Natural history museums grew out of a fruitful marriage between European academia and the, often royal, 'cabinets of curiosities' of the 16th and 17th centuries (Impey & MacGregor 2001). In a taxonomic context, one particularly important product of this time was the system of scientific nomenclature and classification stemming from the scholarly work of Carolus Linnaeus, and much of the modern family-level and genus-level classification of Diptera is still strongly biased from this European origin. The European starting point for the early taxonomic exploration of Diptera has meant a high concentration of types in European museums, but also a considerable impediment from a heavy load of synonyms, old types and mixed type series. The efficiency has therefore been lower in

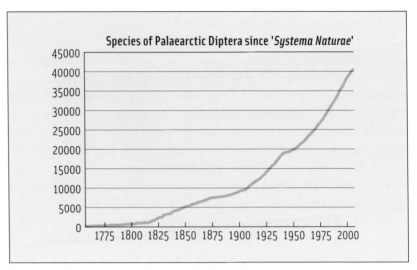

Figure 5.2. Species accumulation curve for Diptera known from the Palaearctic Region. Data from the *BioSystematic Database of World Diptera* (Evenhuis *et al.* 2007).

the Palaearctic when compared with other regions, and 32% of the names proposed for Palaearctic taxa are junior synonyms, junior homonyms or otherwise invalid, while the proportion is only 22% for the Nearctic and 12–14% for the remaining regions (Evenhuis *et al.* 2007).

The taxonomic productivity relating to Palaearctic Diptera has been irregular over time (Fig. 5.1). The year 1830 saw the highest output ever with 1903 nominal species from this region alone (3,080 at a world level) due to large monographic works by especially Robineau-Desvoidy, Meigen and Wiedemann. Mainly because of Robineau-Desvoidy's highly split species concept, less than a third (546) of these names are considered valid as of today. Another peak appeared shortly after, in 1838, where a total of 1,048 names were proposed for Palaearctic species, with almost half of these (469) being valid today. The majority of these names were proposed by Meigen, Macquart and Zetterstedt. The posthumously published '*Memoires*' of Robineau-Desvoidy (1863) resulted in a peak almost as high as that of 1830, this time with 1,643 nominal Palaearctic species, but because of the already mentioned fine-graded species concept combined with added uncertainties from a posthumous publication, only about a tenth of these are now considered valid. Obviously, the output counted as valid species is much more modest, although the 1830 peak is still high above the yearly output of any other year of that century. Only very re-

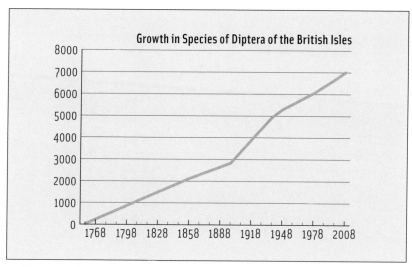

Figure 5.3. Species accumulation curve for Diptera known from the British Isles. Data from Chandler (1998, 2008).

cently, with steadily increasing taxonomic output, have the levels of 1830 and 1838 been surpassed.

Following the crash of the Vienna Stock Exchange in 1873 and aggravated by the general European economic depression and the politically unstable times during the Bismarck period in the latter half of the nineteenth century, taxonomic output was low up to about 1906. From then taxonomic output increased steadily, apparently only mildly affected by the World War I, until it virtually crashed in 1941, crippled by the impact from an intensified World War II. Following the end of the war, the taxonomic output of valid Palaearctic species started to grow back, with pre-WWII levels regained in the early 1970'ies and reaching the highest level ever in the history of Palaearctic dipterology close to the turn of the millennium (730 currently valid species published in 1993). This high output for Palaearctic Diptera is in marked contrast to stagnant or declining levels for the other biogeographic regions and may at least partly be explained from a marked growth in the number of people taking part in describing the Palaearctic Diptera fauna (Fig. 5.4).

Since the all inclusive compilation of 'Systema Naturae' (Linnaeus 1758) and the much less inclusive 'Systema Antliatorum' (Fabricius 1805), a century should pass before the first catalogue specifically on Palaearctic Diptera was produced (Becker *et al.* 1903–1907), and more than three

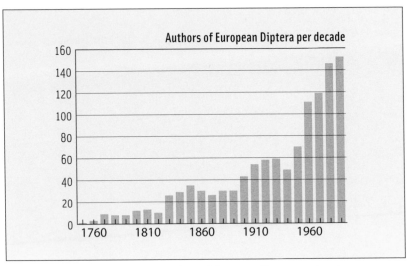

Figure 5.4. Number of authors of European Diptera per decade.
Data from *Fauna Europaea* (Jong 2004, Pape 2004).

quarters of a century should elapse before this catalogue had a successor (Soós & Papp 1984–1993). The latter catalogue, however, was a major undertaking representing the work through more than a decade of a total of 58 taxonomic specialists. Altogether 51,222 names are included in 13 volumes, producing evidence both of the substantial progress made on Palaearctic Diptera in particular, but also of the burden of having to cope with the historical constraints from early taxonomy and nomenclature.

At the country level, Diptera catalogues have been published irregularly at least since Zetterstedt's (1842–1860) encyclopedic monograph '*Diptera Scandinaviae*' and Gobert's (1887) '*Catalogue des diptères de France*', but with an interesting surge in productivity at the turn of the last millennium, seeing country level catalogues for numerous countries: Belgium (Grootaert *et al.* 1991), China (Hua 2006), Czech and Slovak Republics (Ježek 1987, Chvála 1997), Denmark (Petersen & Meier 2001), Finland (Hackman 1980a,b), Germany (Schumann 1991), Great Britain & Ireland (Chandler 1998), Hungary (Papp 2001), Italy (Minelli 1995), Japan (Saigusa & Morimoto 1989), Lithuania (Pakalniškis *et al.* 2000, 2006), Netherlands (Beuk 2002), Poland (Razowski 1991), Spain (Carles-Tolrá Hjorth-Andersen 2002), and Switzerland (Merz *et al.* 1998). With the advent of *Fauna Europaea* (Jong 2004, Pape 2004) at least the part of this information bearing on Europe is now available in updated (and easily updatable!) form from a common portal (see below).

3. Total Estimated Fauna

The Palaearctic Region contains a larger number of named Diptera than any other region, and with about 45,000 species (Evenhuis *et al.* 2007) the region holds 30% of the named world fauna. This may be surprising considering that a large part of the Palaearctic land mass is covered in either boreal forest or arctic tundra. The Palaearctic is indeed the largest of the biogeographic regions, but the relatively high number of species is mainly a reflection of the much longer taxonomic history, the much higher concentration of taxonomists (including many highly skilled 'private' taxonomists), and an inclination by many dipterists for working on their local fauna. The latter may be due to a combination of earmarked funding, availability of good identification sources and reference collections, and the relative ease of acquiring fresh material and biological data through local field work.

Estimating the total number of species for the Palaearctic Region is far from trivial, but a 'guesstimate' of a 2–3 fold increase may not be unrealistic as the species accumulation curve for the Palaearctic shows no sign of levelling off (Fig. 5.2). Even considering the data for the much better studied European fauna with some 19,000 species (Jong 2004, Pape 2004) will give the same picture of a steadily increasing species accumulation curve. Most surprisingly, however, even the comparatively well-studied Diptera fauna of the British Isles produces a steadily growing accumulation curve with no sign of approaching the asymptote (Fig. 5.3).

4. Special Features of the Fauna

4.1 A historically biased family composition

The Palaearctic contains 128 families, which amounts to 79% of all recognised Diptera families (as of Evenhuis *et al.* 2007, which has been the source of the following statistics). The Nearctic Region has a similar coverage, while the other regions are slightly (Neotropical, Oriental, Australasian) or markedly (Afrotropical) lower in their representation of families. It may be surprising that a continental fauna with a history of extensive recent glaciations may have such high taxonomic coverage at the family level. However, family diversity in itself is little informative due to the artificiality of the family rank as currently applied, and the high family coverage is best seen as an artefact stemming from the European bias in classification mentioned earlier.

4.2 Holarctic elements

The lower dipteran family of fungus gnats Bolitophilidae and the calyptrate families Anthomyiidae and Scathophagidae, which both contain a large number of plant-mining species, are markedly Holarctic. For the former, only 2 of the 59 species are found outside the Holarctic, and these are not surprisingly being found in the fungus-gnat rich Oriental region. However, only some 5–15% of the species are shared between the Nearctic and the Palaearctic. Other families with a marked richness of species in the Holarctic are Cecidomyiidae, Chironomidae, Empididae, Fanniidae, Heleomyzidae, Pediciidae, Scenopinidae, Sciomyzidae, and Spaniidae, although for at least the Cecidomyiidae this is most probably an artefact due to lack of revisionary studies of tropical faunas.

4.3 Palaearctic dominance

For a total of twenty families, 50% or more of their known species are found in the Palaearctic Region, here listed alphabetically (with ratio of Palaearctic species to world total): Anthomyiidae (1,158/1,896), Atelestidae (6/10), Axymyiidae (5/6), Braulidae (4/7), Carnidae (67/90), Cecidomyiidae (3,275/6,051), Chamaemyiidae (175/349), Chironomidae (3,579/6,951), Chyromyidae (82/106), Dryomyzidae (15/25), Hesperinidae (4/6), Heterocheilidae (1/2), Lonchopteridae (29/58), Opetiidae (5/5), Opomyzidae (48/61), Pachyneuridae (4/5), Pallopteridae (40/66), Phaeomyiidae (3/3), Rhinophoridae (101/167), Scathophagidae (273/392), Thaumaleidae (87/173).

4.4 Families endemic to the Palaearctic

Considering the tectonic history, the impact by the recent glaciation, and the broad connection with the Oriental region, which in its eastern part would seem to present no significant barriers to dispersal, it should come as no surprise that the Palaearctic Region contains few endemic families. Under the current classification (Evenhuis *et al.* 2007), only the following two families are Palaearctic endemics:

Opetiidae: Only one genus, *Opetia* Meigen, containing small, black, slender-bodied flies. The genus contains a total of three species, including two Japanese and one widespread continental species. Although one species has been reared from rotten wood, the immature stages remain unknown (Chandler 1998). The family is of particular interest as it is the possible sister group of the extremely diverse Cyclorrhapha (Wiegmann *et al.* 1993, Sinclair & Cumming 2006), and as such would represent a lineage dating back to around the K-T boundary some 65 million years ago (Grimaldi & Engel 2005).

Phaeomyiidae: This family was previously treated as a subfamily within the Sciomyzidae, but Griffiths' (1972) proposal of elevating the subfamily to family rank is now followed by most authors. Vala *et al.* (1990) gave further evidence to a status as a separate family. The three known species of this family are accommodated in the single genus *Pelidnoptera* Rondani, and they are all found in the western Palaearctic, with no records east of southern Caucasus (Rozkošný 1998). The biology is known only for a single species, *Pelidnoptera nigripennis* (Meigen), which is a parasitoid of millipedes, and for this reason was recently purposefully introduced to Australia as a means of controlling an invasive European millipede (Vala *et al.* 1990).

A number of small, lower dipteran families are near-endemics. The best example is the Axymyiidae, with only one eastern Nearctic and five Palaearctic species. The Pachyneuridae have one Nearctic, one Oriental and four Palaearctic species, and the Hesperinidae contain one Nearctic and one Neotropical species as well as four Palaearctic species (Krivosheina 1997).

4.5 Large families

The four largest families in the Palaearctic Region are the Chironomidae (3,579 spp.), Cecidomyiidae (3,275 spp.), Tachinidae (3,051 spp.) and Syrphidae (2,058 spp.). These are also in the upper range in other regions, or for the Cecidomyiidae at least expected to get there when this family receives more attention in the tropical parts of the world. The Palaearctic shows most similarity to the Nearctic, which would be expected from the similarity in habitats and climate. The dominance of Chironomidae is remarkable, and this family has more than three times as many species in the Palaearctic than the next region in line, which is the Nearctic with 1,111 species. A very similar pattern is seen for the Cecidomyiidae, which have 3,275 Palaearctic species but only 1,247 Nearctic species. The pattern holds even at the subfamily level, where 278 species of the non-galling Lestremiinae are recorded from the Palaearctic region, contrasting with 92 Nearctic species (Jaschhof 1998). The explanation may be similar in both cases and relate to a combination of longer taxonomic history and much larger area.

Other large families in the Palaearctic region are those represented with a species count from 2,000 and down to 1,000, here listed in descending order of species richness: Dolichopodidae (1,716), Asilidae (1,673), Limoniidae (1,625), Mycetophilidae (1,549), Ceratopogonidae (1,537), Muscidae (1,502), Empididae (1,425), Bombyliidae (1,370), Tipulidae (1,303), Agromyzidae (1,274), Anthomyiidae (1,158) and Sarcophagidae (1,013). Several of these families are rich in species in all remaining regions, with some notable exceptions. The Anthomyiidae are highly concentrated to the Holarctic, and the Empididae have half of their about 3,000 species in the Palaearctic. The Mycetophilidae with more than 4,000 species globally have 1,549 species in the Palaearctic, and this number will most probably increase significantly when the large boreal fauna is better sampled and more thoroughly studied. Kjærandsen *et al.* (2007) made a detailed study of the Swedish fauna of fungus gnats, by which the tally for Sweden increased with 30%. The Sciaridae contain more than 2,000 species globally, and a little less than half of these are found in the Palaearctic. While the geographic distribution and interests of taxonomists may account for part of this Palaearctic dominance, the relative richness of fungus gnats in the boreal parts of the Palaearctic may to some extent be a real phenomenon (Økland *et al.* 2004, Kjærandsen *et al.* 2007), although proper attention to particularly the Sciaridae in many moist tropical areas may change this pattern. Especially the highland and mountainous parts of the Oriental Region appear to contain a very large fauna of mycetophiloids (Hippa & Vilkamaa 2007).

4.6 Least known and best known families

The Diptera family with the proportionally largest number of undescribed species in the Palaearctic is probably the Phoridae, which at this time contain just above 900 Palaearctic species. The phorid fauna is remarkably understudied practically everywhere on the planet, which above all holds for the immensely complex 'super genus' *Megaselia* Rondani. A very thorough study in a Swedish national park close to Stockholm has revealed a surprisingly rich phorid fauna of 330 species, of which a third or more could not be assigned to a named taxon, and several of these may be undescribed (Weber *et al.* 2007, Bonet *et al.* 2006). Using non-parametric estimators, Bonet *et al.* (2006) obtained estimates for a total species richness of Phoridae in the 50 km^2 national park to be in the range of 357 to 491. This compares with the approximately 400 recorded species of European phorids currently present in the *Fauna Europaea* database (Prescher & Weber 2004). While the morphospecies approach used by Bonet *et al.* (2006) may be suspected to inflate species numbers, it should be mentioned that the estimators never reached asymptotic levels. Also, the study included only the genus *Megaselia*. The Phoridae undoubtedly represent one of the last major frontiers in Palaearctic dipterology.

Assessing which families are overall best known with regard to taxonomy, morphology and biology in the Palaearctic Region is not a straightforward measure and has to be considered relative to their size, but Oestridae and Culicidae, which have attracted much interest due to their medical and veterinary importance and for which larvae are not too difficult to obtain, are overall very well known families. The huge family Chironomidae is a taxonomic challenge (and a nomenclatural quagmire), but it is probably the best known of the larger families of Diptera in terms of the morphology of adults, pupae and larvae and with regard to the ecological requirements and habitat preferences of the immatures (Sæther *et al.* 2000). The large family Syrphidae is receiving much attention because of the immediate appeal of most species as well as their importance for pollination and biocontrol, and this family has been particularly favoured by amateur dipterists, with several important treatments published recently (Torp 1994, Ball & Morris 2001, Stubbs & Falk 2002, van Veen 2004, Haarto & Kerppola 2007), and even larvae are becoming still better known (e.g., Rotheray 1993).

4.7 Families with marginal representation in the Palaearctic

A number of Diptera families are represented in the Palaearctic with only a single or a few species. In the lower Diptera, the Lygistorrhinidae is a small family of fungus gnats with only a single Palaearctic species out of a world fauna of 30. In the lower Brachycera, the Hilarimorphidae include four Palaearctic species, one Oriental and 27 Nearctic species. In the Acalyptrata, the small, mainly Neotropical family Tanypezidae has a single

Palaearctic species out of a world fauna of 21. The mainly Afrotropical and Oriental Diopsidae include three Palaearctic species out of a world fauna of 183, and a similar pattern is seen for the mainly tropical families Neriidae with two Palaearctic species out of 111 and the Curtonotidae with three Palaearctic species out of 61.

4.8 Families absent from the Palaearctic Region

Only one family of lower Diptera is absent in the Palaearctic, the Valese-guyidae, which is a very small, scatopsoid family with one extant (Australia) and two extinct (West Indies, Oriental region) species (Amorim & Grimaldi 2006). In the lower Brachycera, the monotypic families Apystomyiidae (Nearctic), Evocoidae (Neotropical) and Oreoleptidae (Nearctic) are absent in the Palaearctic, as are the small families Austroleptidae (Neotropical/Australasian), Pantophthalmidae (Neotropical) and Apsilocephalidae (Nearctic/Neotropical and Australasian). Particularly noteworthy is the absence of the family Apioceridae, as this family is present in all other biogeographical regions (although with only a single species in the Oriental Region). The family Homalocnemiidae is a small clade of empidoid species found in the Neotropical, Afrotropical and Australasian Regions. The aschizan family Ironomyiidae has been found in several extinct species in Upper Jurassic and Cretaceous deposits in the Holarctic, but the family is now confined to Australia (J.F. McAlpine 1973, Zhang 1987, Mostovsky 1995, Grimaldi & Cumming 1999). In the Calyptratae, the monotypic calyptrate family Mystacinobiidae (New Zealand) is absent from the Palaearctic, as are the Glossinidae, although the latter have been found in Miocene European deposits (Wedmann 2000). Some 21 families of acalyptrate families are absent in the Palaearctic, all rather small (<50 spp.) except for the New World Richardiidae, and confined to the warmer areas of one or two biogeographical regions except for the Ctenostylidae (present in the Neotropical, Afrotropical and Oriental regions) and Neurochaetidae (present in the Afrotropical, Oriental and Australasian regions).

4.9 Infraregional patterns

The Palaearctic is remarkable by harbouring the most species rich (in terms of species of trees) temperate forests of the Northern Hemisphere in its eastern parts, and those of the lowest diversity in the western parts (Latham & Ricklefs 1993, MacKinnon 2008). For insects in general, Gressitt (1974: 302) made the broad statement that 'The eastern and western ends of the Palearctic are richer and more varied than in between'. Whether this holds for Diptera is in need of testing through more thorough faunistic studies, but a general explanation for Gressitt's postulated bimodality could be the near absence of the temperate, broadleaf forests in Central Asia, where the relatively species-poor boreal taiga is separated from the extensive and much drier temperate grasslands by a narrow strip of temperate coniferous forest (Olsson *et al.* 2001). The large expanses of mainly lowland boreal forests stretching throughout the Palaearctic were generally established relatively recently during the climatic deterioration from middle to late Miocene (Matthews 1979). Being of relatively recent origin and presenting few geophysical barriers, it is only to be expected that the degree of endemism appears to be low. Much of the Palaearctic Diptera diversity is to be found south of the taiga. An exception is the family Scathophagidae, where the highest species richness is found around the timberline in the forest-tundra ecotone and in subarctic and montane habitats (Gorodkov 1986).

Oosterbroek & Arntzen (1992) and Jong (1998) found that for certain groups of Tipulidae in the Mediterranean, the oldest faunal elements are found in the western part, with a consecutive origin of younger lineages eastwards.

The small calyptrate family Rhinophoridae was poorly circumscribed until recently (Pape & Arnaud 2001), and although now documented from all biogeographic regions, the Mediterranean still contains the majority of the known species of the family. While this is probably exaggerated due to much more intensive collecting in Europe, there is nowhere that rhinophorids are encountered in such abundance as in the Mediterranean (T. Pape, unpubl.). The eastern Palaearctic has remarkably few species (Herting 1961, Pape & Kurahashi 1994), and the marked difference between western and eastern Palaearctic must be considered a real phenomenon that may possibly be explained by the large difference in number of available hosts: The Mediterranean region, and the Balkan Peninsula in particular, is known to be the epicenter of terrestrial isopod diversity (Sfenthourakis 1996, Schotte 2006).

In the Sarcophagidae, the large genus *Sarcophaga* Meigen (*sensu lato*) contains two major species swarms with a marked concentration of species in western Europe. *Sarcophaga* (*sensu stricto*) with some 35 species is almost entirely found in Europe, with only two species reaching eastwards to Inner Mongolia in China and one of these extending its distribution into Siberia (Pape 1996 & unpublished). The subgenus *Sarcophaga* (*Heteronychia*) has a remarkable diversity in the Mediterranean biome (Pape 1996, D. Whitmore, unpubl.), although the subgenus extends into the eastern Palaearctic with several species. The 100+ species may all be predators or parasitoids of shell-bearing pulmonate snails, which is a diverse group in the generally limestone-rich Mediterranean habitats (Cameron 1995).

A possible western relict is found in one of the only two species of the genus *Heterocheila* Rondani, sole genus of the acalyptrate family Heterocheilidae. The Nearctic species is found along the Pacific coasts of North America, while the Palaearctic member is living on the beaches of northern and western Europe (D.K. McAlpine 1998). Both species are associated with stranded seaweed, and especially with *Laminaria* spp. (Egglishaw 1960). The family has an undoubted affiliation with the superfamily Sciomyzoidea (D.K. McAlpine 1991) and may be a close relative of the Phaeomyiidae, which are known only from the western Palaearctic.

The 'special' faunal composition of the far eastern Palaearctic, often considered a special subregion, has been known at least since Wallace (1878), who stated that 'The Manchurian sub-region has ... a very beautiful and varied fauna', although Wallace himself considered 'Europe being the richest and most varied portion' of the Palaearctic. Séguy (1950) operated with a dipterological 'province mandchourienne', including Russian and Chinese Manchuria plus all of Palaearctic Japan. The eastern richness is shown by many groups of animals (Sanmartín *et al.* 2001, MacKinnon 2008) and holds for Diptera as well. A good example is shown by the Drosophilidae (Toda *et al.* 1996), where there is a distinct surge in species richness when going eastwards into the central and southern parts of the Sikhote-Alin Mountains (Primorsky Krai). The rather small lower brachyceran families Xylophagidae and Xylomyidae have a high concentration of species in the easternmost part of the Palaearctic region. Species of both families have larvae that are generally found under the bark of trunks and roots of mature (dying or recently dead) trees or in humid wood detritus of rot holes and similar habitats. The species seem to be rather catholic in their preferences for host trees, although species of Xylomyidae have never been found in conifers (Krivosheina 1988). A calyptrate example is the cone-associated genus *Strobilomyia* Michelsen (Anthomyiidae), which not surprisingly has its center of diversity in the conifer-rich eastern Palaearctic (Michelsen 1988). However, thorough infra-regional comparisons are still immature as the dipterological exploration of the eastern Palaearctic in general is lagging behind that of Europe in terms of both collecting and revisionary studies. Pietsch *et al.* (2003) list 8,000 species of Diptera from the entire Russian Far East, which is about 40% of the tally for the European fauna, although the Far East is about 60% as large as Europe.

The eastern Palaearctic contains a number of relict forms, particularly among the lower Diptera. The species-poor family Tanyderidae has six Palaearctic species, which are found either along the south-central border or in the eastern parts (Wagner 1992). The relict nature of this distribution is corroborated by the presence of this family in Baltic amber (Evenhuis 1994). Other apparently relict distributions with a small number of eastern Palaearctic representatives are the four small families Axymyiidae, Hesperinidae, Deuterophlebiidae and Nymphomyiidae (Courtney 1994, Courtney & Jedlička 1997, Jedlička & Courtney 1997, Krivosheina 1997).

4.10 *Ecological characteristics of arctic Diptera*

The Palaearctic Region shares the large, circumpolar arctic area with the Nearctic Region. With increasing latitude, the diversity of available biotopes decreases, culminating in the extreme high arctic, where suitable habitats for Diptera are strongly restricted and mainly includes a few fresh water biotopes, soil covered by litter or stunted vegetation, plus more specialised habitats like seaweed beds, nests of birds and mammals, and carrion and excrements from larger vertebrates. In the arctic the dominating trophic niche is the saprophagous one, and dipteran larvae are here the dominant group of organisms along with collembolans and mites (Gressitt 1974, Danks 1981). While the arctic is generally characterised by low biomass, the density of midge larvae in eutrophic arctic ponds is generally about 6,000–9,000 or even up to 27,000 per m^2, and more than 2,000 dipteran larvae per square meter were recorded from wet moss tundra in Spitsbergen (Bengtson *et al.* 1974).

Close to the ground is a layer where the microclimate offers reduced wind speeds, higher temperatures (benefitting from the midnight sun!), and higher humidity. This relatively benign layer, which may have a vertical extent limited to a few centimeters, is of crucial importance to the adults of arctic Diptera. In general it can be stated that with increasing latitude, the abiotic factors acquire increasing importance, so that in the most extreme high arctic situations the biotic interactions probably play only an insignificant role. Accordingly, parthenogenesis, which may be regarded as a reduction of the male sex, is fairly common among arctic Chironomidae, Simuliidae and Tipulidae.

Arctic Diptera often have a potentially perennial life cycle with a larval stage of indefinite length, growth simply depending on whether conditions are favourable or not. Thus, the species *Tanytarsus gracilentus* Holmgren completes two generations per year in Iceland, but its larval development in the high arctic is perennial. In extreme cases the larval life of Chironomidae may last up to seven years (Danks 1981).

5. Collecting Palaearctic Diptera

The assembly of Diptera specimens, whether into royal 'cabinets of curiosities' or private natural history collections, grew out of human curiosity and the intellectual surplus of the Renaissance to dwell on the fascination for nature, and as such started even before Linnaeus and the advent of scientific naming. Up through the 19th and 20th centuries, large collections of Diptera were built through the efforts of particularly devoted people. Labelling was often sparse or entirely absent, as the associated information was not an issue for the early collectors, but with a growing knowledge of the immense diversity came an interest in capturing more detail as well as distributional patterns. Early collections of Diptera were usually not built in a particularly structured way, growing through various combinations of personal collecting efforts and the influx of specimens brought by others or through purchase or exchange. With a growing scientific approach and better organised natural history museums, coupled with a world opening up through better means of long-distance transportation, the scientific collecting of entomological specimens became more organised and was often implemented as part of expeditions with a different main purpose. Many of these expeditions headed off to tropical destinations, but a few had a Palaearctic focus.

The exploration of the Palaearctic Diptera was set off with Linnaeus, who travelled to Swedish Lappland and made observations on, among others, the reindeer dermal bot fly (later named by Linnaeus as *Hypoderma tarandi*) and the myriads of biting flies (Linnaeus 1732). Zetterstedt's (1838–1840) *Insecta Lapponica* and his magnum opus *Diptera Scandinaviae* (1842–1860), based partly on extensive collections done by himself throughout Sweden, were essentially a portal to the fauna of boreal and subarctic Diptera. Several early Swedish expeditions to Spitsbergen, led by explorers like C.J. Sundevall, A.E. Nordenskiöld, A.J. Malmgren and A.E. Holmgren, brought back some of the first truly arctic Diptera, which were identified by Boheman (1865) and Holmgren (1869), and most of which were described as new.

The more eastern parts of the Palaearctic were in the early 19th century *terra incognito* with regards to Diptera (indeed as to its insect fauna at large). In 1843–1845, the Baltic-German A.T. von Middendorff led an expedition funded by the St. Petersburg Academy of Sciences to the very superficially known Taimyr Peninsula in Siberia, from where a total of 14 species of Diptera were brought back (Aurivillius 1883). Later, and to demonstrate the feasibility of waterborne commerce between Siberia and Europe, the Finnish-born explorer and scientist A.E. Nordenskiöld headed expeditions to the Yenisey estuary and the Taimyr Peninsula in 1875 and 1876, during which a large material of Diptera was collected. Material from the islands of Novaya Zemlya and Vaygatch was treated by Holmgren (1883), who documented 81 species of Diptera, most of them described as new (but without illustrations as the plates, which were sent to Paris for engraving, were lost in a fire). The Diptera collected along the shores of the estuaries of Yenisey, Ob and Irtysch, plus earlier material collected from Novaya Zemlya, was described by Becker (1897, 1900).

The easternmost coast of arctic Siberia remained entomologically unexplored until A.E. Nordenskiöld in his remodelled whaling-ship *Vega* led his most famous expedition in 1878–1880 from northern Norway to the Bering Strait through the North-East Passage. A small collection of insects from *Vega*'s winter station at Pitlekai included 'about 20 Brachycera, 10 Nematocera' (Aurivillius 1883), but these were never identified. Since then, numerous Russian teams have made expeditions into the Palaearctic, but the vast tundra is still far from well known for its Diptera fauna. As recent as 1994, the Swedish-Russian Tundra Ecology-Expedition was carried out along the entire Russian arctic coast. As a new concept in ecological field research, a ship was used as a mobile base from which researchers were put ashore by helicopters and boats (Goryachkin *et al.* 1994). An extensive material of Diptera was brought back, but the material has remained largely unpublished (S.A. Bengtsson, pers. comm.). However, a material of 565 specimens of Scathophagidae was sorted into 20 species, two of which were described as new to science (Engelmark 1999). An important find of this first trans-Palaearctic arctic coastal transect was that apparent discontinuous distributions may well be an artefact from insufficient collecting in the more remote areas.

The *Sino-Swedish Expedition*, sponsored in part by the Swedish and German governments and led by the Swedish explorer, geographer and geopolitician, Sven Hedin (Wallström 1983), travelled in a series of trips 1927–1930 through the Chinese provinces Xinjiang, Gansu, Sichuan and Nei Menggu (Inner Mongolia) and brought a large material of terrestrial arthropods back to the Swedish Museum of Natural History. The Diptera were described in 17 papers appearing in *Arkiv för Zoologi* (vols 25–27; 1933–1936).

A particularly important entomological collecting effort in the Central Asian steppe biome south of the nemoral forest was that of the Hungarian coleopterist Zoltan Kaszab, who planned and conducted six expeditions to Mongolia in the years 1963–1968 (Kaszab 1964, 1965, 1966, 1967, 1968, 1969). Kaszab's Mongolian expeditions brought back approximately half a million specimens of invertebrates, and of these 29,728 were Diptera (Kaszab 1969). Of these, 8,391 identified specimens are deposited in the Budapest museum (L. Peregovits, unpubl.), with thousands of others scattered among most of the larger natural history collections of the world. Out of 511 publications stemming directly from the Mongolian material collected by Kaszab, 61 deal with Diptera. At least 52 honorific names were proposed on flies collected during the Kaszab expeditions (Evenhuis *et al.* 2007).

A number of Japanese Overseas Grant-in-Aid research projects on the taxonomy of medically important Diptera, mainly calyptrates, were carried out under the leadership of Rokuro Kano. While inventory work was mainly carried out in various core Oriental localities (India, Indonesia, Malaysia, Singapore, Thailand), several important collections were made in Pakistan (Sugiyama 1989) and Nepal (Shinonaga *et al.* 1994) and as such close to the Palaearctic.

A major insect inventory using light traps and yellow pans was initiated by Antonius van Harten in the United Arab Emirates in 2004, with the first volume of at least three planned for recently printed (Harten 2008).

Inventories directed specifically towards the Diptera were recently carried out in the Czech Repulic, one in an industrially affected region (Barták & Vaňhara 2000, 2001) and another in one of the smaller Czech national parks (Barták & Kubik 2006).

The perhaps most efficient mass-sampling device for the collecting of Diptera is the Malaise trap, named after the Swede René Malaise, who invented the trap (Malaise 1937) and was the first to use it during his extensive collecting in Kambaiti, Myanmar, near the border to Yunnan, China. Malaise traps are today routinely deployed in larger insect inventories targeting the actively flying part of the Diptera fauna, although usually without an explicit sampling protocol. Two recent projects differ by their scale and by being designed and implemented with an explicit quantitative approach. The *Diptera Stelviana* project (Ziegler 2006) is an altitudinal transect in the biotically rich Stilfserjoch National Park, which with a size of approximately 133,000 ha is one of the largest and most important nature reserves in the entire European Alpine region. Malaise traps were set up along a transect at five different altitudes, thereby covering all of the significant habitats of the region, from the lowest submontane area of the National Park at Prad (940 m) to the alpine altitude of the Stilfser Joch (2,315 m). Diptera material from these traps are forming the basis for an investigation of the current diversity of Diptera in this region. The three lower traps in the montane zone were operating from mid-May to mid-October 2005, while the two upper traps in the subalpine and alpine zones, were deployed after snowmelt ultimo June and operated until end of August. Building on a history of extensive collecting, the present project carries the potential of a monitoring component.

An even larger project, the *Swedish Malaise Trap Project* (Karlsson *et al.* 2005), has a focus on the least known and most diverse insect groups, i.e., the Hymenoptera and Diptera. A total of 60 Malaise traps were deployed throughout Sweden and were running virtually year round for three consecutive years. About half of the estimated 40 million insects collected are probably Diptera, and in parallel with the initial sorting to order, Diptera are being separated into lower Diptera (or 'Nematocera') and Asilidae, Phoridae, Syrphidae, Sepsidae and 'remaining Brachycera'. Currently, all lower Diptera are being further sorted to family, and the Mycetophiloidea are being processed in specific projects under the Swedish Taxonomy Initiative (Ronquist & Gärdenfors 2003). The 'remaining Brachycera' will be sorted to family when funds are secured (K. Glemhorn & D. Karlsson, pers. comm.).

At the European level, the larger taxonomic institutions have been engaging in a still more closely knit network, starting with the formation of the '*Consortium of European Taxonomic Facilities*' (CETAF; http://www.

cetaf.org/). As an offspring project of CETAF, 27 European taxonomic institutions are involved in the five year project 'European Distributed Institute of Taxonomy' (EDIT; http://www.e-taxonomy.eu/), where one of the work packages is working to apply taxonomy to the conservation of biodiversity by, among other activities, developing and promoting standards, techniques and methodologies for modernized and cost-efficient biodiversity assessments. The first All Taxa Biodiversity Inventory project started in 2007 in the Alpi Marittime and Mercantour nature reserves along the French-Italian border (http://www.atbi.eu/mercantour-maritime), and a second site is currently starting in Slovakia in three nature reserves in the Gemer Area: Slovensky raj, Slovensky kras and Muranska Planina (http://www.atbi.eu/gemer). At this early stage, no dipterological results have been published.

The initiation of large-scale canopy arthropod studies some 25 years ago clearly demonstrated that forest canopies had been a neglected component of particularly the tropical forest ecosystems. Backed by a widely popularised concept of exploring the last biological frontier (e.g., Mitchell 1986, Lowman & Wittman 1995), a lot of arthropod canopy research was done in tropical areas, while very few studies have been carried out using canopy fogging or similar techniques in the Palaearctic Region (e.g., Southwood *et al.* 1982a,b; Ozanne *et al.* 2000), and data on Diptera are particularly sparse. A noteworthy exception is the study by Thunes and coworkers on the arthropod canopy community of Scots pine (*Pinus sylvestris*) in Norway (Thunes *et al.* 2003, 2004), where sampling the canopy of 24 pine trees for two consecutive seasons gave 512 arthropod species of which 210 were Diptera. Among these, 79 species were first records from Norway, and five even considered to be species new to science.

6. Means of Identification

The Palaearctic Diptera fauna is well known compared to that of other regions. A recent and near-comprehensive information source is the 'Manual of Palaearctic Diptera' (Papp & Darvas 1997–2000), which has general introductory chapters and family treatments with keys to genera. A few families, however, were left uncovered, including such important families as the Tipulidae (*sensu lato*), Dolichopodidae, Empididae (*sensu lato*), Tephritidae and Muscidae.

DIPTERA DIVERSITY: STATUS, CHALLENGES AND TOOLS
(EDS T. PAPE, D. BICKEL & R. MEIER), © 2009 KONINKLIJKE BRILL NV.

A wealth of species-level identification sources exist for Palaearctic Diptera. At the family level, the keys to adults (Papp & Schumann 2000) and larvae (Smith & Ferrar 2000) cover the entire Palaearctic Region, while Oosterbroek (2006) produced a key specifically for separating adults of the Diptera families found in Europe. At the species level, the most prominent resource is the series '*Die Fliegen der Palaearktischen Region*' initiated by Erwin Lindner, who also wrote the first volume (Lindner 1924–1993). With the ambition of covering all families in twelve volumes, the series has been quite productive with more than 16,000 pages, although the Cecidomyiidae are left untreated and several families are incompletely covered (Limoniidae, Psychodidae, Chironomidae, Dolichopodidae, Phoridae, Sarcophagidae and Tachinidae). Lieferungen are still being produced, although with still longer intervals, and the last two decades have only seen the publication of a single Lieferung in partial completion of the Sarcophagidae. Several other series of identification keys exist, mostly much broader in taxonomic scope and covering a much smaller geographic region. Too many to deal with separately, and several no longer active, the major series with a considerable dipterological content are: *Die Tierwelt Mitteleuropas, Fauna Entomologica Scandinavica, Fauna Helvetica, Fauna Japonica, Fauna of Saudi Arabia, Fauna Sinica Insecta, Faune de France, Handbooks to the Identification of British Insects, Insects of Mongolia, Insects of the USSR, Keys to the fauna of USSR, Nationalnyckeln* (*Encyclopedia of the Swedish Flora and Fauna*).

7. Recent Bioinformatics Initiatives

The European Commission has an explicit '*Community Biodiversity Strategy*' to provide the framework for development of community policies and instruments in order to comply with the Convention of Biological Diversity. The Strategy recognises the current incomplete state of knowledge at all levels concerning biodiversity, which is a constraint on the successful implementation of the Convention. At the turn of the millennium, the European Commission funded a four year project (2000–2004) aiming as assembling a database of the scientific names and distribution of all multicellular European non-marine animals (*Fauna Europaea*, http://www.faunaeur.org/). Coordinated by the University of Amsterdam, and assisted by the University of Copenhagen and the National Museum of Natural History in Paris, separate group-coordinators for Nematocera

and Brachycera and a total of 58 taxonomic specialists worked together
to assemble data for almost 20,000 valid European species (Jong 2004,
Pape 2004). *Fauna Europaea* contributes to the European Community
Biodiversity Strategy by supporting one of its main themes: to identify
and catalogue the components of European biodiversity into a database
to serve as a basic tool for science and conservation policies. The *Fauna
Europaea* has recently been re-activated through incorporation into a new
EU-funded three-year project called '*A Pan-European Species-directories
Infrastructure*' (PESI; http://www.eu-nomen.eu/pesi/). It is intended that
PESI forms a component of a broader initiative to be known as '*EU-no-
men*' that will service the long-term needs of the biodiversity community
in Europe on taxonomic data standards and ensure an integrated access
to European, and over time Palaearctic, authoritative taxonomic digital
resources. The Diptera component of PESI has not yet been activated.

Communication and data-sharing is gradually moving into the virtual domain, as for example *Diptera.info* (http://www.diptera.info/news.
php — maintained by Paul Beuk), which is an interactive site open to
all dipterists with the broad scope of facilitating the communication and
sharing of knowledge about Diptera. The *Dipterists Forum* (http://www.
dipteristsforum.org.uk/ — maintained by Stuart Ball) was initiated in
1993 with the purpose of promoting the study and conservation of Diptera of the British Isles. The *Dipterists Forum* is associated with the Biological Records Centre (http://www.brc.ac.uk/default.htm), which is the
UK national focus for terrestrial and freshwater species recording (other
than birds), and it volunteers to coordinate the gathering of records for
several different fly families. Particularly elaborate is the '*Hoverfly Recording Scheme*' (http://www.hoverfly.org.uk/), which since it was launched in
1976 has provided a total of 535,921 records (as of 11 January 2008).

The '*Swedish Species Gateway*' (http://www.artportalen.se/) developed
as a bird recording scheme with public data entry under a peer review
system, but it has since expanded gradually and was recently prepared
for accepting data for several families of Diptera: Chironomidae, Asilidae, Xylophagidae, Xylomyidae, Stratiomyidae, Bombyliidae, Therevidae,
Mythicomyiidae, Syrphidae, Conopidae and Tachinidae. Out of the total of 10,802 Diptera records (including observations), 8,922 are for syrphids.

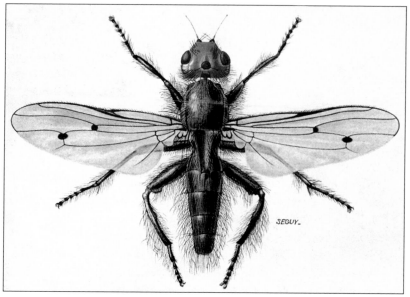

Figure 5.5. *Thyreophora cynophila* (Panzer). Some 50 years after its description in 1794, this remarkable, Central European species was no longer collected and may be extinct as a result of the European megafauna impoverishment. Reproduced from Séguy (1950).

8. Conservation

Diptera are rich in stenotopic species with very particular microhabitat or breeding site requirements, which provides them with a high potential for habitat quality assessment. Chironomidae and Chaoboridae are the key taxa for diagnosing limnic habitats (Bazzanti & Seminara 1986, Real *et al.* 2000, Mousavi 2002), and Haslett (1988) used Syrphidae as bioindicators of environmental stress on ski slopes in Austria. Many saproxylic and fungivorous species of the Syrphidae and the mycetophiloid families have an association with old-growth forests and contain a considerable potential as indicators of woodland quality, which may help designing and implementing management strategies like forest cutting regimes and tree species composition (Speight 1986; Good & Speight 1996, Økland 1994, 1996, 2000; Kjærandsen & Jordal 2007; Økland *et al.* 2004, 2008). The strict preferences for particular habitats that convey their potential use in conservation planning also call for action where such habitats are on the retreat. Special measures have been taken to conserve saproxylic Diptera in England (Rotheray & MacGowan 2000, Rotheray *et al.* 2001), and increase in forest cover and changes in their management in the Netherlands since the 1950'ies have meant that the saproxylic Syrphidae in general are on the increase (Reemer 2005). A similar situation has been indicated for Germany (Ssymank & Doczkal 1998). While this is certainly positive, there is no doubt that the increasing urbanisation and the ever more efficient land-use in most of Europe is having an overall negative impact on the Diptera fauna. An indication of this is that still more species of Diptera are finding their way into European 'Red Lists', e.g., Falk (1991), Stark (1996), Binot *et al.* (1998) and Pollet (2000).

Permanent anthropogenic extinctions of dipteran species must have occurred as soon as early man started to eradicate the megafauna. The first Palaearctic dipteran to become extinct possibly through the activities of man may have been the stomach bot fly of the woolly mammoth, *Cobboldia russanovi* (Grunin), larval remains of which have been found in mammoths excavated from the Siberian permafrost (Grunin 1973). A more recent case is that of the European bone skipper *Thyreophora cynophila* (Panzer) (Fig. 5.5). This conspicuous, read-headed fly was described at the end of the 18th century, and at that time it could be observed on big cadavers like dead dogs, mules and horses in the early spring (Robineau-Desvoidy 1830). Museum specimens exist from Germany, France and Austria, all taken in the first half of the 19th century, but suddenly, 50 years after its discovery, *T. cynophila* disappeared and has never been collected again. This disappearance may be due to changes in livestock management and improved carrion disposal following the Industrial Revolution in Europe, but the underlying scenario probably is the reduction of the megafauna, including the near absence of large predators to leave large carcasses with partly crushed long-bones to give access to the medullar canal and bone marrow, which may have been the favoured breeding habitat for *T. cynophila*.

Acknowledgements

Thanks are due to Verner Michelsen and Toyhei Saigusa for critical comments and valuable suggestions.

REFERENCES

Amorim, D.S. & Grimaldi, D. (2006) Valeseguyidae, a new family of Diptera in the Scatopsoidea, with a new genus in Cretaceous amber from Myanmar. *Systematic Entomology* 31: 508–516.

Aurivillius, C. (1883) Insect life in arctic lands. Pages 403–459 *in*: Nordenskiöld, A.E. (ed.), *Studier och forskningar föranledda af mina resor i höga Norden. Ett populärt vetenskapligt bihang till "Vegas färd kring Asien och Europa".* F. & G. Beiers Förlag, Stockholm. [Translated in 1970 by the Boreal Institute, University of Alberta.]

Ball, S.G. & Morris, R.K.A. (2001) *British hover flies (Diptera, Syrphidae).* Biological Records Centre, Huntingdon, 167 pp.

Barták, M. & Kubik, S. (2006) *Diptera of Podyjí National Park.* Fakulta Agrobiologie, Praha, 432 pp.

Barták, M. & Vaňhara, J. (eds) (2000) Diptera in an industrially affected region (North-Western Bohemia, Bílina and Duchcov environs), I. *Folia Facultatis Scientiarum Naturalium Universitatis Masarykianae Brunensis, Biologia* 104: 1–204.

Barták, M. & Vaňhara, J. (eds) 2001) Diptera in an industrially affected region (North-Western Bohemia, Bílina and Duchcov environs), II. *Folia Facultatis Scientiarum Naturalium Universitatis Masarykianae Brunensis, Biologia* 105: 241–514.

Bazzanti, M. & Seminara, M. (1986) Profundal macrobenthos structure as a measure of long-term environmental stress in a polluted lake. *Water, Air, & Soil Pollution* 33: 435–442.

Becker, T. (1897) Beitrag zur Dipteren-Fauna von Nowaja-Semlja. *Annuaire du Musée Zoologique de l'Académie Impériale des Sciences, St.-Petersburg* 2: 396–404.

Becker, T. (1900) Beiträge zur Dipteren-Fauna Sibiriens. *Acta Soc. Scient. Fenn.* 26: 1–66.

Becker, T., Bezzi, M., Kertèsz, K. & Stein, P. (eds) (1903–1907) *Katalog der Paläarktischen Dipteren.* Vol. 1: iii + 382 + [1] pp.; Vol. 2: 396 pp.; Vol. 3: 828 pp.; Vol. 4: i + 328 pp.; Budapest.

Bengtson, S.-A., Fjellberg, A. & Solhöy, T. (1974) Abundance of tundra arthropods in Spitsbergen. *Entomologica scandinavica* 5: 137–142.

Beuk, P.L. (2002) *Checklist of the Diptera of the Netherlands.* KNNV Uitgeverij, Utrecht, 448 pp.

Binot, M., Bless, R., Boye, P., Grüttke, H. & Prescher, P. (eds) (1998) Rote Liste gefährdeter Tiere Deutschlands. *Schriftenreie für Landschaftspflege und Naturschütz* 55: 1–434.

Boheman, C.H. (1865) Spetsbergens insekt-fauna. *Öfversigt af Kungliga Svenska Vetenskaps-Akademiens Förhandlingar* 22: 563–577, 1 pl.

Bonet, J., Ulefors, S.-O., Viklund, B. & Pape, T. (2006) Mass sampling, species richness and distribution of scuttle flies (Diptera: Phoridae, *Megaselia*) in a wildfire affected hemiboreal forest. Page 27 *in*: Suwa, M. (ed.), Abstract volume, 6th International Congress of Dipterology, 23–28 September 2006, Fukuoka, Japan.

Cameron, R.A.D. (1995) Patterns of diversity in land snails: The effects of environmental history. Pp. 187–200 *in*: Bruggen, A.C. van, Wells, S.M. & Kemperman, T.C.M. (eds), *Biodiversity and conservation of the Mollusca*. Backhuys, Leiden.

Carles-Tolrá Hjorth-Andersen, M. (ed.) 2002) Catálogo de los Diptera de España, Portugal y Andorra. *Boletin de la Sociedad Entomologica Aragonesa; Monografia* 8: 1–323.

Chandler, P.J. (ed.) 1998. Checklists of insects of the British Isles (New Series). Part 1: Diptera. *Handbooks for the Identification of British Insects* 12(1): 1–234. Royal Entomological Society of London.

Chandler, P.J. 2008. Corrections and changes to the Diptera Checklist (19). *Dipterists Digest* 15: 16–19.

Chvála, M. (1997) *Check list of Diptera (Insecta) of the Czech and Slovak Republics*. 1st edition. Karolinum, Charles University Press, Praha, 130 pp.

Courtney, G.W. (1994) Biosystematics of the Nymphomyiidae (Insecta: Diptera): life history, morphology and phylogenetic relationships. *Smithsonian Contributions to Zoology* 550: 1–41.

Courtney, G.W. & Jedlička, L. (1997) Family Nymphomyiidae. Pages 21–27 *in*: Papp, L. & Darvas, B. (eds), *Contributions to a manual of Palaearctic Diptera (with special reference to flies of economic importance)*, Vol. 2, Nematocera and Lower Brachycera. Science Herald, Budapest.

Crosskey, R.W. (ed.) 1980. *Catalogue of the Diptera of the Afrotropical Region*. British Museum (Natural History), London, 1437 pp.

Danks, H.V. (1981) *Arctic Arthropods. A review of systematics and ecology with particular reference to the North American fauna*. Entomological Society of Canada, Ottawa, 608 pp.

Delany, M.J. (1989) The zoogeography of the mammal fauna of southern Arabia. *Mammal Review* 19: 133–152.

Egglishaw, H.J. (1960) Studies on the family Coelopidae (Diptera). *Transactions of the Royal entomological Society* 112: 109–140.

Engelmark, R. (1999) Dungflies (Diptera: Scathophagidae) collected by the Swedish-Russian tundra ecology expedition '94, with the description of two new species *Nanna indotatum* and *Cochliarium sibiricum*. *Entomologisk Tidskrift* 120: 157–167.

Evenhuis, N.L. (1994) *Catalogue of the fossil flies of the world (Insecta: Diptera)*. Backhuys Publishers, Leiden, 600 pp.

Evenhuis, N.L., Pape, T., Pont, A.C. & Thompson, F.C. (eds) (2007) *BioSystematic Database of World Diptera*, Version 10. Available at http://www.diptera.org/biosys.htm, accessed March 2008.

Fabricius, J.C. (1805) *Systema antliatorum secundum ordines, genera, species, adiectis synonymis, locis, observationibus, descriptionibus*. C. Reichard, Brunsvigae [= Brunswick], xiv + 15–372 + [1] + 30 pp.

Falk, P. (1991) A review of the scarce and threatened flies of Great Britain (part I). *Research and survey in nature conservation* 39: 1–194. The Joint nature Conservation Committee, Peterborough.

Gobert, E. (1887) *Catalogue des diptères de France.* Imprimerie Henri Delesques, Caen, 87 pp. [+ 1 unnumbered page with "Statistique des genres et des Familles".]

Good, J.A. & Speight, M.C.D. (1996) *Saproxylic Invertebrates and Their Conservation Throughout Europe.* Convention on the Conservation of European Wildlife and Natural Habitats, Council of Europe, Strasbourg, 58 pp.

Gorodkov, K.B. (1986) Family Scathophagidae. Pages 11–41 *in*: Soós, Á. & Papp, L. (eds), *Catalogue of Palaearctic Diptera, Vol. 11, Scathophagidae—Hypodermatidae.* Akadémiai Kiadó, Budapest.

Goryachkin, S.V., Zlotin, R.I. & Tertitsky, G.M. (1994) *Russian-Swedish Expedition "Tundra Ecology-94": Diversity of natural ecosystems in the Russian Arctic, A guidebook.* Lund University, Lund, 98 pp.

Gressitt, J.L. (1974) Insect biogeography. *Annual Review of Entomology* 19: 293–321.

Griffiths, G.C.D. (1972) The phylogenetic classification of Díptera Cyclorrhapha with special reference to the structure of the male postabdomen. *Series Entomologica* 8: 1–340.

Grimaldi, D.A. & Cumming, J. (1999) Brachyceran Diptera in Cretaceous ambers and Mesozoic diversification of the Eremoneura. *Bulletin of the American Museum of Natural History* 239: 1–124.

Grimaldi, D.A. & Engel, M.S. (2005) *Evolution of the Insects.* Cambridge University Press, Cambridge, 755 pp.

Grootaert, P., De Bruyn, L. & De Meyer, M. (eds) (1991) Catalogue of the Diptera of Belgium. *Studiendocumenten van het K.I.B.N.* 70: 1–338.

Grunin, K.J. (1973) The first discovery of larvae of the mammoth bot-fly *Cobboldia* (*Mamontia*, subgen. n.) *russanovi* sp.n. (Diptera, Gasterophilidae). *Entomologicheskoe Obozrenie* 52: 228–233. [In Russian with English subtitle; English translation in *Entomological Review, Washington* 52: 165–169.]

Haarto, A. & Kerppola, S. (2007) [*Finnish hoverflies and some species in adjacent countries.*] Otava, Keuruu, 647 pp. [In Finnish.]

Hackman, W. (1980a) A check-list of the Finnish Diptera. I. Nematocera and Brachycera (s. str.). *Notulae Entomologicae* 60: 17–48.

Hackman W. (1980b) A check-list of the Finnish Diptera. II. Cyclorrhapha. *Notulae Entomologicae* 60: 117–162.

Harten, A. van (ed.) (2008) *Arthropod fauna of the United Arab Emirates.* Vol. 1. Dar Al Ummah Printing, AbuDhabi, 754 pp.

Haslett, J.R. (1988) Qualitätsbeurteilung alpiner Habitate: Schwebfliegen (Diptera: Syrphidae) als Bioindikatoren für Auswirkungen des intensiven Skibetriebes auf alpinen Wiesen in Österreich. *Zoologischer Anzeiger* 220: 179–184.

Herting, B. (1961) Rhinophorinae. *In*: Lindner, E. (ed.), *Die Fliegen der palaearktischen Region* 64e (Lieferung 216): 1–36. E. Schweizerbart'sche Verlagsbuchhandlung, Stuttgart.

Hippa, H. & Vilkamaa, P. (2007) The *flagria* group of *Keilbachia* Mohrig (Diptera, Sciaridae) in a biodiversity hot spot: nine new sympatric species from Kambaiti, Myanmar. *Zootaxa* 1556: 31–50.

Holmgren, A.E. (1869) Bidrag till kännedomen om Beeren eilands och Spetsbergens insekt-fauna. *Kungliga Svenska VetenskapsAkademiens Handlingar, n. ser.* [=ser. 4] 8(5): 3–55.

Holmgren, A.E. (1883) Insecta a viris doctissimis Nordenskiöld illum ducem sequentibus in insulis Waigatsch et Novaja Semlia anno 1875 collecta. Hymenoptera et Diptera. *Entomologisk Tidskrift* 3: 162–190.

Hua L.-z. (2006) *List of Chinese Insects*, Vol. 4, Sun Yat-sen University Press, Guangzhou, 540 pp.

Impey, O. & MacGregor, A.C. (eds) (2001) *The Origins of Museums: The Cabinet of Curiosities in Sixteenth and Seventeenth-century Europe*. House of Stratus, London, i–xx + 431 pp.

Jaschhof, M. (1998) Revision der Lestremiinae (Diptera: Cecidomyiidae) der Holarktis. *Studia dipterologica Supplement* 4: 1–552.

Jedlička, L. & Courtney, G.W. (1997) Family Deuterophlebiidae. Pages 13–19 *in*: Papp, L. & Darvas, B. (eds), *Contributions to a manual of Palaearctic Diptera (with special reference to flies of economic importance)*, Vol. 2, Nematocera and Lower Brachycera. Science Herald, Budapest.

Ježek, J. (ed.) (1987) Enumeratio Insectorum Bohemoslovakiae. Check list of Czechoslovak Insects II (Diptera). *Acta Faunistica Entomologica Musei Nationalis Pragae* 18: 1–341.

Jong, H. de (1998) In search of historical biogeographic patterns in the western Mediterranean terrestrial fauna. *Biological Journal of the Linnean Society* 65: 99–164.

Jong, H. de (2004) Diptera: Nematocera. *Fauna Europaea*, Version 1.1. Available at http://www.faunaeur.org, accessed April 2006.

Karlsson, D., Pape, T., Johanson, K.A., Liljeblad, J. & Ronquist, F. (2005) The Swedish Malaise trap project, or how many species of Hymenoptera and Diptera are there in Sweden? *Entomologisk Tidskrift* 126: 43–53. [In Swedish with English abstract.]

Kaszab, Z. (1964) Ergebnisse der zoologischen Forschungen von Dr. Z. Kaszab in der Mongolei. 1. Reisebericht der I. Expedition. *Annales Historico-Naturales Musei Nationalis Hungarici* 56: 229–240.

Kaszab, Z. (1965) Ergebnisse der zoologischen Forschungen von Dr. Z. Kaszab in der Mongolei. 26. Reisebericht der II. Expedition. *Annales Historico-Naturales Musei Nationalis Hungarici* 57: 203–215.

Kaszab, Z. (1966) Ergebnisse der zoologischen Forschungen von Dr. Z. Kaszab in der Mongolei. 69. Reisebericht der III. Expedition. *Annales Historico-Naturales Musei Nationalis Hungarici* 58: 243–258.

Kaszab, Z. (1967) Ergebnisse der zoologischen Forschungen von Dr. Z. Kaszab in der Mongolei. 114. Reisebericht der IV. Expedition. *Annales Historico-Naturales Musei Nationalis Hungarici* 59: 191–210.

Kaszab, Z. (1968) Ergebnisse der zoologischen Forschungen von Dr. Z. Kaszab in der Mongolei. 166. Reisebericht der V. Expedition. *Annales Historico-Naturales Musei Nationalis Hungarici* 60: 109–129.

Kaszab, Z. (1969) Ergebnisse der zoologischen Forschungen von Dr. Z. Kaszab in der Mongolei. 193. Reisebericht der VI. Expedition. *Annales Historico-Naturales Musei Nationalis Hungarici* 61: 189–209.

Kjærandsen, J., Hedmark, K., Kurina, O., Polevoi, A., Økland, B. & Götmark, F. (2007) Annotated checklist of fungus gnats from Sweden (Diptera: Bolitophilidae, Diadocidiidae, Ditomyiidae, Keratoplatidae and Mycetophilidae). *Insect Systematics & Evolution Supplement* 65: 1–128.

Kjærandsen, J. & Jordal, J.B. (2007) Fungus gnats (Diptera: Bolitophilidae, Diadocidiidae, Ditomyiidae, Keroplatidae and Mycetophilidae) from Møre og Romsdal. *Norwegian Journal of Entomology* 54: 147–171.

Krivosheina, N.P. (1988) Family Xylomyidae. Pages 38–42 *in*: Soós, Á. & Papp, L. (eds), *Catalogue of Palaearctic Diptera, Vol. 5, Athericidae—Asilidae*. Akadémiai Kiadó, Budapest.

Krivosheina, N.P. (1997) Family Hesperinidae. Pages 35–39 *in*: Papp, L. & Darvas, B. (eds), *Contributions to a manual of Palaearctic Diptera (with special reference to flies of economic importance), Vol. 2, Nematocera and Lower Brachycera*. Science Herald, Budapest.

Latham, R.E. & Ricklefs, R.E. (1993) Continental comparisons of temperate-zone tree species diversity. Pages 294–314 *in*: Ricklefs, R.E. & Schluter, D. (eds), *Species diversity in ecological communities: historical and geographical perspectives*. University of Chicago Press, Chicago.

Lindner, E. (ed.) (1924–1993) *Die Fliegen der paläarktischen Region. Band I–XII.* Schweizerbart'sche Verlagsbuchhandlung, Stuttgart.

Linnaeus, C. (1732) *Carl Linnaeus, Lapplandsresa år 1732. Caroli Linnaei, Iter Lapponicum 1732*. Wahlström & Wikstrand, Stockholm, 277 pp. [1975]

Linnaeus, C. (1758) *Systema naturae per regna tria naturae*. 10th ed., Vol. 1. Laurentii Salvii, Holmiae [=Stockholm], 824 pp.

Lowman, M.D. & Wittman, P.K. (1995) The last biological frontier? Advancements in research on forest canopies. *Endeavour* 19: 161–165.

MacKinnon, J. (2008) Species richness and adaptive capacity in animal communities: lessons from China. *Integrative Zoology* 3: 95–100.

Malaise, R. (1937) A new insect trap. *Entomologisk Tidskrift* 58: 148–160.

Matthews, J.V. (1979) Tertiary and Quaternary environments: historical background for an analysis of the Canadian insect fauna. *Memoirs of the entomological Society of Canada* 108: 31–86.

McAlpine, D.K. (1991) Relationships of the genus *Heterocheila* (Diptera: Sciomyzoidea) with description of a new family. *Tijdschrift voor Entomologie* 134: 193–199.

McAlpine, D.K. (1998) Family Heterocheilidae. Pages 345–347 *in*: Papp, L. & Darvas, B. (eds), *Contributions to a manual of Palaearctic Diptera (with special reference to flies of economic importance), Vol. 3, Higher Brachycera*. Science Herald, Budapest, 880 pp.

McAlpine, J.F. (1973) A fossil ironomyiid fly from Canadian amber (Diptera: Ironomyiidae). *Canadian Entomologist* 105: 105–111.

Merz, B., Bächli, G., Haenni, J.-P. & Gonseth, Y. (eds) (1998) Diptera — Checklist. *Fauna Helvetica* 1: 1–369.

Michelsen, V. (1988) A world revision of *Strobilomyia* gen.n.: the anthomyiid seed pests of conifers (Diptera: Anthomyiidae). *Systematic Entomology* 13: 271–314.

Minelli, A., Ruffo, S. & La Posta, S. (eds) (1995) *Checklist delle specie della fauna Italiana*, Fascicles 63–78. Calderini, Bologna. [Online version at: http://www.faunaitalia.it/checklist/.]

Mitchell, A.W. (1986) *The enchanted canopy. A journey of discovery to the last unexplored frontier, the roof of the world's rainforests*. Macmillan Publishing Company, New York, 255 pages.

Mostovsky, M.B. (1995) New taxa of ironomyiid flies (Diptera, Phoromorpha, Ironomyiidae) from Cretaceous deposits of Siberia and Mongolia. *Paleontologicheskii Zhurnal* 4: 86–103.

Mousavi, S.K. (2002) Boreal chironomic communities and their relations to environmental factors – the impact of lake depth, size and acidity. *Boreal environment research* 7: 63–75.

Økland, B. (1994) Mycetophilidae (Diptera), an insect group vulnerable to forest practices? A comparison of clearcut, managed and semi-natural spruce forests in southern Norway. *Biodiversity and Conservation* 3: 68–85.

Økland, B. (1996) Unlogged forests: Important sites for preserving the diversity of mycetophilids (Diptera, Sciaroidea). *Biological Conservation* 76: 297–310.

Økland, B. (2000) Management effects on the decomposer fauna of Diptera in spruce forests. *Studia Dipterologica* 7: 213–223.

Økland, B., F. Götmark, B. Nordén, N. Franc, O. Kurina & A. Polevoi. (2004) Regional diversity of mycetophilids (Diptera: Sciaroidea) in Scandinavian oak-dominated forests. *Biological Conservation* 121: 9–20.

Økland, B., Götmark, F. & Nordén, B. (2008) Oak woodland restoration: testing the effects on biodiversity of mycetophilids in southern Sweden. *Biodiversity and Conservation* 17: 2599–2616.

Olson, D.M., Dinerstein, E., Wikramanayake, E.D., Burgess, N.D., Powell, G.V.N., Underwood, E.C., D'amico, J.A., Itoua, I., Strand, H.E., Morrison, J.C., Loucks, C.L., Allnutt, T.F., Ricketts, T.H., Kura, Y., Lamoreux, J.F., Wettengel, W.W., Hedao, P. & Kassem, K.R. (2001) Terrestrial ecoregions of the world: a new map of life on Earth. *BioScience* 51: 933–938.

Oosterbroek, P. (2006) *The European families of the Diptera, identification, diagnosis, biology*. Utrecht, KNNV Publishing, 204 pp.

Oosterbroek, P. & Arntzen, J.W. (1992) Area-Cladograms of Circum-Mediterranean Taxa in Relation to Mediterranean Palaeogeography. *Journal of Biogeography* 19: 3–20.

Ozanne, C.M.P., Speight, M.R., Hambler, C. & Evans, H.F. (2000) Isolated trees and forest patches: patterns in canopy arthropod abundance and diversity in *Pinus sylvestris* (Scots pine). *Forest Ecology and Management* 137: 53–63.

Pakalniškis, S., Bernotienė, R., Lutovinovas, E., Petrarca, V., Podėnas, S., Rimšaite, J., Saether, O.A. & Spungis, V. (2006) Checklist of Lithuanian Diptera. *New and rare for Lithuania insect species, records and descriptions* 18: 16–154.

Pakalniškis, S., Rimšaitė, J., Sprangauskaitė-Bernotienė, R., Butautaitė, R., Podėnas, S. (2000) Checklist of Lithuanian Diptera. *Acta Zoologica Lituanica* 10(1): 3–58.

Pape, T. (1996) Catalogue of the Sarcophagidae of the world (Insecta: Diptera). *Memoirs of Entomology, International* 8: 1–558.

Pape, T. (ed.) (2004) Diptera: Brachycera. *Fauna Europaea*, Version 1.1. Available at http://www.faunaeur.org, accessed April 2006.

Pape, T. & Arnaud, P.H., Jr. (2001) *Bezzimyia* — a genus of New World Rhinophoridae (Insecta, Diptera). *Zoologica Scripta* 30: 257–297.

Pape, T. & Kurahashi, H. (1994) First records of Rhinophoridae (Insecta: Diptera) from Japan. *Japanese Journal of Entomology* 62: 475–481.

Papp, L. (ed.) (2001) *Checklist of the Diptera of Hungary.* Hungarian Natural History Museum, Budapest, 550 pp.

Papp, L. & Darvas, B. (eds) (1997–2000) *Contributions to a manual of Palaearctic Diptera (with special reference to flies of economic importance)*, Vols. 1–4. Science Herald, Budapest.

Papp, L. & Schumann, H. (2000) Key to families — adults. Pages 163–200 *in*: Papp, L. & Darvas, B. (eds), *Contributions to a manual of Palaearctic Diptera (with special reference to flies of economic importance), Vol. 1, General and applied dipterology.* Science Herald, Budapest.

Petersen, J.F.T. & Meier, R. (eds) (2001) A preliminary list of the Diptera of Denmark. *Steenstrupia* 26: 1–276.

Pietsch, T.W., Bogatov, V.V., Amaoka, K., Zhuravlev, Y.N., Barkalov, V.Y., Gage, S., Takahashi, H., Lelej, A.S., Storozhenko, S.Y., Minakawa, N., Bennett, D.J., Anderson, T.R., Ôhara, M., Prozorova, L.A., Kuwahara, Y., Kholin, S.K., Yabe, M., Stevenson, D.E. & MacDonald, E.L. (2003) Biodiversity and biogeography of the islands of the Kuril Archipelago. *Journal of Biogeography* 30: 1297–1310.

Pollet, M. (2000) A documented Red List of the dolichopodid flies (Diptera: Dolichopodidae) of Flanders. *Communications of the Institute of Nature Conservation* 8: 1–190. [In Dutch with English summary.]

Prescher, S. & Weber, G. (2004) Phoridae. *In*: Pape, T. (ed.) 2004. *Fauna Europaea*, Version 1.1. Available at http://www.faunaeur.org, accessed April 2006.

Puff, C. & Nemomissa, S. (2001) The Simen Mountains (Ethiopia): comments on plant biodiversity, endemism, phytogeographical affinities and historical aspects. *Systematics and Geography of Plants* 71: 975–991.

Razowski, J. (ed.) (1991) *Checklist of animals of Poland.* Vol. 2, Part XXXII/28. Diptera: 77–269. Polska Akademia Nauk, Instytut Systematyki i Ewolucji Zwierzat.

Real, M., Rieradevall, M. & Prat, N. (2000). *Chironomus* species (Diptera: Chironomidae) in the profundal benthos of Spanish reservoirs and lakes: factors affecting distribution patterns. *Freshwater Biology* 43: 1–18.

Reemer, M. (2005) Saproxylic hoverflies benefit by modern forest management (Diptera: Syrphidae). *Journal of Insect Conservation* 9: 49–59.

Robineau-Desvoidy, J.B. (1830) Essai sur les myodaires. *Mémoires présentés par divers Savans à l' Académie Royal des Sciences de l'Institut de France* 2: 1–813.

Robineau-Desvoidy, J.B. (1863) *Histoire naturelle des diptères des environs de Paris.* Oeuvre posthume du Dr Robineau-Desvoidy. Publiée par les soins de sa famille, sous la direction de M.H. Monceaux. V. Masson et Fils. Tome premier: xvi + 1143 pp.; Tome second, 920 pp.

Ronquist, F. & Gärdenfors, U. (2003) Taxonomy and biodiversity inventories: time to deliver. *Trends in Ecology & Evolution* 18: 269–270.

Rotheray, G.E. (1993) Colour guide to hoverfly larvae (Diptera, Syrphidae). *Dipterists Digest* 9: 1–156.

Rotheray, G.E., Hancock, G., Hewitt, S., Horsfield, D., MacGowan, I., Robertson, D. & Watt, K. (2001) The biodiversity and conservation of saproxylic Diptera in Scotland. *Journal of Insect Conservation* 5: 77–85.

Rotheray, G.E. & MacGowan, I. (2000) Status and breeding sites of three presumed endangered Scottish saproxylic syrphids (Diptera, Syrphidae). *Journal of Insect Conservation* 4: 215–223.

Rozkošný, R. (1998) Family Phaeomyiidae. Pages 377–382 *in*: Papp, L. & Darvas, B. (eds), *Contributions to a manual of Palaearctic Diptera (with special reference to flies of economic importance), Vol. 3, Higher Brachycera.* Science Herald, Budapest, 880 pp.

Sæther, O.A., Ashe, P. & Murray, D.A. (2000) Family Chironomidae. Pages 113–334 *in*: Papp, L. & Darvas, B. (eds), *Contributions to a manual of Palaearctic Diptera (with special reference to flies of economic importance), Appendix, Nematocera and Lower Brachycera.* Science Herald, Budapest.

Saigusa, T. & Morimoto, K. (1989) Diptera. Pages 699–873 *in*: Morimoto, K., Tadauchi, O. & Saito, H. (eds), *A check list of Japanese insects.* Entomological Laboratory, Faculty of Agriculture, Kyushu University, Fukuoka.

Sanmartín, I., Enghoff, H. & Ronquist, F. (2001) Patterns of animal dispersal, vicariance and diversification in the Holarctic. *Biological Journal of the Linnean Society* 73: 345–390.

Schotte, M. (2006) Roly-poly lifestyles. *Wings — Essays on Invertebrate Conservation* 29: 22–27.

Schumann, H., Bährmann, R. & Stark, A. (eds) (1991) Check-Liste der Dipteren Deutschlands. *Studia Dipterologica Supplement* 2: 1–354.

Séguy, E. (1950) La biologie des diptères. *Encyclopedie Entomologique, Série A* 26: 5–609.

Sfenthourakis, S. (1996) The species-area relationship of terrestrial isopods (Isopoda; Oniscidea) from the Aegean archipelago (Greece): a comparative study. *Global Ecology and Biogeography Letters* 5: 149–157.

Shinonaga, S., Kurahashi, H. & Iwasa, M. (eds) (1994) Studies on the taxonomy, ecology and control of the medically important flies in India and Nepal. *Japanese Journal of Sanitary Zoology (Supplement)* 45: 1–316.

Sinclair, B.J. & Cumming, J.M. (2006) The morphology, higher-level phylogeny and classification of the Empidoidea (Diptera). *Zootaxa* 1180: 1–172.

Sjöstedt, Y. & Hummel, D. (1933) Schwedisch-chinesische wissenschaftlische Expedition nach den nordwestlichen Provinzen Chinas, unter Leitung von Dr. Sven Hedin und Prof. Sü Ping-chang. Insekten gesammelt von schwedischen Arzt der Expedition Dr. David Hummel 1927–1930. *Arkiv för Zoologi* 25A(3): 1-16, 2 maps + plates 1-9 [following page 34].

Smith, K.G.V. & Ferrar, P. (2000) Key to families – larvae. Pages 201–239 *in*: Papp, L. & Darvas, B. (eds), *Contributions to a manual of Palaearctic Diptera (with special reference to flies of economic importance), Vol. 1, General and applied dipterology.* Science Herald, Budapest.

Soós, Á. & Papp, L. (1984–1993) *Catalogue of Palaearctic Diptera.* Vols 1–14. Akadémiai Kiadó, Budapest.

Southwood, T.R.E., Moran, V.C. & Kennedy, C.E.J. (1982a) The assessment of arboreal insect fauna: comparisons of knockdown sampling and faunal lists. *Ecological Entomology* 7: 331–340.

Southwood, T.R.E., Moran, V.C. and Kennedy, C.E.J. (1982b) The richness, abundance and biomass of the arthropod communities on trees. *The Journal of Animal Ecology* 51: 635–649.

Speight, M.C.D. (1986) Attitudes to insects and insect conservation. *Proceedings of the 3rd European Congress of Entomology*: 369–385.

Ssymank, A. & Doczkal, D. (1998) Rote Liste der Schwebfliegen (Diptera: Syrphidae). Pages 65–72 *in*: Binot, M., Bless, R., Boye, P., Grüttke, H. & Prescher, P. (eds), *Rote Liste gefährdeter Tiere Deutschlands.* Schriftenreihe für Landschaftspflege und Naturschutz 55.

Stark, A. (1996) Besonderheiten der Dipterenfauna Sachsen-Anhalts — eine Herausforderung für den Natur- und Umweltschutz. *Berichte des Landesamtes für Umweltschutz Sachsen-Anhalt, Halle* [1996](21): 100–108.

Stubbs, A.E. & Falk, S.J. (2002) *British hoverflies. An illustrated identification guide,* 2nd ed. British Entomological and Natural History Society, Reading, 469 pp., 12 plates.

Sugiyama, E. (1989) Sarcophagine flies from Pakistan (Diptera: Sarcophagidae). *Japanese Journal of Sanitary Zoology* 40 (Supplement): 113–124.

Thunes, K.H., Skartveit, J. & Gjerde, I. (2003) The canopy arthropods of old and mature pine *Pinus sylvestris* in Norway. *Ecography* 26: 490–502.

Thunes, K.H., Skartveit, J., Gjerde, I., Starý, J., Solhøy, T., Fjellberg, A., Kobro, S., Nakahara, S., Zur Strassen, R., Vierbergen, G., Szadziewski, R., Hagan, D.V., Grogan,

W.L., Jonassen, T., Aakra, K., Anonby, J., Greve, L., Aukema, B., Heller, K., Michelsen, V., Haenni, J.-P., Emeljanov, A.F., Douwes, P., Berggren, K., Franzen, J., Disney, R.H.L., Prescher, S., Johanson, K.A., Mamaev, B., Podenas, S., Andersen, S., Gaimari, S.D., Nartshuk, E., Søli, G.E.E., Papp, L., Midtgaard, F., Andersen, A., Tschirnhaus, M. von, Bächli, G., Olsen, K.M., Olsvik, H., Földvari, M., Raastad, J.E., Hansen, L.O. & Djursvoll, P. (2004) The arthropod community of Scots pine (*Pinus sylvestris* L.) canopies in Norway. *Entomologica Fennica* 15: 65–90.

Toda, M.J., Sidorenko, V.S., Watabe, H.-a., Kholin, S.K. & Vinokurov, N.N. (1996) A Revision of the Drosophilidae (Diptera) in East Siberia and Russian Far East: Taxonomy and Biogeography. *Zoological Science* 13: 455–477.

Torp, E. (1994) *Danmarks svirrefluer*. Danmarks Dyreliv, Vol. 6. Apollo Books, Stenstrup, 490 pp.

Vala, J.-C., Bailey, P.T. & Gasc, P.T. (1990) Immature stages of the fly *Pelidnoptera nigripennis* (Fabricius) (Diptera: Phaeomyiidae), a parasitoid of millipedes. *Systematic Entomology* 15: 391–399.

van Veen, M.P. (2004) Hoverflies of Northwest Europe. Identification keys to the Syrphidae. KNNV Publishing, Utrecht, 245 pp.

Wagner, R. (1992) Family Tanyderidae. Pages 37–39 *in*: Soós, Á. & Papp, L. (eds), *Catalogue of Palaearctic Diptera, Vol. 1, Trichoceridae—Nymphomyiidae*. Akadémiai Kiadó, Budapest

Wallace, A.R. (1878) Distribution (part 1). Pages 267–286 *in*: Baynes, T.S. (ed.), *The Encyclopædia Britannica*, 9th ed., Vol. 7; A.C. Black, Edinburgh.

Wallström, T. (1983) *Svenska upptäckare*. Bra böcker, Höganäs, 294 pp. [In Swedish.]

Weber, G., Prescher, S., Ulefors, S.-O. & Viklund, B. (2007) Fifty-eight species of Scuttle flies (Diptera, Phoridae: *Megaselia* spp.) new to Sweden from the Tyresta National Park and Nature Reserve. *Studia dipterologica* 13: 231–240.

Wedmann, S. (2000) Die Insekten der oberoligizänen Fossillagerstätte Enspel (Westerwald, Deutschland): Systematik, Biostrationomie und Paläoökologie. *Mainzer Naturwissenschaftliches Archiv* 23: 1–154.

Wiegmann, B.M., Mitter, C. & Thompson, F.C. (1993) Evolutionary origin of the Muscomorpha (Diptera): tests of alternative morphological hypotheses. *Cladistics* 9: 41–81.

Zetterstedt, J.W. (1838–1840) *Insecta Lapponica*. L. Voss, Lipsiae (Leipzig), vi + 1139 pp.

Zetterstedt, J.W. (1842–1860) *Diptera Scandinaviae disposita et descripta*. Vols 1–14: 1–6609. Officina Lundbergiana, Lundae [=Lund].

Zhang J.-f. (1987) Four new genera of Platypezidae. *Acta Palaeontologica Sinica* 26: 595–603.

Ziegler, J. (2006) The "Diptera stelviana" project. A dipterological perspective on a changing alpine landscape. *Studia dipterologica* 13: 1–8.

CHAPTER SIX

AFROTROPICAL DIPTERA
– RICH SAVANNAS, POOR RAINFORESTS

ASHLEY H. KIRK-SPRIGGS[1] & BRIAN R. STUCKENBERG[2]
[1]Albany Museum, Grahamstown, South Africa;
[2]Natal Museum, Pietermaritzburg, South Africa

INTRODUCTION

Africa (Fig. 6.1) is a hot continent that was levelled by prolonged Creta-ceous and early Tertiary erosion, during which time little upheaval took place. There was no Pleistocene Ice Age in Africa, so aside from the equa-torial peaks of mounts Ruwenzori, Kenya and Kilimanjaro, and the Atlas Mountains of North Africa, there are no recent glacial landforms. In trop-ical sub-Saharan Africa generally there are two vast biomes — equatorial rainforest (usually termed the Guineo-Congolian rainforest, Fig. 6.1: 1), mostly in the western lowlands — and savanna (Fig. 6.1: 2, 3) that occu-pies the greater part of the continent.

The dipteran fauna of the region bears the imprint of vast savanna evolution in the tropics and a complex vegetational biome history in the subtropical south (Mucina & Rutherford 2006). Biogeographically, the Afrotropics as currently perceived are limited northwards by the young Sahara Desert (Fig. 6.1: 17), which developed in the Pliocene and became hyper-arid in the Pleistocene (Maley 1996). The Sahara was previously an enormous savanna and grassland, which supposedly extended the range of Afrotropical Diptera much closer to the Mediterranean (e.g., Adams & Faure 1997). Vermileonidae in the Atlas Mountains and the Canary Islands (Stuckenberg 2000, Stuckenberg & Fisher 1999) and species of the asilid genus *Habropogon* Loew in countries bordering the Mediterranean (Londt 2000) are examples of isolated relicts of the fauna that predated aridification.

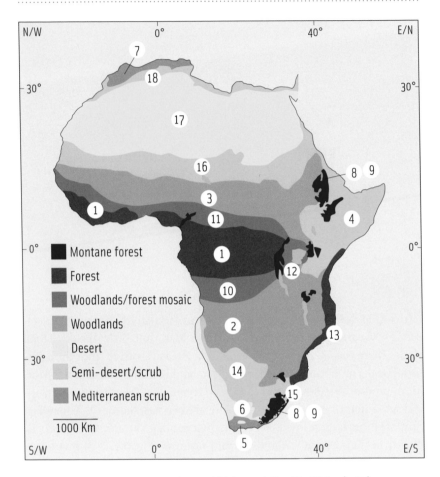

Figure 6.1. The phytochoria of Africa as defined by White (1983)
(after Clarke 2000a).

Regional centres of endemism:

1. Guineo-Congolian, 2. Zambezian, 3. Sudanian, 4. Somalia-Masai, 5, Cape,
6. Karoo-Namib, 7. Mediterranean, 8. Afro-montane archipelago-like
regional centre of endemism.

Regions of extreme floristic impoverishment:

9. Afro-alpine archipelago-like region of extreme floristic impoverishment.

Regional transition zones:

10. Guineo-Congolian/Zambezia, 11. Guineo-Congolian/Sudania,
14. Kalahari-Highveld, 16. Sahel, 17. Sahara. 18. Mediterranean/Sahara.

Regional Mosaics:

12. Lake Victoria, 13. Zanzibar-Inhambane, 15. Tongoland-Pondoland.

1. Extent of the Afrotropical Region

The historical and contemporary use of the term 'Afrotropical Region' is worthy of comment, as is the true extent of this region. The concept of an 'Ethiopian Region' was first proposed by Sclater (1858), based on the distribution of birds, and was later adopted by Alfred Russell Wallace in his enduring work *The Geographical Distribution of Animals*, published in 1876. Wallace defined the Ethiopian Region as the area of continental Africa, Madagascar and its islands, and the Arabian Peninsula south of the Tropic of Cancer and further divided this region into four sub-regions: the West African, East African, South African and Malagasy sub-regions. Although our concept of these sub-regions has altered somewhat as more floristic and faunistic information has become available (Werger 1978 for review), the perceived extent of the region as a whole has little changed. Following the renaming of the political state formerly known as Abyssinia to Ethiopia in 1941, a great deal of confusion in the application of the term 'Ethiopian Region' ensued in the literature and numerous alternative terms began to appear, e.g., sub-Saharan Africa, Afrotropical Realm, etc. As a result of this, Crosskey & White (1977) proposed the replacement name 'Afrotropical Region' corresponding exactly to the Ethiopian Region as it was then known, and this term has become widely accepted and adopted by systematists and biogeographers alike.

Later, for convenience, Crosskey (1980: 27, 32) used the northern borders of the states of Mauritania, Mali, Niger, Chad and Sudan as the regional boundary between the Afrotropical and Palaearctic parts of continental Africa, and for the Arabian Peninsula the northern boundaries of the modern state of Yemen. Crosskey also included Madagascar in his concept of the Afrotropical Region, regarded as a separate zoogeographical region by some biogeographers, and the South Atlantic islands of Ascension, St. Helena, Tristan da Cunha and Gough, as well as the Cape Verde Islands, the Gulf of Guinea islands and the islands of the western Indian Ocean.

Recently, more extensive sampling of Diptera in the Arabian Peninsula, especially in Yemen, Oman and the United Arab Emirates, together with published studies of the Rhopalocera (Larsen 1984) and the Neuroptera (Hölzel 1998), have raised some interesting open questions in respect to the true extent of the Afrotropical Region in this important transitional zone encompassing elements of Afrotropical, Palaearctic and Oriental faunas.

2. South African Complexities

South Africa has the most diverse and distinctive Diptera fauna in the Afrotropics. That country is ecologically complex, with 24 bioclimatic regions (Phillips 1959). South Africa has Africa's oldest mountains — the Cape Fold Mountains, which are part of a Triassic orogeny that predates the break-up of Gondwana — and the Great Escarpment, which was initiated in the east by the separation of Antarctica and southeastern Africa in the Early Jurassic (*ca.* 200 Mya). These mountains preserve endemic Gondwanan Blephariceridae, Psychodidae, Empididae, Africa's only tanyderid, and close-to-basal Chironomidae (Stuckenberg 1962). The Cape Floral Kingdom, *Capensis*, occurs in two famously diverse biomes with about 13,000 endemic plant species — these are the Fynbos shrubland of the Cape Fold Mountains and the Succulent Karoo (Taylor 1978). So much topographic, climatic and floristic diversity promoted radiation among the Diptera, and there are species-rich, systematically complex faunas of the families Asilidae, Bombyliidae, Empididae, Limoniidae, Mydidae, Nemestrinidae, Tabanidae, Therevidae and Vermileonidae. Recent studies prove flies to be important as pollinators in the Cape flora, and there are numerous and often remarkable examples of convergent adaptations of the mouth-parts for feeding in co-adapted flowers (e.g., Barraclough 2006b; Manning & Goldblatt 1995; Struck 1992, 1994).

3. Rostrum Elongation as a Notable Adaptation in the Diptera of the Cape Flora

The Fynbos flora is notable for its great taxonomic diversity and profuse flowering of nectar-bearing plants, many of which have nectaries recessed in tubular corollas. This resource evidently has produced a co-adaptive response among Diptera in that elongation of the proboscis has evolved. While such an adaptation is frequent in families such as Bombyliidae, Nemestrinidae and Tabanidae, other families among the Fynbos flies also present a long proboscis, which in some of these cases is unique in the families represented (Stuckenberg 1998). The following are notable examples:

Arthroteles Bezzi (Rhagionidae). The only rhagionid genus whose species have an elongate, slender proboscis, formed by lengthening of the labium. Its sister-group is the locally represented genus *Atherimorpha* White, in which the proboscis has the short, stout form normal in Rhagionidae.

Peringueyomyina barnardi Alexander (Tanyderidae). This endemic, monotypic genus has a very limited distribution in the Fynbos (Duxbury & Barraclough 1994); it is unique in the family in its elongate, slender head form; there is ventral, tubular extension of the head capsule, bearing small mouth-parts terminally.

Rhynchoheterotricha stuckenbergae Freeman (Sciaridae). The only Afrotropical sciarid with a very long proboscis. It is formed by a slender, tubular extension of the head capsule, bearing the mouth-parts at the apex.

Forcipomyia subgenus *Rhinohelea* de Meillon & Wirth (Ceratopogonidae). This contains two species of small midges with uniquely elongate mouth-parts, one of these collected from a flowering *Erica* species (B.R. Stuckenberg unpubl.).

Many other cases could be cited of Fynbos flies with less remarkable, but nevertheless unusual mouth-part elongation. In the rhinophorid genus *Phyto* Robineau-Desvoidy, progressive lengthening of the proboscis occurs in a group of closely-related species (Pape 1997) found in the southwestern Cape mountains.

In the montane environment of the Drakensberg escarpment in the east of the country, another rich flora is present, where a large variety of nectar-feeding flies occurs. Among these is a species of *Arthroteles* Bezzi, obviously derived from the main occurrence of the genus in the Cape mountains. Proboscis elongation is conspicuous among the Drakensberg dexiine Tachinidae. In this grassland flora are many species of *Helichrysum* (Compositae), the conspicuous flowers of which attract these flies.

4. Namibia: Deserts and Arid Savannas

Extending along the Atlantic coast of Namibia and southern Angola is the spectacular Namib Desert, the oldest in Africa, the aridification of which began in the early Middle Miocene (*ca.* 16 Mya) (Barnard 1998). The aridification of the Namib Desert also impacted on the adjacent mountains (Namibian Escarpment and desert inselbergs). Among the Diptera are

Diptera Diversity: Status, Challenges and Tools
(eds T. Pape, D. Bickel & R. Meier). © 2009 Koninklijke Brill NV.

Figure 6.2. The Gondwanan empidid genus *Homalocnemis* Philippi occurs in Chile and New Zealand; one species is recorded on the hyper-arid Namib Desert coast.

peculiar, highly-adapted desert mydids that survive through autogeny; the flies have vestigial mouthparts, and their larvae store nutrients for oogenesis (Wharton 1982). Adaptations to extreme xeric conditions are also demonstrated in the camillid genus *Katacamilla* Papp, recorded as breeding in dung in rock hyrax abodes and in arid bat caves in Namibia; eggs have been shown to survive extended periods of desiccation in a viable state, larval development being triggered by seasonal precipitation or the urine of bats and other cave-dwelling mammals (Barraclough 1998, Kirk-Spriggs *et al.* 2002). Some extraordinary discoveries of old lineages of Diptera were unexpected in the Namib Desert: a species of the empidid genus *Homalocnemis* Philippi (Fig. 6.2) was collected on a flowering succulent

Figure 6.3. The Brandberg Massif is the most spectacular topographical feature of the Namibian landscape; covering an area of 650 km² and rising 1.8 km above the Namib peneplain it predates the break-up of Gondwana. With its largely succulent plateau flora and orographic rainfall, it has acted as a refugium for arid Gondwanan insects *(NASA)*.

between the desert dunes and the beach — the other species of the genus occur in humid forests in Chile and New Zealand (Chvála 1991), and two species of the psychodid genus *Nemapalpus* Macquart were found in rock hyrax abodes in the arid highlands (Bethanie District and the Khomas Hochland) — elsewhere, the species of the genus occur in humid forests (Stuckenberg 1978). These flies must be solitary survivors of a Namibian mid-Tertiary woodland fauna. Other notable examples of flies restricted to the hyper-arid region of Namibia include *Orthactia deserticola* Lyneborg in the Therevidae (M. Hauser, pers. comm.), *Zumba antennalis* (Ville-neuve) in the Calliphoridae *sensu lato* (Kurahashi & Kirk-Spriggs 2006).

On the edge of the Namib Desert is the magnificent Brandberg Massif (Fig. 6.3); Namibia's highest mountain (highest peak Königstein 2,575 m.a.s.l.), which comprises a massive inselberg 650 km^2 in size, rising 1.8 km above the Namib peneplain. The Brandberg is a granitic ring complex, which pre-dates the break-up of Gondwana and thus also the change in continental climatic and environmental conditions that prevailed during the Plio-Pleistocene (Marais & Kirk-Spriggs 2000). Geologically, it consists of a series of alkali granites that intruded into the throat of an active volcano in the Early Cretaceous (*ca.* 300 Mya) (Miller 2000). The extensive undulating upland plateau (*ca.* 2000 m) exhibits a winter rainfall climate and associated flora, and shares floral elements with the Succulent Karoo biome of southern Namibia (Kirk-Spriggs 2003). The orographic rainfall, and vegetation of the Brandberg, coupled with its long isolation, has created refugia for Gondwanan faunal elements, and it has a relatively high proportion of endemic species as a result. Gondwanan elements have been identified in the coleopterous family Cerambycidae (Adlbauer 2000), and most strikingly, in the recent discovery of the genus *Alavesia* Waters & Arillo (Empidoidea), previously only known from Cretaceous amber (B. Sinclair, unpubl.). The Brandberg has been the subject of a dedicated floral and faunal biodiversity study and contributions on numerous dipterous families have appeared in the volume *Dâures — Biodiversity of the Brandberg Massif, Namibia* (Kirk-Spriggs & Marais 2000). Notable endemic species on the Brandberg are: the dolichopodid *Schistostoma brandbergensis* (Shamshev & Sinclair 2006), the vermileonid *Leptyoma* (*Perianthomyia*) *monticola* (Stuckenberg 2000), the mythicomyiid genus *Hesychastes* (Evenhuis 2001) and species *Psiloderoides dauresensis* (Kirk-Spriggs & Evenhuis 2008).

Namibia has two clearly-defined areas of endemism, the arid northwestern escarpment, the origin of which was initiated by the slow continental break-up of west Gondwana 130–145 Mya (Barnard 1998), and parts of which still retain relict Gondwana land surfaces, and the southern winter-rainfall zone (Succulent Karoo biome, Gariep Centre), that is restricted to the southern coastal region bordering the Orange River (Irish 1994). Endemism is tangibly demonstrated in numerous groups of flora and fauna in these two areas (Barnard 1998, Simmons *et al.* 1998); studies of the Diptera occurring on the Namibian Escarpment have been few however. An example of an escarpment-restricted species is the endemic *Thoracites nigrifacies* Kurahashi (Calliphoridae *sensu lato*) (Kura-

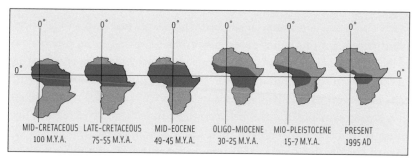

Figure 6.4. Inferred changes to the extent and distribution of forest cover in Africa since the mid-Cretaceous. Shoreline changes are not represented. Changes in the position and orientation of the African continent are due to continental drift
(after Clarke 2000b).

hashi & Kirk-Spriggs 2006). Examples of species apparently restricted to the Succulent Karoo of southern Namibia are more numerous and occur in the Vermileonidae (Stuckenberg 2000), Mythicomyiidae (Evenhuis 2000), Dolichopodidae (Grichanov *et al.* 2006), Calliphoridae *sensu lato* (Kurahashi & Kirk-Spriggs 2006), and Tephritidae (Hancock *et al.* 2001).

While the 'arid' savanna biome, which covers the greater part of Namibia, undoubtedly has a rich Diptera fauna and includes many species more widely distributed in the temperate south and more northerly tropical parts of Africa (e.g., Bombyliidae — Greathead 2000, 2006; Dolichopodidae — Grichanov *et al.* 2006), it has now been clearly demonstrated that the most species-rich area for Namibian Diptera is the 'mesic' savanna of the northeast of the country (Kavango and Caprivi Regions) (e.g., Calliphoridae *sensu lato* — Kurahashi & Kirk-Spriggs 2006; Tephritidae — Hancock *et al.* 2001, 2003; Lonchaeidae — McGowan 2005). Sampling in this region using Malaise trapping and other techniques has resulted in the single greatest increase in dipterous species now known to occur in the country (A.H. Kirk-Spriggs, pers. obs.).

5. The Rainforests

Africa is unusual in that its equatorial rainforests have had a history of radical disturbance due to Neogene climatic change (Maley 1996). Forest cover has changed dramatically since the Mid-Cretaceous (100 Mya), when forest covered the greater part of West Africa and what is today the western Sahara Desert (Fig. 6.4). By the Oligo-Miocene (30–25 Mya) these forests formed a continuous forest belt which stretched across the African continent and from which forest cover has regressed to its present day extent (Clarke & Burgess 2000). This explains why many taxa occurring in the Eastern Arc Mountains of Kenya and Tanzania share affinities with taxa in the Guineo-Congolian rainforest (Clausnitzer 2003), and there are also strong floristic affinities between these now disjunctive regions that must pre-date the mid-Tertiary uplift of these mountains. The Eastern Arc Mountains do, however, exhibit interesting affinities with other regions, particularly Madagascar and the mountains of West Africa (Burgess *et al.* 2007). Much of the vast Congo Basin forest is rooted in sand (*sables ocres*) that was a dune desert in the Miocene, and it has been hypothesised that the present range of this forest is, therefore, relatively recent (Clarke 2000b). This instability is reflected in the equatorial rainforest fauna — it is remarkably low in diversity, and there is no evidence of a highly adapted canopy fauna (Meadows 1996). Consequently, there are few lowland rainforest endemics of systematic significance among the Diptera.

The current state of knowledge of canopy invertebrates in tropical forests was reviewed by Basset (2001). He lists eight studies specifically related to canopy invertebrates in the Guineo-Congolian rainforest: three studies in Cameroon (Basset *et al.* 1992, Dejean *et al.* 1992, Watt *et al.* 1997), two in Uganda (Corbet 1961; Wagner 1998, 1999, 2000), two in the Democratic Republic of Congo (as Zaïre) (Sutton & Hudson 1980, Wagner 1997), and one in Rwanda (Wagner 1997). Twenty percent of these studies have focus on the Coleoptera, and only one (Corbet 1961) is specifically related to the Diptera. Basset (2001) notes, however, that although often neglected in taxonomic studies, Diptera play a much more significant role in arboreal community interactions than implied by their usual designation as 'tourists' in the canopy. Indeed, if biomass and density of canopy invertebrates is considered worldwide, the Diptera rank fourth after the Blattodea, Hymenoptera and Coleoptera (Basset 2001).

In addition to the canopy studies noted above, Moran *et al.* (1994) studied the herbivorous insects on twelve species of evergreen broadleaved trees in a small relict forest in the Pondoland Centre in South Africa (Fig. 6.1: 15). They found the fauna to also be markedly depauperate as compared to that reported from native broadleaved trees from other parts of the world, despite the fact that the sampled forest included numerous endemic and rare tree species. In the more temperate south, Moran & Southwood (1982) undertook canopy fogging of trees in the Grahamstown and

Hogsback area of the Amathole Mountains of the Eastern Cape, South Africa. This study was primarily concerned with guild composition rather than an estimation of biodiversity. The late Harold Oldroyd (1957) observed that the Congo forest Tabanidae were inhabitants mainly of forest margins and the transitional zones between tropical forest and savanna (Fig. 6.1: 10, 11).

6. Afromontane Biome

Afromontane forest occurs in South Africa, on the eastern escarpment and other sites with orographic rain (Partridge & Maud 1987). In successive countries to the north, similar forests are scattered along the rift valley escarpments and on uplands and isolated mountains. The Ethiopian Highlands, which began to rise in the Tertiary (*ca.* 75 Mya), form a rugged mass of mountains in Ethiopia, Eritrea and northern Somalia in north-eastern Africa, reaching altitudes of 1,500–4,600 m.a.s.l. The opening of the series of rift valleys in the closing stages of the Pliocene divided these highlands, thus creating Africa's great salt lakes (e.g., Fig. 6.1: 12). This rifting gave rise to large, alkali basalt shield volcanoes in the Ethiopian and Virunga regions beginning about 25–29 Mya. The associated forests share a characteristic dipteran fauna whose patterns of endemism and cladogenesis suggest that the apparent ecological gaps between these forests may not in fact invariably be barriers to dispersal and faunal exchange. Species of the chamaemyiid genus *Leucopis* subgenus *Leucopella* Malloch have been shown to occur in the Great Rift Valley forests and the Ethiopian Highlands, but one species is restricted to the Arabian Peninsula (Gaimari & Raspi 2002). A similar distribution is exhibited by true examples of the curtonotid genus *Cyrtona* Séguy, species of which appear to have radiated along the Great Rift Valley forests and dispersed into coastal areas of South Africa (A.H. Kirk-Spriggs, unpubl.).

There may be a localised montane hot spot in the Cameroon area, where a still undescribed blepharicerid occurs (B.R. Stuckenberg, pers. obs.). This is supported by the fact that the Cameroon forests have been demonstrated as an area of endemicity for birds (de Jong & Congdon 1993) and butterflies (Stuart *et al.* 2003). The Ruwenzori Massif in Uganda also has some remarkable Diptera, such as the endemic, monotypic psychodid genus *Eutonnoiria* Alexander, one of only three endemic genera of Psychodidae in Africa (Duckhouse & Lewis 1980).

7. Arid Coasts

While most of Africa has undergone ceaseless climate fluctuations, generating the expansion and contractions of forests and savannas over millennia, the continent's arid coasts have remained relatively stable (Barnard 1998). A study of the marine-littoral biogeography of the Diptera of the south-western and southern African seaboard (Kirk-Spriggs *et al.* 2001) has indicated that the distribution of coastal Diptera is largely influenced by the effects of the cold Benguela and warm Agulhas currents, and the associated primary production of kelp, and some interesting distributional patterns are illustrated for the Canacidae (as Tethinidae) and Sarcophagidae, as discussed by Kirk-Spriggs *et al.* (2001).

8. Savanna

Savanna (termed the Sudano-Zambezian Region, Fig. 6.1: 2 & 3) characterises much of the vast expanse of the flat landscape of tropical Africa around the Guineo-Congolian Region (Werger & Coetzee 1978). It varies from grassland to woodland with grasses, and it is an old biome with a highly adapted invertebrate fauna, which forms complex mosaics as exhibited, for example, in the chrysomelid beetle genus *Monolepta* (Kirk-Spriggs 2003, Wagner 2001). With the limited rainfall confined to the warmest months, the Diptera of the savanna are highly seasonal and also fire-adapted (Phillips 1965). The vast extent of this relatively homogeneous grassland is reflected in the wide distribution of many of the flies (e.g., Asilidae, Bombyliidae), with even small graminivorous acalyptrates, e.g., Chloropidae, ranging from Ethiopia to South Africa. Although great expanses of the African savanna remain poorly sampled and our knowledge is largely based on scattered records and type localities, centres of endemicity in the grasslands are apparent. Poor sampling of African grasslands in general may be largely due to the misconception that such habitats are largely monotonous.

The environment has produced convergent adaptations of body form, colouring and behaviour among the smaller flies (B.R. Stuckenberg, pers. obs.). Ismay (2000) has noted, for example, the number of small species of Chloropidae in the Afrotropics that are bright yellow in colour with a distinct black, shining pleural spot. He notes that this similarity occurs in the chloropid genera *Arctuator* Sabrosky, *Conioscinella* Duda,

Oscinimorpha Lioy and *Pselaphia* Becker as occurring in Namibia, and in some Milichiidae, Phoridae and Hybotidae. Water-retaining cavities and rot holes in savanna trees are important breeding sites for taxa, and synchronous, wet-season flowering of trees provides a critical resource of nectar and pollen for many Diptera. The role of Diptera as pollinators of flowering trees is surely underestimated in Africa. Fogging of flowering *Acacia* trees in Namibia, for example, has revealed an extensive associated Diptera fauna, which is predominated by the Calliphoridae *sensu lato* (A.H. Kirk-Spriggs, pers. obs.). Radiation of the larger savanna mammals prompted diversification of the Oestridae, the greatest development having been in the Afrotropics and the Palaearctic Region — the two African rhinoceros species are hosts to the larvae of the two magnificent species of *Gyrostigma* Brauer, the future of which is looking increasingly precarious with the dwindling numbers of their hosts (Barraclough 2006a).

There have been few studies of insect canopy faunas in African savannas using modern fogging and mist-blowing techniques. The only published study is that of Krüger & McGavin (1997), who studied the insect fauna associated with *Acacia* species in Mkomazi Game Reserve in northeast Tanzania using a mist blower. They found canopy faunas to be highly diverse, and that the Coleoptera diversity paralleled that of other tropical areas. The Diptera component represented 1.0% of the biomass share. The National Museum of Namibia has also recently undertaken extensive canopy fogging of individual trees in the arid and mesic savanna biome of Namibia, but these studies remain unpublished.

9. Madagascar

Madagascar is an ancient fragment of Gondwana which separated from Pangea from the mid-Jurassic to the Early Cretaceous. The dominant biome during the Tertiary was probably monsoon rainforest, with drier forests and scrub in the west (de Wit 2003). The Madagascan Diptera encompass a remarkable mix of relationships. Two endemic Gondwanan relicts are the blepharicerid genus *Paulianina* Alexander and the acrocerid genus *Parahelle* Schlinger. Oriental and African relationships occur in many families, though few with South Africa. Some large families are poorly represented — there are few Bombyliidae (18 species in eight genera, 16 of which are endemic) and Empididae (although not officially recorded from Madagascar there are numerous species), and the Nemestrinidae are represented by only one species. Much of the fauna could still be unknown — the first vermileonid was discovered recently by Mike Irwin (Stuckenberg 2002), two species of Rhinophoridae await description (A.H. Kirk-Spriggs, pers obs.; T. Pape, unpubl.), at least 100 undescribed species of Lauxaniidae have been collected (B.R. Stuckenberg, pers. obs.), and additional families are likely to be recorded. With fewer than 2,500 recorded species, considerably more collection and study of Madagascar Diptera is obviously required, and is now being undertaken on a coordinated basis (see below).

10. Afrotropical Faunistics

The size, taxonomic composition and characteristics of the Afrotropical Diptera have received growing attention in recent decades. With the publication of the first comprehensive systematic synopsis of this fauna — *Catalogue of the Diptera of the Afrotropical Region* (Crosskey 1980) — it was possible to obtain statistical data on the taxa described or recorded to that date, as well as general information on distribution and biology. This was one of the best catalogues of its kind, and immediately became the basis for further dipterological studies. The total number of Afrotropical species recorded in that work was 16,318, in 95 families, of which 1,944 species were known from Madagascar and 14,669 species from the continental Afrotropical Region.

Growing concerns about the developing biodiversity crisis led to the publication of various synoptic overviews directed at assessments of the extent to which knowledge of the world's fauna in general was known. The first contemporary prediction of the probable number of insect species on Earth was that of Erwin (1982). He gave a figure of 30 million, based on his pioneering work on the canopy fogging of tropical rainforest trees. This prediction has since been substantially downscaled. Gaston (1991) took a 'taxonomist-based' approach, in which specialist systematists were asked to provide predictions of the probable number of described and undescribed species. From these tentative but possibly more reliable data he came up with a tangible figure of five million. Contributions relating to the Afrotropics began to appear. Estimations in respect to the South African insect fauna were provided for each insect order in the textbook *Insects of Southern Africa* (Scholtz & Holm 1985). The chapter on Diptera by Barraclough & Londt (1985) gives a total of 6,243 recorded species, although the origin of these data remains unclear. Scholtz & Chown (1995) and Scholtz (1999) reviewed the same data for all the insect orders, and a synoptic analysis by Miller & Rogo (2001) covered a wide range of associated topics. It had been estimated by Barraclough that the true size of the dipteran fauna in the Afrotropics could be double what had already become known by 1985 (Scholtz & Chown 1995). Some of this literature was discussed by Kirk-Spriggs (2003) in an introductory study of African biogeographical patterns.

The most recent published review of the Afrotropical Diptera, by Irwin *et al.* (2003), appeared in the stupendous book *The Natural History of Madagascar* (Goodman & Benstead 2003). Drawing on the computerised *BioSystematic Database of World Diptera* online from the USDA/National Museum of Natural History, Smithsonian Institution (Thompson 2006), it could be reported that by early 2001, 19,051 valid fly species had been recorded in the Afrotropical Region. This represented about 13% of the world's fly species. Of these, 1,796 were known from Madagascar — a smaller total than that obtained from data in the 1980 *Catalogue*.

The present study attempts to revise data for the Afrotropical Diptera (Table 6.2). For this, the excellent resources of the *BioSystematic Database of World Diptera* have been invaluable. It must be stressed, however, that the accumulated taxonomic literature deals almost entirely with morphospecies, which can be recognised by traditional, usually morphological criteria (but with some interesting exceptions among black flies and mosquitoes, see below), and that no calculated projections of faunal size through standardised collecting techniques or rates of accumulation of taxa is available (but see Dikow *et al.* 2009). The true number of existing species, including cryptic and sibling ones, the recognition of which requires genetic or molecular studies, cannot even be guessed at. It is salutary to recall, for example, that whereas just over 120 species of *Anopheles* Meigen mosquitoes were recognised in 1987, later genetic studies showed the existence of species complexes within every taxon examined, leading to the estimation that the African anopheline fauna may actually attain 600 species (Coetzee 1999). Another startling example is provided by the vector of onchocerciasis in West Africa, once considered to be the species *Simulium damnosum* Theobald, which has since been shown by cytogenetic studies to be a complex of at least 40 species (Coetzee 1999). Such situations may be far more prevalent among Diptera than has been realised. A case in point is the tephritid genus *Ceratitis* MacLeay; it has been reported that close study of apparently polyphagous species revealed the existence of species complexes involving undescribed stenophagous species (De Meyer 2001).

In our study we follow the approach of Gaston (1991) to estimate the number of undescribed species from the Afrotropical Region. Practising systematic authorities on each fly family were contacted and asked to confirm the number of described species provided through the *BioSystematic Database of World Diptera*, and also to predict the probable number of undescribed species likely to occur in the region. In the case of families for which no systematists are currently engaged in research, competent systematists, well-versed in the Afrotropical fauna, were asked to make such predictions. In the case of families for which the fauna is extremely poorly known, e.g., Cecidomyiidae, Phoridae and Sphaeroceridae, it was only possible to provide a highest and lowest estimation; this is why two figures must be provided for the Diptera of the Afrotropical Region as a whole — a lowest and highest estimation.

11. Estimate of Total Afrotropical Diptera Fauna

Irwin *et al.* (2003) discuss what they term the 'discovery phase', being the part of a timeline during which most species in a given environment are discovered and described. They note that this is quite advanced for most of the North American insect fauna and is virtually complete in the case of the best-documented insect fauna of Great Britain. The situation in the Afrotropics is markedly different. Table 6.2 provides details of the number of described species for each family known to occur in the Afrotropical Region, plus specialist estimates of the number of undescribed species. Figures presented in the table indicate the following:

> It can be concluded that 19,689 species of Diptera are currently known from the Afrotropical Region, this representing an overall increase of 3,371 species in the past 26 years, since the publication of Crosskey's (1980) *Catalogue*.

> Specialist estimates of the overall number of undescribed species range from 29,961–37,930 (*mean*: 33,946). The range difference of 7,969 is largely due to the range-estimates given for three of the least known families, the Cecidomyiidae (5,000–10,000), Phoridae (2,000–3,000), and Sphaeroceridae (1,000–2,000). A round figure of 30,000 species at the lower end is therefore reasonable for general purposes.

DIPTERA DIVERSITY: STATUS, CHALLENGES AND TOOLS
(EDS T. PAPE, D. BICKEL & R. MEIER). © 2009 KONINKLIJKE BRILL NV.

Table 6.1. Numbers of described species of Diptera from the six zoogeographical regions of the world and the percentage of the world fauna. Data from Thompson (2006), unless otherwise stated.

Zoogeographical region	Number of known species	Percentage of known world fauna
Afrotropical Region*	19,689	13%
Palaearctic Region	40,291	27%
Oriental Region	21,559	14%
Australasian/Oceanian Region	17,948	12%
Nearctic Region	21,356	14%
Neotropical Region	29,783	20%
Totals	150,626	100%

*Numerical data based on familial information provided in Table 6.2.

This estimate implies that ⅔ to ½ of the dipteran species might be known. If the total number of species added to the regional list from 1981–2006 is considered (3,371), this gives an average of 129 new species per year. Hypothetically, at that average rate of taxonomic growth it would take upwards of three centuries (231–289 years) to reach the end of the discovery phase.

12. Comparison to Other Zoogeographic Regions

The number of species known to occur in the six zoogeographical regions is here briefly reviewed, based on the revised figure for the Afrotropical Region presented above, and data from the *BioSystematic Database of World Diptera* (Thompson 2006) for the other regions. The number of genera from the six regions is not considered, as that taxonomic ranking is not necessarily definitive between regions. From data presented in Table 6.1 the Afrotropical Region accounts for a mere 13% of the species of Diptera known worldwide. This is rather a reflection of taxonomic effort than of actual numbers. Also, the geographical extent of each of the five regions varies enormously, with the Palaearctic comprising a considerably greater proportion of landmass than the other five; the Oriental and Australasian/Oceanian regions comprising the least. It is not surprising, therefore, that the Palaearctic Region accounts for 27% of the world spe-

Figure 6.5. The remarkable *Mormotomyia hirsuta* Austen, the only known species of the endemic family Mormotomyiidae, associated with bats in Kenya.

cies, being as it is the largest of the zoogeographic regions and having as it does the best-studied fauna. Predictive figures for the six regions may present a more plausible estimation of the world Diptera fauna.

13. Endemic Afrotropical Families

Four families of Diptera are now known to be endemic to the Afrotropical Region, *viz.* the Mormotomyiidae, Marginidae, Natalimyzidae and Glossinidae. A brief résumé of what is known for each of these families is provided below. McAlpine (1991) has intimated that aside from the Marginidae and Natalimyzidae to which he refers (as unknown family in the latter case), it may be assumed that further new dipterous families of limited diversity and distribution await discovery in the Afrotropics. Given the extremely limited distribution of the family Mormotomyiidae and the meagre distribution of the Marginidae in wet forests, these are the families of Diptera most at risk of extinction (McAlpine 1991).

Mormotomyiidae (Fig. 6.5) — This family of truly remarkable flies comprises a single species, *Mormotomyia hirsuta* Austen, which was described from a single locality, Ukazzi Hill, Garissa District, Kenya. Flies were found to occur in a cave-like cleft about a yard wide, with its horizontal and oblique side-cracks inhabited by unidentified bats. Larvae were found to develop in the accumulated dung of bats, and to be saprophagous, either on the decaying organic matter itself or predacious on other inhabitants. Upon entering the cave at Ukazzi, H.B. Sharpe noted that the flies *'came floating down from above like feathers'* (Austen 1936: 430); a behaviour later noted by van Emden (1950), who remarked that the fall of the flies is apparently much slowed by the long, shaggy, hair-like body setae and occurred in a slight spiral motion. As well as the disproportionately long legs, which abound in numerous 'mummy-brown' hair-like setae, the wings are drastically reduced to dysfunctional thin straps, and the halteres are entirely absent. Austen (1936) discussed the relationship of the family based on morphological characters. He believed it to be an acalyptrate fly 'perhaps' most closely related to the Sphaeroceridae (as Borboridae). Later van Emden (1950) described the immature stages (egg, third instar larva and puparium) and discussed relationships based on larval and adult morphology. He considered that the genus represents a well-founded family intermediate between the Scathophagidae (as Cordyluridae) and the Acalyptratae.

Natalimyzidae (Fig. 6.6) — This family was only formally described in 2006 (Barraclough & McAlpine 2006), although it has been recognised as a distinct family for some forty years. The family is apparently widespread in Africa, with current records from South Africa (Western Cape, Eastern Cape, Kwazulu-Natal, Mpumalanga, Limpopo), Zimbabwe, Kenya and Nigeria. Although regarded as predominantly associated with grassland in South Africa, recent Malaise trapping in indigenous forest of the Eastern Cape has indicated that the family is also common in forests (A.H. Kirk-Spriggs, pers. obs.). Larvae have not yet been described, but appear to be micro-floral grazers in decaying grass (Miller 1984). Some 20 species are known from South Africa alone. The family is currently ascribed to the subfamily Sciomyzoidea and molecular studies are currently underway as part of the *FlyTree* Project (B. Wiegmann, unpubl.). The family was included in an identification key to the acalyptrate flies of southern Africa as 'New family' by Barraclough (1995).

Marginidae — This family is one of the two endemic Afrotropical acalyptrate families described in the past few decades (McAlpine 1991). The description was based on two species of the monotypic genus *Margo*, namely: *M. aperta* from Chirinda Forest, near Mount Selinda in Zimbabwe, and *M. clausa* from near the coastal town of Manakara in Mada-

Figure 6.6. An undescribed species of the endemic family Natalimyzidae from the Eastern Cape, South Africa. The family was only described in 2006.

gascar. McAlpine (1991) discussed the systematic relationships of the family in some detail and suggested its placement in the Opomyzoidea. The occurrence of the species in both the continental Afrotropical Region and Madagascar suggests that additional species may await discovery in humid African forests. The immature stages remain unknown and molecular work is required to correctly determine superfamilial placement.

Glossinidae — *Tsetse* require no introduction and are well known due to their status as vectors of pathogens such as African trypanosomiasis (sleeping sickness) and its animal equivalent *Nagana*. Although the family is restricted to tropical Africa today, two Oligocene fossil species, G. †*oligocenus* Scudder and G. †*osborni* Cockerell, are known from the United States and are frequently used as examples of regional extinction events (Grimaldi 1992, Evenhuis 2006). The family was also recently found in Miocene European deposits (Wedmann 2000). Twenty-three species of the genus are extant in tropical Africa today and an extensive literature has developed related to their taxonomy, behaviour, habitat preferences, development, and role as disease vectors. They are readily distinguished from other blood-sucking flies by the hatchet-shaped discal cell of the wing and the multi-plumed arista.

14. Some Absent Families

Notable families absent from the Afrotropical fauna are the Atelestidae, Hilarimorphidae, Pseudopomyzidae, Spaniidae, Trichoceridae (Africa is the only continent without this family of lower Diptera) and Xylophagidae.

15. Largest Afrotropical Families

The three largest families of Diptera in the region are the Asilidae (1,643), the Bombyliidae (1,425) and the Limoniidae (1,034). These three families alone constitute almost 21% of the fauna – none of which is likely to double in numbers through future research. Other families with >500 described species are the Ceratopogonidae (913), Chironomidae (556), Culicidae (752), Dolichopodidae (738), Muscidae (932), Syrphidae (562), Tabanidae (815), Tachinidae (1,013), and Tephritidae (947). The total number of species for all 12 families mentioned here is 11,330, which is more than half (57.5%) of the fauna. Many of the smaller families could, therefore, be doubled by new species without markedly affecting the total estimate.

16. Least Known Families in the Afrotropics

The following families are regarded as least known (>1000 undescribed species predicted), and are ranked in order of importance, i.e., least known first. Figures for Madagascar are derived from Irwin *et al.* (2003), with Phoridae updated from Disney (2005).

> Cecidomyiidae — gall midges are by far the least known family of Afrotropical Diptera. They are under-collected and under-studied. Most research has been directed to those species of agricultural importance or to species associated with particular plant hosts. Currently only 213 described species (3 in Madagascar!), representing only an estimated 2–4% of the true number of species (5,000–10,000).

> Phoridae — scuttle flies are extremely diverse in the region and remain under-studied. Currently only 418 described species (18 in Madagascar), representing only an estimated 12–17% of the estimated number of species (2,000–3,000).

> Sphaeroceridae — lesser dung flies are an enormous group which has received only limited attention. Currently only 321 described species (11

in Madagascar), representing only an estimated 14–24% of the estimated number of species (1,000–2,000). These estimations may be rather high, given that the distribution of dung-dependant flies in savannas is hardly known, particularly the dependence of species associated with the dung of indigenous species as opposed to domesticated species.

Ceratopogonidae — biting midges are also poorly known. Most recent research has focused on the *Culicoides* Latreille vectors of veterinary arboviruses. Currently 913 described species (12 in Madagascar), representing 31% of the estimated number of species (2000).

Mycetophilidae — fungus gnats are extremely prolific in tropical forests and in savanna in the wet season. Currently only 249 described species (a single endemic species in Madagascar), representing 11% of the estimated number of species (2000).

Sciaridae — dark-winged fungus gnats have been very poorly studied, but are known to be prolific in wet forest and savannas. Currently only 71 described species (6 in Madagascar), representing 3% of the predicted number of species (2000).

Dolichopodidae — Currently 738 described species (52 in Madagascar), representing 33–42% of the predicted number of species (1,000–1,500).

Tachinidae — Currently 1,013 described species (187 in Madagascar), representing 51% of the predicted number of species (2,000).

17. Notable Taxonomic Growth

Examples of productivity since 1980 by some individual specialists:

Asilidae: the largest Afrotropical family. Jason Londt has described 25% of the species. He enlarged the genus *Neolophonotus* Engel by 192 new species, or 76% of the present total of 253 species — this is currently the largest Afrotropical genus of Diptera.

Camillidae: Originally considered to be a minor Palaearctic family. David Barraclough increased the number of Afrotropical camillids from 1 genus and 2 species to 4 genera and 20 species, showing it to be primarily an African family.

18. Best Known and Collected Countries

The best known dipterous faunas are those for the modern states of South Africa, Namibia, Kenya, and Nigeria. This assumption is based on the

Diptera Diversity: Status, Challenges and Tools
(eds T. Pape, D. Bickel & R. Meier). © 2009 Koninklijke Brill NV.

extent of Diptera material from these countries in European and African museums.

19. Possible Gondwanan Elements in the Afrotropical Diptera

It has been long known that South Africa has the most distinctive invertebrate fauna in the Afrotropics. Included are various taxa the phylogenetic relationships of which indicate them to be of ancient occurrence in this region (Kirk-Spriggs 2003, Stuckenberg 1962). The explanation for their presence has been that they are remnants of a fauna that diversified and dispersed across the Gondwanan landmass before its prolonged break-up into separate continental masses. These Palaeogenic elements thus have been termed 'Gondwanan', and their presence in South Africa is of great interest, especially as their distribution pattern concurs with a biogeographical situation also involving South America and Australia. Each of these three continents has essentially two insect faunas — a southern one, mostly associated with relatively temperate environments, the other mainly in more northerly, warmer or even tropical latitudes. These austral insect faunas have taxa in common and appear to share an evolutionary history that reflects continental drift. Two areas of such putative Gondwanan insects occur in the Afrotropics — namely, in South Africa and in Madagascar. They need to be considered separately.

South Africa has Africa's oldest mountains. They are of two kinds, with completely different origins. In the south of the country, extending more or less east-west, with a smaller interlocking north-south section in the west, is a series of elongate ranges, constituted by similar sedimentary rocks, known as the Cape Fold Mountains. They are part of an ancient orogeny that predated the break-up of Gondwana. At that time they were continuous with old mountains in south-eastern Australia, with the trans-Antarctic ranges, and even with a small range in the Buenos Aires Province of Argentina known as the Sierra de la Ventana. Recognition of this once enormous orogeny extending across Gondwana, was an insight by the South African geologist Alex L. Du Toit, whose landmark book *Our wandering continents; an hypothesis of continental drifting* (1937) provided the first elaboration of continental relationships and drift.

Also in South Africa is the eastern Great Escarpment, called the Drakensberg over much of its length. This was initiated as a result of the separation of Antarctica from southeastern Africa in the Jurassic, when a

Figure 6.7. Africa's only tanyderid, *Peringueyomyina barnardi* Alexander from the Cape Fold Mountains of South Africa.

new drainage system formed in the hinterland of the new South African coastline, flowing eastwards towards the expanding Indian Ocean. Extremely prolonged water erosion established by this drainage, operating throughout the Mesozoic and twice rejuvenated by Cenozoic episodes of continental uplift, created this escarpment in eastern South Africa. It was progressively eroded westward, until the presence of a massive, almost horizontal sequence of hard basaltic rocks retarded the rate of erosion and resulted in steep exposure of an underlying, very thick sequence of sediments. With permanent benefit of summer rains derived from the expanding, warm Indian Ocean, this escarpment could acquire and retain freshwater and terrestrial invertebrate faunas during much of the Mesozoic and Cenozoic (Partridge & Maud 1987).

All these mountains are preserved ancient landforms with a characteristic biota, which include putative Gondwanan Blephariceridae, Thaumaleidae, Psychodidae, Empididae, Africa's only tanyderid (Fig. 6.7), near-basal Chironomidae, and other flies with possible austral relationships. Among the more convincing cases are the following taxa:

Blephariceridae — all the South African species are in the genus *Elporia* Edwards, which belongs to the tribe Paltostomatini, the other genera of which are all Neotropical (Stuckenberg 2004). *Elporia* is probably the sister-group of the genus *Kelloggina* Williston, which is limited to the old coastal highlands of central and southern Brazil. Because of the strong cladistic support for blenpharicerid classification, a good case can be made for a Gondwanan origin of *Elporia*.

Thaumaleidae — only two species of this small nematoceran family associated with mountain streams have been described from the Afrotropics (Sinclair & Stuckenberg 1995). One occurs in the Natal Drakensberg, the other in the Cape Fold Mountains. They constitute the endemic South African genus *Afrothaumalea* Stuckenberg, which is part of a monophylum of genera also occurring in Australasia and temperate South America. The closest relative of *Afrothaumalea* appears to be the Australian and southern Chilean genus *Niphta* Theischinger (Sinclair & Stuckenberg 1995).

Psychodidae — the distinctive genus *Gondwanotrichomyia* Duckhouse was erected for two species limited to montane evergreen forests of eastern South Africa (Duckhouse 1980). Related species occur in Australia and Chile.

Rhagionidae — a genus with species of 'archaic' habitus, *Atherimorpha* White, has been recorded as well represented in South Africa (Nagatomi & Nagatomi 1990), inhabiting mesic montane grasslands and the fynbos flora of the Cape Fold Mountains. Although these species are classified as congeneric with evidently Gondwanan clades of *Atherimorpha* in southern South America and eastern Australia, this relationship has not yet been confirmed through morphological study. It was proposed first by Bezzi (1926), but may have been based on symplesiomorphies.

Empididae — the near-basal genus *Homalocnemis* Philippi (Fig. 6.2) was recorded from the edge of the Namib Desert by Chvála (1991); the genus is also recorded from Chile and New Zealand (see above). Other possible Gondwanan genera in the Empididae are discussed by Sinclair (2003).

Tabanidae — the endemic genus *Stuckenbergina* Oldroyd (1962) is the only Afrotropical member of the tribe Pangoniini. This tribe otherwise has a notable austral distribution, involving elements shared between South America and Australia. The two described South African species are associated with the Cape Fold Mountains.

Chironomidae — putative cases of Palaeogenic genera are discussed by Sæther & Ekrem (2003).

Simuliidae — distribution patterns and possible Gondwanan subgenera are discussed by Miranda-Esquivel & Coscarón (2003).

During the Gondwanan history of Madagascar, when this island was still united with Africa in the west and with India in the east — and when the compound Indian-Madagascan landmass was contiguous with the still united Australian + Antarctic continental sectors of Gondwana, which also had contact through western Antarctica to southernmost South America — the possibility of an evolution of a pan-austral insect fauna which dispersed variably over this huge area is entirely plausible (Wells 2003). Particularly significant in this regard is the recent remarkable expansion of knowledge of the palaeobotany of Antarctica (Hill & Scriven 1995). At the time when that continental area had a central position with contact to all of the other austral areas which later attained independent continental status with the break-up of Gondwana, Antarctica was not a frigid wilderness with scanty flora. Hill & Scriven stated that Antarctica was a key area in the development of the extant vegetation of the Southern Hemisphere, and that ancestors of the present austral flora were established there by the end of the Cretaceous. Angiosperms were present by the Mid-Cretaceous, and may have been invasive elements in Antarctica, possibly coming from South America *via* the Antarctic Peninsula. Madagascar + India could thus have shared faunal components that may have evolved and diversified in southern South America and Antarctica.

Expeditions to Madagascar in the late 1950s (by B.R. Stuckenberg) were directed at collecting Diptera in general, but with a special objective to search for what seemed to be the Gondwanan elements that were already known in South Africa. None of them could be found. Indeed, among the Diptera in Madagascar it was a blepharicerid genus well developed there, *Paulianina*, that was the only taxon for which a Gondwanan origin could at that time be plausibly postulated. This genus is the sister-group of the austral Neotropical — Australian genus *Edwardsina* Alexander, and the two genera together constitute the near-basal subfamily Edwardsininae. The biogeography of *Edwardsina* had long attracted attention, the genus having been considered a likely Gondwanan relict by earlier dipterists, such as R.J. Tillyard, A.L. Tonnoir, I.M. Mackerras and D.H.D. Edwards. *Paulianina* is classified in a different subfamily to that of the South African genus *Elporia* (Blepharicerinae), the sister-group of which may be the Brazilian *Kelloggina* (see above), so a separate explanation for the presence of Edwardsininae in Madagascar could be expected.

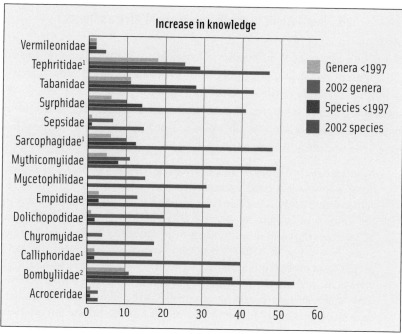

Figure 6.8. Examples of the increase in knowledge of 14 families of Diptera occurring in Namibia from 1997 to 2002. <1997 figures from Crosskey (1980). This increase has been due to dedicated sampling efforts using Malaise traps and other techniques, the field preparation of good quality material, and taxonomic efforts by specialist systematists.
[1]species numbers reduced by 50%
[2]species numbers reduced by 75%

There is still one other record of blepharicerids in sub-Saharan Africa — two finds of immature stages of a still undescribed genus and species apparently in the blepharicerine tribe Paltostomatini, in West Africa. This taxon was recorded as a species of *Elporia* (Germain *et al.* 1967), but that assessment is probably erroneous, as the larval stages differ in having one additional pair of prolegs. Also in Madagascar is an undescribed blepharicerid not related to Edwardsininae (Paulian 1954). It is known only from river systems draining off the Andringitra Massif, and it was once thought to be possibly a member of the tribe Apistomyiini. Adults were found in 1958, but determination of its relationships among the Blepharicerinae has been inconclusive (Stuckenberg 2004). This species may have been derived from tropical sub-Saharan Africa, outside the range of *Elporia*.

DIPTERA DIVERSITY: STATUS, CHALLENGES AND TOOLS
(EDS T. PAPE, D. BICKEL & R. MEIER). © 2009 KONINKLIJKE BRILL NV.

20. Innovations in the Bioinventory of Africa's Diptera

If our knowledge of the Diptera of Africa is to improve, despite the all too apparent constraints of the new millennium, a new approach to the sampling and study of the African Diptera fauna must be adopted. To assess such changes in approach that need to be considered, it is useful to examine and assess some of the more recent innovative approaches that have been undertaken both within and from outside Africa.

Namibian bioinventory — as a developing African country, the Republic of Namibia is a good example of a sub-Saharan country whose recent bioinventory initiatives have resulted in a substantial increase in the number of Diptera known from that country (see Fig. 6.8). The use of Malaise traps and other modern techniques to sample the dipterous fauna in a range of habitat types and biomes, together with the field preparation of good quality material, has encouraged international specialists to work on the Namibian fauna, and numerous faunal reviews have, and continue to, appear as a result. Targeted biodiversity studies, such as the National Museum of Namibia's Marine-littoral Survey 1998 and Brandberg Biodiversity Project have also been widely regarded as benchmark studies.

An Arthropod Survey of Madagascar's Protected Areas (1998–2009) — initiated by Mike Irwin (University of Illinois of Urbana-Champaign) and chiefly funded through the Schlinger Foundation, this project involves the long-term monitoring of Diptera by use of Malaise traps in protected areas throughout the island of Madagascar. Traps are serviced by locals on a monthly basis and most traps being continually deployed for at least one year, thus enabling seasonal data to be analysed. The project has trained local Madagascans as sorters, but sorted material is managed and distributed by the California Academy of Sciences in San Francisco. At the termination of the sampling phase (2009), it may be several decades before this material can be fully processed, but it is already turning up numerous families not previously recorded from Madagascar and is making great strides in our knowledge of the Malagasy fauna.

United Arab Emirates Insect Project — initiated by Sheikh Tahnoon bin Zayed al Nahyan from Abu Dhabi, and run by Antonius van Harten, this project is restricted to the United Arab Emirates, but also includes available material from Yemen and Oman, collected by van Harten and others (see above). The project is now in its third year, and the first of a series of volumes entitled *The Arthropod Fauna of the United Arab Emir-*

ates was recently published (van Harten 2008), with contributions on 17 families of Diptera by specialist researchers. This coordinated project is doing much to better our understanding of this fascinating transitional region.

The following projects, programmes and initiatives are here suggested as especially suitable for bringing the bioinventory of Africa's Diptera fauna forwards.

A Manual of Afrotropical Diptera — the publication of such a manual, providing generic keys, taxonomic information and notes on immature stages and biology, in line with manuals produced for the other zoo-geographical regions of the world, would do much to stimulate future interest in Afrotropical Diptera.

Catalogue of Afrotropical Diptera — it has been over 26 years since the publication of Crosskey's (1980) *Catalogue*. Although catalogues to various families of Diptera have appeared, which deal with the fauna on a world basis, there is now an urgent need for a new, fully up-dated Afrotropical catalogue, preferably with an on-line version.

Country faunal bioinventories — there is now a disproportionate amount of information available on the Diptera of South Africa and Namibia. If we are to gain better insight into the biological diversity and biogeography of Afrotropical Diptera it is necessary to undertake coordinated, long-term sampling using Malaise traps and specialist collection in other under-collected African countries.

Targeted biodiversity studies — multi-disciplinary studies of specific habitat types and biodiversity hotspots are urgently required, especially in threatened habitats. Examples of high priority areas are West African and Congo rainforests (especially Cameroon, Sierra Leone, Liberia, Côte d'Ivoire, and Ghana), the Eastern Arc Mountains, the Albertine Rift, the arid north-eastern regions of the Horn of Africa, the escarpments of South Africa, Namibia and Angola, the isolated inselbergs of Namibia and Angola, the Hoggar and Tibesti Mountains of the central Sahara and the Ethiopian Highlands. Recent steps towards an inventory of the species-rich and highly endemic biota of the Eastern Arc Mountains are currently being taken by the EU-funded network of excellence, *European Distributed Institute of Taxonomy*, Workpackage 7 (T. Pape, pers. comm.).

Table 6.2. Systematic list of known species of Afrotropical Diptera by family and specialist predictions of the number of undescribed species. Numbers of described species based on *BioSystematic Database of World Diptera* (Thompson 2006); emended by specialists [in brackets]. Figures and predictions based on data collected in 2006.

	Estimations of undescribed species			
Family	Known species	Estimated No. undescribed species	Percentage of species known	Percentage of fauna awaiting description
Acroceridae[6]	67	10+	87%	13%
Agromyzidae[28]	273	400+	41%	59%
Anisopodidae[45]	16	2+	89%	11%
Anthomyiidae[44]	66	20–25+	77%	23–27%
Anthomyzidae[7]	23	100+	19%	81%
Apioceridae[45]	4	1+	80%	20%
Asilidae[23]	1,533[1643]	1,900+	46%	54%
Asteiidae[24]	15	15+	50%	50%
Athericidae[45]	22	8+	73%	27%
Aulacigastridae[6]	4	2–3+	67%	33–43%
Bibionidae[45]	71	6+	92%	8%
Blephariceridae[45]	28	8+	78%	22%
Bombyliidae[15]	1,425	300+	83%	17%
Braulidae[6]	3	0	100%	0%
Calliphoridae[37]	207	50+	81%	19%
Camillidae[6]	22	20–25+	52%	48–53%
Canacidae[41]	25	42+	37%	63%
Carnidae[25]	6	5+	55%	45%
Cecidomyiidae[26]	213	5,000–10,000+	4%	96–98%
Celyphidae[42]	13[12]	7+	63%	37%
Ceratopogonidae[5]	913	2,000+	31%	69%
Chamaemyiidae[32]	15[10]	40–50+	20%	80–83%
Chaoboridae[45]	8	6+	57%	43%
Chironomidae[43]	556	374+	60%	40%
Chloropidae[4]	401	700+	36%	64%
Chyromyidae[9]	13[10]	70–100+	13%	88–91%
Clusiidae[6]	10	15–20+	40%	60–67%
Coelopidae[6]	5	0	100%	0%
Conopidae[6]	160	40+	80%	20%

Estimations of undescribed species			
Family	Known species	Estimated No. undescribed species	Percentage of species known	Percentage of fauna awaiting description
Corethrellidae[45]	3	3+	50%	50%
Cryptochetidae[6]	12	2+	86%	14%
Ctenostylidae[6]	2	0	100%	0%
Culicidae[38]	752	752+	50%	50%
Curtonotidae[35]	25	50+	33%	67%
Diadocidiidae[45]	1	2+	33%	67%
Diastatidae[6]	12	2–3+	86%	14–20%
Diopsidae[6]	137	1233+	90%	10%
Dixidae[31]	8	16+	33%	67%
Dolichopodidae[36]	738	1,000–1,500+	42%	58–67%
Drosophilidae[6]	451	250+	64%	36%
Empididae[18]	367	640+	36%	64%
Ephydridae[41]	339	578+	37%	63%
Fanniidae[8]	13	3–6+	80%	20–30%
Glossinidae[45]	23	0	100%	0%
Heliomyzidae[12]	62	80+	44%	56%
Hippoboscidae[45]	130	0	100%	0%
Keroplatidae[2]	170	500+	25%	75%
Lauxaniidae[32]	91[90]	400+	18%	82%
Limoniidae[27]	1,034	1,500+	41%	59%
Loncheidae[29]	64	250+	20%	80%
Lonchopteridae[6]	6	0	100%	0%
Lygistorrhinidae[46]	7	20+	26%	74%
Marginidae[6]	3	1–2+	75%	25–40%
Micropezidae[6]	65	15+	81%	19%
Milichiidae[13]	63[67]	192–203+	26%	74–75%
Mormotomyiidae[35]	1	0	100%	0%
Muscidae[8]	932	233–466+	80%	20–40%
Mycetophilidae[2]	249	2,000+	11%	89%
Mydidae[3]	200	200+	50%	50%
Mythicomyiidae[19]	55	220+	20%	80%
Natalimyzidae[6]	1	30–40	3%	97–98%
Nemestrinidae[6]	51	30+	63%	37%

Estimations of undescribed species			
Family	Known species	Estimated No. undescribed species	Percentage of species known	Percentage of fauna awaiting description
Neminidae[6]	7	2+	78%	23%
Neriidae[6]	20	5+	80%	20%
Neurochaetidae[6]	12	1–2+	92%	8–14%
Odiniidae[6]	8	8–10+	50%	50–60%
Oestridae[16]	35	8+	81%	19%
Opomyzidae[6]	5	0	100%	0%
Periscelididae[6]	10	5+	67%	33%
Phoridae[34]	418	2,000–3,000+	17%	83–88%
Piophilidae[6]	7	0	100%	0%
Pipunculidae[1]	152	300+	34%	66%
Platypezidae[2]	41	40+	41%	49%
Platystomatidae[40]	241[280]	70+	80%	20%
Psilidae[6]	52	10+	84%	16%
Psychodidae[45]	299	199+	60%	40%
Ptychopteridae[6]	9	1–2+	90%	10–18%
Pyrgotidae[30]	141	10+	93%	7%
Rhagionidae[45]	54	70+	80%	20%
Rhiniidae[37]	151	30+	83%	17%
Rhinophoridae[16]	28	50+	36%	64%
Sarcophagidae[16]	394	330+	54%	46%
Scathophagidae[8]	4[5]	0	100%	0%
Scatopsidae[47]	43[45]	400+	10%	90%
Scenopinidae[11]	69[68]	46+	60%	40%
Sciaridae[20]	71	2,000+	3%	97%
Sciomyzidae[33]	64	35+	65%	35%
Sepsidae[14]	131	131+	50%	50%
Simuliidae[21]	210	60+	78%	22%
Sphaeroceridae[7]	321	1,000–2,000+	24%	76–86%
Stratiomyidae[17]	388	100–150+	80%	20–28%
Syrphidae[40]	593[562]	241+	80%	30%
Tabanidae[39]	815	50+	94%	6%
Tachinidae[6]	1,013	1,000+	51%	49%
Tachiniscidae[6]	2	0	100%	0%

Estimations of undescribed species			
Family	Known species	Estimated No. undescribed species	Percentage of species known	Percentage of fauna awaiting description
Tanyderidae[45]	1	0	100%	0%
Tephritidae[22]	947	100–200+	90%	10–17%
Tethinidae[41]	23	41+	36%	64%
Thaumaleidae[45]	2	2+	50%	50%
Therevidae[10]	147[160]	50+	76%	24%
Tipulidae[27]	374	200+	65%	35%
Ulididiidae[6]	24	5+	83%	17%
Vermileonidae[45]	29	12	71%	29%
Xenasteiidae[6]	2	0	100%	0%
Xylomyidae[45]	6	0	100%	0%

Sources of information:

[1] Marc De Meyer & Mihály Földvári (pers. comms.)
[2] Peter Chandler (pers. comm.)
[3] Torsten Dikow (pers. comm.)
[4] John Deeming (pers. comm.)
[5] Rudy Meiswinkel (pers. comm.)
[6] David Barraclough (pers. comm.)
[7] Jindrich Roháček (pers. comm.)
[8] Adrian Pont (pers. comm.)
[9] Martin Ebejer (pers. comm.)
[10] Martin Hauser (pers. comm.)
[11] Shaun Winterton (pers. comm.)
[12] Andrzej Woźnica (pers. comm.)
[13] John Swann (pers. comm.)
[14] Andrey Ozerov (pers. comm.)
[15] David Greathead (pers. comm.)
[16] Thomas Pape (pers. comm.)
[17] Norm Woodley (pers. comm.)
[18] Brad Sinclair (pers. comm.)
[19] Neal Evenhuis (pers. comm.)
[20] Pekka Vilkamaa (pers. comm.)
[21] Douglas Craig (pers. comm.)
[22] David Hancock (pers. comm.)
[23] Jason Londt (pers. comm.)
[24] Amnon Freidberg (pers. comm.)
[25] David Barraclough & Amnon Freidberg (pers. comms.)
[26] Netta Dorchin & Keith Harris (pers. comms.)
[27] Chen Young (pers. comm.)
[28] Michael von Tschirnhaus (pers. comm.)
[29] Iain MacGowan (pers. comm.)
[30] Valery Korneyev (pers. comm.)
[31] Henry Disney (pers. comm.)
[32] Stephen Gaimari & Ray Miller (pers. comm.)
[33] Lloyd Knutson (pers. comm.)
[34] Mikhail Mostovski (pers. comm.)
[35] Ashley Kirk-Spriggs (pers. obs.)
[36] Igor Grichanov (pers. comm.)
[37] Hiromu Kurahashi (pers. comm.)
[38] Maureen Coetzee (pers. comm.)
[39] John Chainey (pers. comm.)
[40] Andrew Whittington (pers. comm.)
[41] Wayne Mathis (pers. comm.)
[42] Ray Miller (pers. comm.)
[43] Pete Cranston (pers. comm.)
[44] Michael Ackland (pers. comm.)
[45] Brian Stuckenberg (pers. obs.)
[46] Heikki Hippa (pers. comm.)
[47] Jean-Paul Haenni (pers. comm.)

Acknowledgements

We wish to thank specialists listed in Table 6.2, who confirmed the numbers of described species and provided estimations of the numbers of undescribed species. Chris Thompson kindly provided extracts and statistics from the BDWD, and other information was provided by Neal Evenhuis, Martin Hauser, David Furth, Mike Irwin, Nigel Wyatt, Marc De Meyer and Bradley Sinclair, and literature was provided by David Barraclough, Steve Gaimari, Maureen Coetzee and Scott Miller. The specimens illustrated in Figs 6.5 and 6.7 were loaned by Mikhail Mostovski and Fig. 6.2 was provided by Bradley Sinclair. Eugéne Marais and Susan Abraham assisted with the preparation of Fig. 6.8.

REFERENCES

Adams, J.M. & Faure, H. (eds) (1997) QEN members. *Review and atlas of palaeoveg-etation: preliminary land ecosystem maps of the world since the Last Glacial Maxi-mum.* Oak Ridge National Laboratory, TN, USA. Available at http://www.esd.ornl.gov/ern/qen/adams1.html, accessed 24 August 2007.

Adlbauer, K. (2000) Neue Bockkäfer vom Brandberg in Namibia (Coleoptera: Ceram-bycidae). *Cimbebasia* 16: 137–142.

Austen, E.E. (1936) A remarkable semi-apterous fly (Diptera) found in a cave in East Africa, and representing a new family, genus and species. *Proceedings of the Zoo-logical Society of London* [1936]: 425–431.

Barnard, P. (1998) Overview of Namibia and its biological diversity. Pages 15–55 *in*: Barnard, P. (ed.), *Biological diversity in Namibia — a country study.* Namibian National Biodiversity Taskforce, Windhoek.

Barraclough, D.A. (1995) An illustrated identification key to the acalyptrate fly fam-ilies (Diptera: Schizophora) occurring in southern Africa. *Annals of the Natal Museum* 36: 97–133.

Barraclough, D.A. (1998) *Katacamilla* Papp, 1978, a genus of Camillidae (Diptera: Schizophora) associated with the dung of birds, bats and hyraxes in Africa and the Arabian Peninsula. *African Entomology* 6: 159–176.

Barraclough, D.A. (2006a) Bushels of bots: Africa's largest fly is getting a reprieve from extinction. *Natural History* [2006] (June): 18–21.

Barraclough, D.A. (2006b) An overview of the South African tangle-veined flies (Diptera: Nemestrinidae), with an annotated key to the genera and a checklist of species. *Zootaxa* 1277: 39–63.

Barraclough, D.A. & Londt, J.G.H. (1985) Order Diptera (flies). Pages 283–321 *in*: Scholtz, C.H. & Holm, E. (eds), *Insects of southern Africa.* Butterworths, Durban.

Barraclough, D.A. & McAlpine, D.K. (2006) Natalimyzidae, a new African family of acalyptrate flies (Diptera: Schizophora: Sciomyzoidea). *African Invertebrates* 47: 117–134.

Basset, Y. (2001) Invertebrates in the canopy of tropical rain forest. How much do we really know? *Plant Ecology* 153: 87–107.

Basset, Y., Aberlenc, H.-P. & Delvare, G. (1992) Abundance and stratification of foli-age arthropods in a lowland rain forest of Cameroon. *Ecological Entomology* 17: 310–318.

Bezzi, M. (1926) South African Rhagionidae (Diptera) in the South African Museum. *Annals of the South African Museum* 23: 297–324.

Burgess, N.D., Butynski, T.M., Cordeiro, N.J., Doggart, N., Fjeldså, J., Howell, K.M., Kilahama, F., Loader, S.P., Lovett, J.C., Menegon, M., Moyer, D.C., Nashanda, E., Perkin, A., Rovero, F., Stanley, W.T. & Stuart, S.N. (2007) The biological impor-tance of the Eastern Arc Mountains of Tanzania and Kenya. *Biological Conserva-tion* 134: 209–231.

Chvála, M. (1991) First record of the Gondwanan genus *Homalocnemis* Philippi (Diptera: Empididae) from Namibia and the Afrotropical Region. *Annals of the Natal Museum* 32: 13–18.

Clarke, G.P. (2000a) 1.2 Defining the eastern African coastal forests. Pages 9–26 *in*: Burgess, N.D. & Clarke, G.P. (eds), *Coastal forests of eastern Africa*. IUCN, Gland, Switzerland and Cambridge.

Clarke, G.P. (2000b) Climate and climatic history. Pages 47–67 *in*: Burgess, N.D. & Clarke, G.P. (eds), *Coastal forests of eastern Africa*. IUCN, Gland, Switzerland and Cambridge.

Clausnitzer, V. (2003) Odonata of African forests – a review. *Cimbebasia* 18: 173–190.

Clarke, G.P. & Burgess, N.D. (2000) Geology and geomorphology. Pages 29–39 *in*: Burgess, N.D. & Clarke, G.P. (eds), *Coastal forests of eastern Africa*. IUCN, Gland, Switzerland and Cambridge.

Crosskey, R.W. (ed.) (1980) *Catalogue of the Diptera of the Afrotropical Region*. British Museum (Natural History), London; 1437 pp.

Crosskey, R.W. & White, G.B. (1977) The Afrotropical Region. A recommended term in zoogeography. *Journal of Natural History* 11: 541–544.

Coetzee, M. (1999) The historical and current state of medical entomology systematics in South Africa. *Transactions of the Royal Society of South Africa* 54: 65–73.

Corbet, P.S. (1961) Entomological studies from a high tower in Mpanga Forest, Uganda. VI. Nocturnal flight activity of Culicidae and Tabanidae as indicated by light traps. *Transactions of the Royal Entomological Society of London* 113: 301–314.

Dejean, A., Belin, M. & McKey, D. (1992) Les relations plantes-fourmis dans la canopée. Pages 76–80 *in*: Hallé, F. & Pascal, O. (eds), *Biologie d'une canopée de forêt équatoriale – II*. Rapport de Mission: radeau des cimes octobre novembre 1991, Réserve de Campo, Cameroun. Foundation Elf, Paris.

de Jong, R. & Congdon, T.C.E. (1993) The montane butterflies of the eastern Afrotropics. Pages 133–172 *in*: Lovett, J.C. & Wasser, S.K. (eds), *Biogeography and ecology of the rainforests of eastern Africa*. Cambridge University Press, Cambridge.

De Meyer, M. (2001) Distribution patterns and host-plant relationships within the genus *Ceratitis* MacLeay (Diptera: Tephritidae) in Africa. *Cimbebasia* 17: 219–228.

De Wit, M.J. (2003) Madagascar: Heads it's a continent, tails it's an island. *Annual Review of Earth and Planetary Sciences* 31: 213–248.

Dikow, T., Meier, R., Vaidya, G.G. & Londt, J.G.H. (2009) Biodiversity Research Based on Taxonomic Revisions — A Tale of Unrealized Opportunities. Pages 323-345 *in*: Pape, T., Bickel, D. & Meier, R. (eds), *Diptera Diversity: Status, Challenges and Tools*. Brill, Leiden.

Disney, R.H.L. (2005) Phoridae (Diptera) of Madagascar and nearby islands. *Studia dipterologica* 12: 139–177.

Duckhouse, D.A. (1980) *Trichomyia* species (Diptera: Psychodidae) from southern Africa and New Zealand, with a discussion of their affinities and of the concept of monophyly in Southern Hemisphere biogeography. *Annals of the Natal Museum* 24: 177–191.

Duckhouse, D.A. & Lewis, D.J. (1980) 3. Family Psychodidae. Pages 93–105 *in*: Crosskey, R.W. (ed.), *Catalogue of the Diptera of the Afrotropical Region*. British Museum (Natural History), London.

Du Toit, A.L. (1937) *Our wandering continents; an hypothesis of continental drifting*. Oliver and Boyd, Edinburgh and London, 366 pp.

Duxbury, K.J. & Barraclough, D.A. (1994) Rarely encountered Diptera families in southern Africa: an introductory conservation perspective. *Annals of the Natal Museum* 35: 25–43.

Emden, F.I. van (1950) *Mormotomyia hirsuta* Austen (Diptera) and its systematic position. *Proceedings of the Royal Entomological Society of London* (B) 19: 121–128.

Erwin, T.L. (1982) Tropical forests: their richness in Coleoptera and other arthropod species. *Coleopterist's Bulletin* 36: 74–75.

Evenhuis, N.L. (2000) A revision of the 'microbombyiid' genus *Doliopteryx* Hesse (Diptera: Mythicomyiidae). *Cimbebasia* 16: 117–135.

Evenhuis, N.L. (2001) A new 'microbombyiid' genus from the Brandberg Massif, Namibia (Diptera: Mythicomyiidae). *Cimbebasia* 17: 137–141.

Evenhuis, N.L. (2006) *Catalogue of the fossil flies of the world (Insecta: Diptera)*. Online version, available at http://hbs.bishopmuseum.org/fossilcat/fosscurto.html, accessed 18 April 2007.

Gaimari, S.D. & Raspi, A. (2002) The species of *Leucopis*, subgenus *Leucopella* Malloch (Diptera: Chamaemyiidae), from northeastern Africa and Yemen. *African Entomology* 10: 241–264.

Gaston, K.J. (1991) The magnitude of global insect species richness. *Conservation Biology* 5: 183–196.

Germain, M., Grenier, P. & Mouchet, J. (1967) Présence de Blepharoceridae (Diptera, Nematocera) au Cameroun. *Cahiers Office de la Recherche Scientifique et Technique Outre-Mer* (Série Entomologie Medicale et Parasitologie) 5: 133–139.

Goodman, S.M. & Benstead, J.P. (eds) (2003) *The natural history of Madagascar*. University of Chicago Press, Chicago and London, 1159 pp.

Greathead, D.J. (2000) The family Bombyliidae (Diptera) in Namibia, with descriptions of six new species and an annotated checklist. *Cimbebasia* 16: 55–93.

Greathead, D.J. (2006) New records of Namibian Bombyliidae (Diptera), with notes on some genera and descriptions of new species. *Zootaxa* 1149: 1–88.

Grichanov, I.Ya, Kirk-Spriggs, A.H. & Grootaert, P. (2006) An annotated checklist of Namibian Dolichopodidae (Diptera) with the description of a new species of *Grootaertia* and a key to species of the genus. *African Invertebrates* 47: 207–227.

Grimaldi, D.A. (1992) Vicariance biogeography, geographic extinctions, and the North American Oligocene tsetse flies. Pages 178–204 *in*: Novacek, M.J. & Wheeler, Q.D. (eds), *Extinction and phylogeny*. Columbia University Press, New York.

Hancock, D.L., Kirk-Spriggs, A.H. & Marais, E. (2001) An annotated checklist and provisional atlas of Namibian Tephritidae (Diptera: Schizophora). *Cimbebasia* 17: 41–72.

Hancock, D.L., Kirk-Spriggs, A.H. & Marais, E. (2003) New records of Namibian Tephritidae (Diptera: Schizophora), with notes on the classification of subfamily Tephritinae. *Cimbebasia* 18: 49–70.

Hill, R.S. & Scriven, L.J. (1995) The angiosperm-dominated woody vegetation of Antarctica: a review. *Review of Palaeobotany and Palynology* 86: 175–198.

Hölzel, H. (1998) Zoogeographical features of Neuroptera of the Arabian Peninsula. *Acta Zoologica Fennica* 209: 129–140.

Irish, J. (1994) The biomes of Namibia, as determined by objective categorisation. *Navorsinde van die Nasionale Museum, Bloemfontein* 10: 559–592.

Irwin, M.E., Schlinger, E.I. & Thompson, F.C. (2003) Diptera, true flies. Pages 692–702 *in*: Goodman, S.M. & Benstead, J.P. (eds), *The natural history of Madagascar*. University of Chicago Press, Chicago and London.

Ismay, J.W. (2000) Chloropidae (Diptera: Chloropoidae). Pages 269–289 *in*: Kirk-Spriggs, A.H. & Marais, E. (eds), *Dâures – biodiversity of the Brandberg Massif, Namibia*. Cimbebasia Memoir 9, National Museum of Namibia, Windhoek.

Kirk-Spriggs, A.H. (2003) Foreword: African biogeography patterns – an introduction. *In*: Proceedings of the workshop 'African Diptera Biogeography & Gondwanaland', 5th International Congress of Entomology, Brisbane, Australia, 29th September – 4th October 2002. *Cimbebasia* 19: 149–156.

Kirk-Spriggs, A.H., Barraclough, D.A. & Meier, R. (2002) The immature stages of *Katacamilla cavernicola* Papp, the first described for the Camillidae (Diptera: Schizophora), with comparison to other known Ephydroidea larvae, and notes on biology. *Journal of Natural History* 36: 1105–1128.

Kirk-Spriggs, A.H. & Evenhuis, N.L. (2008) A new species of *Psiloderoides* Hesse (Diptera: Mythicomyiidae) from the Brandberg Massif, Namibia. *African Entomology* 16: 122–126.

Kirk-Spriggs, A.H. & Marais, E. (eds) (2000) *Dâures – biodiversity of the Brandberg Massif, Namibia*. Cimbebasia Memoir 9, National Museum of Namibia, Windhoek, 389 pp.

Kirk-Spriggs, A.H., Ismay, J.W., Ackland, M., Roháček, J., Mathis, W.N., Foster, G.A., Pape, T., Cranston, P. & Meier, R. (2001) Inter-tidal Diptera of southwestern Africa (Chironomidae, Canacidae, Chloropidae, Milichiidae, Tethinidae, Ephydridae, Sphaeroceridae, Coelopidae, Sarcophagidae and Anthomyiidae). *Cimbebasia* 17: 85–135.

Krüger, O. & McGavin, G.C. (1997) The insect fauna of *Acacia* species in Mkomazi Game Reserve, north-east Tanzania. *Ecological Entomology* 22: 440–444.

Kurahashi, H. & Kirk-Spriggs, A.H. (2006) The Calliphoridae of Namibia (Diptera: Oestroidea). *Zootaxa* 1322: 1–131.

Larsen, T.B. (1984) The zoogeographical composition and distribution of the Arabian butterflies (Lepidoptera: Rhopalocera). *Journal of Biogeography* 11: 119–158.

Londt, J.G.H. (2000) Afrotropical Asilidae (Diptera) 33. A revision of Afrotropical *Habropogon* Loew, 1847 (Diptera: Asilidae: Stenopogoninae), with the description of four new species. *Annals of the Natal Museum* 41: 139–150.

Maley, J. (1996) The African rainforest – main characteristics of changes in vegetation and climate from the Upper Cretaceous to the Quaternary. *Proceedings of the Royal Society of Edinburgh* 104b: 31–73.

Manning, J.C. & Goldblatt, P. (1995) The *Prosoeca peringueyi* (Diptera: Nemestrinidae) pollination guild in southern Africa: long-tongued flies and their tubular flowers. *Annals of the Missouri Botanical Gardens* 82: 517–534.

Marais, E. & Kirk-Spriggs, A.H. (2000) Inventorying the Brandberg Massif, Namibia. Pages 91–102 *in*: Kirk-Spriggs, A.H. & Marais, E. (eds), *Dâures – biodiversity of the Brandberg Massif, Namibia*. Cimbebasia Memoir 9, National Museum of Namibia, Windhoek.

McAlpine, D.K. (1991) Marginidae, a new Afrotropical family of Diptera (Schizophora: ? Opomyzoidea). *Annals of the Natal Museum* 32: 167–177.

McGowan, I. (2005) New species of Lonchaeidae (Diptera: Schizophora) from central and southern Africa. *Zootaxa* 967: 1–23.

Meadows, M.E. (1996) 10 Biogeography. Pages 161-172 *in*: Adams, W.M., Goudie, A.S. & Orme, A.R. (eds), *The physical geography of Africa*. Oxford University Press, New York.

Miller, R.M. (1984) A new acalyptrate fly from southern Africa, possibly representing a new family. Page 32 *in*: *XVII International Congress of Entomology, 20–26 August, Hamburg, Federal Republic of Germany*, Abstract volume.

Miller, R.McG. (2000) Geology of the Brandberg Massif, Namibia and its environs. Pages 17–38 *in*: Kirk-Spriggs, A.H. & Marais, E. (eds), *Dâures – biodiversity of the Brandberg Massif, Namibia*. Cimbebasia Memoir 9, National Museum of Namibia, Windhoek.

Miller, S.E. & Rogo, L.M. (2001) Challenges and opportunities in understanding and utilisation of African insect diversity. *Cimbebasia* 17: 197–218.

Miranda-Esquivel, D.R. & Coscarón, S. (2003) Distributional patterns of Neotropical, Afrotropical and Australian-Oriental *Simulium* Latreille subgenera (Diptera: Simuliidae). *Cimbebasia* 19: 165–174.

Moran, V.C., Hoffman, J.H., Impson, F.A.C. & Jenkins, J.F.G. (1994) Herbivorous insect species in the tree canopy of a relict South African forest. *Ecological Entomology* 19: 147–154.

Moran, V.C. & Southwood, T.R.E. (1982) The guild composition of arthropod communities in trees. *Journal of Animal Ecology* 51: 239–306.

Mucina, L. & Rutherford, M.C. (eds) (2006) The vegetation of South Africa, Lesotho and Swaziland. *Strelitzia* 19: 748–790.

Nagatomi, A. & Nagatomi, H. (1990) A revision of *Atherimorpha* White, 1915 from southern Africa (Diptera: Rhagionidae). *Annals of the Natal Museum* 31: 33–82.

Oldroyd, H. (1957) *The horse-flies (Diptera: Tabanidae) of the Ethiopian Region. Vol. III subfamilies Chrysopinae, Scepsidinae and Pangoniinae and a revised classification.* British Museum (Natural History), London, xii + 489 pp. + Plates II–XIII.

Oldroyd, H. (1962) South African horseflies of the tribe Pangoniini (Diptera: Tabanidae). *Journal of the Entomological Society of Southern Africa* 25: 51–55.

Pape, T. (1997) Two new species of the *Phyto carinata* species-group (Diptera: Rhinophoridae). *Annals of the Natal Museum* 38: 159–168.

Partridge, T.C. & Maud, R.R. (1987) Geomorphic evolution of Southern Africa since the Mesozoic. *South African Journal of Geology* 90: 179–208.

Paulian, R. (1954) Fauna des eaux douces de Madagascar. II. Larves de Blepharoceridae (Dipt.). *Mémoires de l'Institut scientifique de Madagascar* (E) 4 (1953): 431–441.

Phillips, J.L. (1959) *Agriculture and ecology in Africa: a study of actual and potential development south of the Sahara.* Faber and Faber, London, 424 pp.

Phillips, J.L. (1965) Fire — as master and servant: its influence in the bioclimatic regions of trans-Saharan Africa. Pages 7–109 *in: Proceedings of the 4th Tall Timbers Fire Ecology Conference.* Tall Timbers Research Station, Tallahassee, Florida.

Sæther, O.A. & Ekrem, T. (2003) Biogeography of Afrotropical Chironomidae (Diptera), with special reference to Gondwanaland. *Cimbebasia* 19: 175–191.

Scholtz, C.H. (1999) Review of insect systematics research in South Africa. *Transactions of the Royal Society of South Africa* 54: 53–63.

Scholtz, C.H. & Chown, S.L. (1995) Insects in southern Africa: how many species are there? *South African Journal of Science* 91: 124–126.

Scholtz, C.H. & Holm, E. (eds) (1985) *Insects of southern Africa.* Butterworths, Durban, 502 pp.

Sclater, P.L. (1858) On the general geographical distribution of the members of the class Aves. *Journal and Proceedings of the Linnaean Society. Zoology* 2: 130–145.

Shamshev, I.V. & Sinclair, B.J. (2006) The genus *Schistostoma* Becker from southern Africa, with an evaluation of its generic status (Diptera: Dolichopodidae *s. l.*: Microphorinae). *African Invertebrates* 47: 335–346.

Simmons, R.E., Griffin, M., Griffin, R.E., Marais, E. & Kolberg, H. (1998) Endemism in Namibia: patterns, processes, and predictions. *Biodiversity and Conservation* 7: 513–530.

Sinclair, B.J. (2003) Southern African Empidoidea (Diptera) – phylogenetic patterns and biogeographic implications. *Cimbebasia* 19: 205–213.

Sinclair, B.J. & Stuckenberg, B.R. (1995) Review of the Thaumaleidae (Diptera) of South Africa. *Annals of the Natal Museum* 36: 209–214.

Stuart, S.N., Jensen, F.P., Brøgger-Jensen, S. & Miller, R.I. (1993) The zoogeography of the montane forest avifauna of eastern Tanzania. Pages 203–228 *in:* Lovett, J.C. & Wasser, S.K. (eds), *Biogeography and ecology of the rainforests of eastern Africa.* Cambridge University Press, Cambridge.

Struck, M. (1992) Pollination ecology in the arid winter rainfall region of southern Africa: a case study. *Mitteilungen aus dem Institute für allemeine Botanik* 24: 61–90.

Struck, M. (1994) Flowers and their insect visitors in the arid winter rainfall region of southern Africa: observations on permanent plots. *Journal of Arid Environments* 28: 51–74.

Stuckenberg, B.R. (1962) The distribution of the montane palaeogenic element in the South African invertebrate fauna. *Annals of the Cape Provincial Museums (Natural History)* 2: 190–205.

Stuckenberg, B.R. (1978) Two new species of *Nemopalpus* (Diptera: Psychodidae) found in rock hyrax abodes in South West Africa. *Annals of the Natal Museum* 23: 367–374.

Stuckenberg, B.R. (1998) A new Namibian wormlion species, with an account of the biogeography of *Leptynoma* Westwood *s. str.* and its association with anthophily in the Fynbos and Succulent Karoo Biomes (Diptera, Vermileonidae). *Annals of the Natal Museum* 39: 165–183.

Stuckenberg, B.R. (2000) The Vermileonidae (Diptera) of Namibia, with the description of a new species of *Perianthomyia* Stuckenberg from the Brandberg Massif. *Cimbebasia* 16: 47–54.

Stuckenberg, B.R. (2002) A new genus and species of Vermileonidae (Diptera: Brachycera) from Madagascar. *Tijdschrift voor Entomologie* 155: 1–8.

Stuckenberg, B.R. (2004) Labial morphology in Blephariceridae (Diptera: Nematocera): a new interpretation with phylogenetic implications, and a note on colocephaly. *African Invertebrates* 45: 223–236.

Stuckenberg, B.R. & Fisher, M. (1999) A new species of *Lampromyia* Macquart, from Oman: the first record of Vermileonidae (Diptera) from the Arabian Peninsula. *Annals of the Natal Museum* 40: 127–136.

Sutton, S.L. & Hudson, P.J. (1980) The vertical distribution of small flying insects in the lowland rain forest Zaïre. *Zoological Journal of the Linnaean Society* 68: 111–123.

Taylor, H.C. (1978) Capensis. Pages 171–229 *in*: Werger, M.J.A. (ed.), *Biogeography and ecology of southern Africa.* Vol. 1, Dr. W. Junk Publishers, the Hague.

Thompson, F.C. (ed.) (2006) *BioSystematic Database of world Diptera. Version 8.5.* Available at http://198.77.169.80/Diptera/names/Status/bioregion.htm, accessed 17 April 2007.

van Harten, A. (ed.) (2008) *Arthropod fauna of the United Arab Emirates.* Vol. 1; Dar Al Ummah Publishers, Abu Dhabi; 754 pp.

Wagner, T. (1997) The beetle fauna of different species in forest of Rwanda and east Zaïre. Pages 169–183 *in*: Stork, N.E., Adis, J. & Didham, R.K. (eds), *Canopy arthropods.* Chapman & Hall, London.

Wagner, T. (1998) Influence of tree species and forest type on the chrysomelid community in the canopy of an Ugandan tropical forest. Pages 253–269 *in*: Biondi, M., Daccordi, M. & Furth, D.G. (eds), *Proceedings IV. International Symposium on the*

Chrysomelidae; Proceedings XX. International Congress of Entomology, Firenze 1996, Torino.

Wagner, T. (1999) Arboreal chrysomelid community structure and faunal overlap between different types of forests in Central Africa. Pages 247–270 *in*: Cox, M.L. (ed.), *Advances in Chrysomelidae biology 1*. Backhuys Publishers, Leiden.

Wagner, T. (2000) Influence of forest type and tree species on canopy-dwelling beetles in Budongo Forest, Uganda. *Biotropica* 32: 502–514.

Wagner, T. (2001) Biogeographical and evolutionary aspects of Afrotropical *Monolepta* (Chevrolat), *Afromaculepta* (Hasenkamp & Wagner) and *Bonesioides* (Laboissière) (Coleoptera: Chrysomelidae, Galerucinae). *Cimbebasia* 17: 237–244.

Wallace, A.R. (1876) *The geographical distribution of animals; with a study of the relations of living and extinct faunas as elucidating the past changes of the Earth's surface*. Vol. 1, MacMillan & Co., London, 503 pp.

Watt, A.D., Stork, N.E., McBeath, C. & Lawson, G.L. (1997) Impact of forest management on insect abundance and damage in a lowland tropical forest in southern Cameroon. *Journal of Applied Ecology* 34: 985–998.

Wedmann, S. (2000) Die Insekten der oberoligizänen Fossillagerstätte Enspel (Westerwald, Deutschland): Systematik, Biostrationomie und Paläoökologie. *Mainzer Naturwissenschaftliches Archiv* 23: 1–154.

Wells, N.A. (2003) Some hypotheses on the Mesozoic and Cenozoic paleoenvironmental history of Madagascar. Pages 16–34 *in*: Goodman, S.M. & Benstead, J.P. (eds), *The natural history of Madagascar*. The University of Chicago Press, Chicago and London.

Werger, M.J.A. (1978) Biogeographical division of southern Africa. Pages 301–462 *in*: Werger, M.J.A. (ed.), *Biogeography and ecology of southern Africa*. Vol. 1, Dr. W. Junk Publishers, the Hague.

Werger, M.J.A. & Coetzee, B.J. (1978) The Sudano-Zambezian Region. Pages 465–513 *in*: Werger, M.J.A. (ed.), *Biogeography and ecology of southern Africa*. Vol. 1, Dr. W. Junk Publishers, the Hague.

Wharton, R.A. (1982) Observations on the behaviour, phenology and habitat preferences of Mydas flies in the central Namib Desert (Diptera: Mydidae). *Annals of the Transvaal Museum* 33: 145–151.

White, F. (1983) *The vegetation of Africa. A descriptive memoir to accompany the UNESCO/AETFAT/UNSO vegetation map of Africa* (3 separate map sheets at 1 : 5,000.000, dated 1981, and legend). UNESCO, Paris, 356 pp.

CHAPTER SEVEN

ORIENTAL DIPTERA, A CHALLENGE IN DIVERSITY AND TAXONOMY

Patrick Grootaert

Royal Belgian Institute of Natural Sciences, Brussels, Belgium

INTRODUCTION

If one asks 'what about the diversity of Diptera in the Oriental Region, and where are the gaps in our knowledge on the diversity?' the reply can be very quick by saying that we do not know very much about the fly fauna of this region. What we know is that it is has a fascinating fauna with ancestral groups, with weird endemics and a huge, unknown diversity. Alfred Russel Wallace, the founder of zoogeography, was the first to be amazed by the dazzling fauna of the Orient. Here he gained his insight in the origin and evolution of species (Wallace 1876, Oosterzee 1997); views, which he continuously exchanged in his correspondence with Charles Darwin during his journey in Asia.

The Oriental Region can indeed be considered as a unit. It hosts a fauna that is different from the adjacent Palaearctic fauna in the north and the Australasian fauna or West Pacific fauna on its southern and south eastern border. About 91% of its fly fauna is endemic to the region (Evenhuis *et al.* 2007). Moreover, within the Orient there are various areas of endemicity with a unique and isolated, independent evolution. The fauna of Sulawesi is perhaps the most striking one. It is very difficult to say on a scientific basis what number of species we can expect in this region. In the following paragraphs we will demonstrate with various examples that there are indeed huge gaps in our knowledge and they are not likely to be filled in soon.

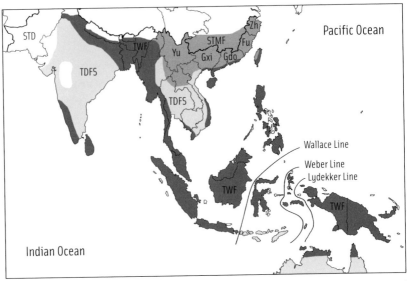

Figure 7.1. Boundaries of the Oriental Region as used for the *BioSystematic Database of World Diptera* (Evenhuis *et al.* 2007) with indication of biomes (after Encarta): STMF: subtropical mixed forest; STD: subtropical desert and semidesert; TDFS: tropical dry forest and savanna; TWF: tropical wet forest. Five provinces of South China are considered to belong to the Oriental Region: YU: Yunnan, Gxi: Guangxi, Gdo: Guangdong, Fu: Fujian, Zh: Zhejiang

1. Boundaries and a Few Definitions

To understand the basics of the Oriental Diptera fauna it is important to place it on the world map, to look to its boundaries, geological origin, climate and biomes.

Fig. 7.1 shows a map of the Oriental Region (or Realm), sometimes also called the Indomalayan Region by botanists. The Palaearctic Region borders it to the west and to the north, and the Australasian Region to the south. The western part of the Oriental region is formed by the Indian subcontinent that is a huge landmass separated in the north from the Palaearctic Region and in the east from Southeast Asia by mountain chains. In the west the Indian subcontinent transits into Palaearctic Arabia, Pakistan still being Oriental.

The border between the Oriental and the Palaearctic Region in China has always been a point of discussion. Sometimes the 30° northern latitude is chosen as an arbitrary northern limit for the Oriental fauna in China (cf. Palaearctic catalogue, Soós & Papp 1991). For the *BioSystematic Database of World Diptera* (BDWD), Evenhuis *et al.* (2007) consider the southern provinces of Yunnan, Guangxi, Guangdong (including Hainan), Fujian and Zhejiang as the best approximation. Coleopterists often consider the whole of China as Palaearctic, ignoring the presence of many Oriental elements in South China. Anyway, for Diptera there is a distinct (or 'typical') tropical Oriental fauna present in the mainland of South China, which is however almost completely endemic and thus different from adjacent Southeast Asia (e.g., Yang *et al.* 2006b).

The famous Wallace Line, considered by many authors as the eastern border with the Australasian Region, is also still a point of strong debate (Oosterzee 1997, Beck *et al.* 2006). Indeed, for mammals and birds there is a quite marked limitation of the Oriental and Australasian regions, but not for plants and insects.

The gradual reduction of Asian faunal influence and increase in Australian faunal prominence occurs as one moves eastward through Wallacea (see below) from Wallace's Line. A number of lines have been drawn in addition to Wallace's that apply to specific animal groups (see Oosterzee 1997). Wallace himself vacillated about to which region Sulawesi truly belongs (Oosterzee 1997). He finally said it would remain a matter of opinion if it would be considered Asian or Australian, it is unique regarding its insect fauna anyway.

Lydekker's Line, which was defined by Lydekker (1896), is perhaps the most significant line except for the one that Wallace himself defined. Corresponding to the edge of the Sahul Continental shelf, it marks the boundary between Australia and the transitional zoogeographic zone of Wallacea.

Finally, Weber's Line, which was originally based on the distribution of freshwater fish, was redefined as the line of faunal balance between Australia and Asia. To the west, the fauna is more than 50% Oriental, while to the east of the line, the fauna is more than 50% Australian in origin. It is important to note that this line does not represent a faunal break because it separates areas that are most similar, the opposite of a break (Oosterzee 1997). The BDWD (Evenhuis *et al.* 2007) considers Weber's Line as the eastern border of the Oriental Region. This concept is followed here, and this has its consequences on the calculation of the number of endemics present in the Orient.

It is a bit unfortunate that a mixture of names is used for the various parts of the Oriental Region and even more confusing is that not all authors use the same boundaries for these areas. The following names for areas are often encountered in the literature.

> *Sundaland* comprises the Malay Peninsula (West Malaysia), the Malay archipelago, the islands of Sumatra, Java, Bali and Borneo. The islands of Sundaland rest on the shallow continental shelf of Asia. The eastern border of Sundaland is formed by Wallace's Line.

> *Wallacea* is the area between Wallace's Line and Lydekker's Line. It includes the islands of Nusa Tenggara or the Lesser Sunda Islands (Lombok, Sumbawa, Komodo, Flores, Sumba and Timor), Sulawesi, the islands of North Maluku, including Halmahera, and most of the province of Maluku, excluding the Aru Islands, which lie on the Australian continental shelf and thus belong to *Australasia*. Some authors include also the Philippines into Wallacea (Beck *et al.* 2006), an opinion that is not followed here. Although the distant ancestors of Wallacea's plants and animals may have been from Asia or Australia-New Guinea, Wallacea is presently home to many endemic species. Because many of the islands are separated from one another by deep water, there is a tremendous species endemism on each island as well.

Botanists often include Sundaland, Wallacea and the Philippines up to New Guinea in the floristic region of *Malesia* based on similarities in their fauna that is predominantly of Asian origin.

2. Origin of the Oriental Region

The flora and fauna of East and Southeast Asia show a complex pattern that is the result of plate movements, shifting of land to sea and from continent to ocean, of shifting coastlines and changing palaeoclimates (Metcalfe 1998). Hence, the geography of the Oriental Region is a gigantic jigsaw puzzle of continental fragments and the question is if the endemics and the distribution of plants and animals are related to that puzzle or to what extent they survived on these fragments? In other words, which terranes did carry the Diptera fauna or the ancestors of the fly fauna that we find today in the Oriental Region? During the drifting of the various terranes it is likely that vicariance speciation took place, but the question is when?

The geology of the area is very complex and the region is composed of three main plates: an Asian, a Pacific and an Australian shelf. Using multidisciplinary techniques (palaeobiogeography, tectonostratigraphy and palaeomagnetism) Metcalfe (1998) showed that all the East and SE Asian continental terranes had their origin on the margin of Gondwanaland. Most probably three continental slivers (terranes) rifted and separated off the Indian-Australian margin of Gondwanaland: the first during the Devonian (409–363 million years ago), the second during the Early-Middle Permian (340 Mya) and the last during Late Triassic to Late Jurassic (210–150 Mya). The northwards drift of these terranes was accompanied by the opening and closing of three successive oceans: the Palaeotethys, the Mesotethys and the Ceno-Tethys.

Simplified we can say that first South China collided with Central China and joined Laurasia some 350 Mya. Indochina joined Laurasia around 200 Mya. The islands of Sumatra, Java, the lesser Sunda and the different parts of Sulawesi joined even later.

Greater India drifted off Gondwana around 148 Mya (Briggs 1989) and travelled rapidly to Eurasia. The general idea is that India was isolated during its journey northwards, but Briggs (2003) argued that it stayed in close contact with Africa instead, allowing faunal exchange. India made its initial contact with Eurasia at the end of the Cretaceous some 52 Mya. The collision caused the uplifting of the Himalayas by the subduction of the northern, leading margin of the Indian plate. It also caused an influx of Gondwana plants and animals, it led to major changes in habitats, climate and drainage systems and promoted dispersal from Gondwana via India into Southeast Asia, but created barriers as well between Southeast Asia and the rest of Asia (Hall 1998). The uplifting of the Himalayas and the recent development of a monsoonal climate in particular had a dramatic effect in shrinking back Sundaic rainforest habitats and causing the origin of Salawin, Chao Praya, Mekong, Red and Yellow rivers drainage systems that actually are the major ecological factors ruling the fauna and flora in the Orient.

Central to the understanding of the origin of the fly fauna is that the terranes of Central and South China and Indochina split off from northern Gondwana more than 200 Mya before all Diptera families had formed. Did Indochina carry its fauna along when it drifted northward and collided with South China? Or did most of the current Oriental fauna originate from the Indian subcontinent? Various authors promote the latter hypothesis, but others question it arguing that most of the Indian subcontinent was submerged during its drifting. Cranston (2005) suggests that maybe the mountains of the Southern Ghats in India retained some fauna that later colonised northern India and eventually the rest of the Orient. In addition, Briggs (1989) claims that Greater India was isolated from 'Africa' and Asia during about 100 million years but did not develop any particular biota with many endemic genera and families in its terrestrial and shallow marine habitats. In a more recent paper, Briggs (2003) hypothesises that India stayed in close contact with the African continent during its journey northwards and that there was continuous exchange of fauna and flora of at least vagile organisms. This would also explain why India did not develop a high endemicity.

Nagatomi (1991) suggested by combining phylogenetic and amber fossils data that most lower Brachycera originated in Gondwana. Wiegmann *et al.* (2003) present a molecular time-scale for Brachycera fly evolution based on 28s ribosomal DNA (Table 7.1). Combining these data it thus seems more likely indeed that the Brachycera dispersed from India into

Table 7.1. Divergence time estimates for brachyceran fly lineages based on 28s rDNA sequences (after Wiegmann *et al.* 2003). Posterior means for Bayesian divergence time estimates are followed in parentheses by 95% credibility intervals.

	Estimates without a clock	Estimates with a clock	Earliest known fossil (Mya)	Hypothesized group age prior to study[b] (Mya)
Brachycera	216 (194, 241)	214 (192, 238)	187 (208)	200
Muscomorpha	216 (194, 241)	214 (192, 238)	144	198
Stratiomyomorpha	204 (176, 232)	199 (172, 228)	187	198
Xylophagomorpha	192 (160, 224)	180 (149, 212)	187	200
Tabanomorpha	192 (160, 224)	180 (149, 212)	187	200
Heterodactyla	197 (177, 223)	195 (176, 221)	144	185
Eremoneura	166 (143, 192)	165 (144, 189)	150[c]	165
Empidoidea	163 (143, 189)	154 (128, 181)	130[d]	150
Cyclorrapha	142 (122, 169)	154 (128, 181)	130[d]	150
Schizophora	84 (70, 113)	107 (75, 142)	80[d]	88
Drosophila/Musca	48 (29, 76)	81 (39, 121)	70[d]	75-100

[b] Evenhuis 1994; [c] Nagatomi & Yang 1998; [d] Grimaldi 1999; Grimaldi & Cumming 1999.

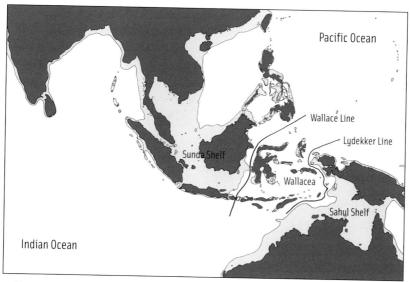

Figure 7.2. Sea level drop to –120 m of the current sea level during the glaciations, showing land connections
(after Voris 2000).

Southeast Asia and not from the slivers splitting off from Australia during the Devonian, Permian and Jurassic periods.

Sea level changes and climatic changes in the Pleistocene, especially during the last ice age, modified the distribution patterns of insects greatly and were determinant for the actual distribution patterns. The sea level dropped to 120 m below the present level, interconnecting many islands like Borneo and the Philippines with mainland Indochina (Fig. 7.2). Dispersal of the terrestrial fauna was facilitated, which explains the wider distribution of some species. At the same time the climate became colder and arid. The tropical rain forests were limited along the large rivers in Sundaland, while most of Sundaland became a savanna-like landscape. After the last ice age, the sea level rose and separated the islands again, and vicariance speciation occurred. Groups that have a weak dispersal power such as cicadas remain restricted to their regions of origin. Groups with a greater dispersal capacity such as hawk moths continue to disperse and show less vicariance speciation resulting in fewer endemics and more species having larger ranges.

DIPTERA DIVERSITY: STATUS, CHALLENGES AND TOOLS
(EDS T. PAPE, D. BICKEL & R. MEIER). © 2009 KONINKLIJKE BRILL NV.

3. Biomes

The major biomes in the Oriental Region are tropical dry forest and savanna, tropical wet forest and subtropical mixed forest (Fig. 7.1). The term tropical dry forest and savanna is perhaps not good for Southeast Asia and should be referred to as monsoonal or tropical monsoonal forest (A. Plant, pers. comm.).

Tropical wet forest is considered as the most speciose biome (Kitching *et al.* 2004), but we have no clue to how many species of Diptera exist in the Oriental region. As an example, one year of collecting with five Malaise traps in secondary rain forest of Nee Soon and Sime forest in Singapore resulted in 93 dolichopodid species, the Sciapodinae not being included. Two days of sweep net collecting in a primary rain forest in Endau Rompin (Malaysia) resulted in 83 species. Endau Rompin and Singapore are low altitude sites separated by about 150 km, but almost all species were different (P. Grootaert, unpubl.). In addition, lowland tropical rain forest is considered as relatively less speciose as compared to median elevation rain forest (e.g., Wolda 1988).

The Indian subcontinent is mainly covered by a dry forest and savanna characterized by a long, dry season. In the northwest there is subtropical desert and semi-desert. The southwestern tip of India and northern India, Bangladesh and Burma was originally covered by tropical wet forest. South China and the northern mountain ranges of Thailand, Laos and Cambodia are mainly covered by subtropical mixed forest. Central Thailand and Cambodia have a tropical dry forest and savanna, characterized by a long dry season. South Thailand, Malaysia and Sundaland have tropical wet forest with short dry season. The eastern part of Java and all the islands to the east are dryer again, with tropical dry forest and savanna.

The general idea that there are no seasons in the tropical rain forests is a misleading myth (Wolda 1988, Frith & Frith 2006). Monsoon regimes affect the phenology of most Diptera. In general, the greatest activity of adult Diptera is seen during the transition from dry to wet season and again during the transition from wet to dry season (P. Grootaert, unpubl.). Aridity is a limitation for the activity of most Diptera and so are long periods of heavy rainfall. In Singapore almost continuous activity of some dolichopodid species could be observed in the sheltered gallery forest bordering small streams, but even in these stable conditions, most

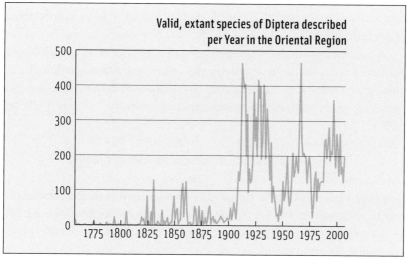

Figure 7.3. Number of Oriental Diptera species described per year since Linnaeus' *Systema Naturae* 10th edition (Evenhuis *et al.* 2007).

species exhibit seasonal activity. Even mangroves with their buffered atmospheric humidity show marked seasonal peaks of activity.

4. History of Taxonomy in the Region

In the early years of taxonomy (1758 to 1908), very few species were described from the Oriental region (Fig. 7.3). From 1908 onwards there is a sudden interest in the Oriental fauna and Dutch and German authors described many species. In the year 1911 there was even a peak of 465 Diptera species being described. With a temporary low during World War I, activity remains high and the bulk of the actually known Oriental species have been described in the period 1908–1940. At the beginning of World War II, taxonomy in the region comes to a halt and it resumes slowly from the 1950's onward. In 1965 there is again a peak of described species, but afterwards the activity drops again. Only from 1988 onward is reached a steady publication rate of at least 200 species a year, which has continued until today. That the publication rate for the region remains low is probably due to less interest in taxonomy by decision makers, regional as well as global, and that the activities of the existing dipterists are saturated and that these activities have gradually turned to molecular studies and other topics that have higher chances of funding than the more classic alpha taxonomy.

It is certainly not due to poor diversity in the region, as will be shown further. Except for some economically important groups such as Culicidae and Tephritidae, taxonomy is not really goal driven. Taxonomists mainly work on groups for purely personal curiosity or interest (Bickel 2009).

5. Diversity and the Number of Species per Family

As of November 2007, 152,715 valid fly species have been recorded in the world (Evenhuis *et al.* 2007, Appendix). The Oriental Region hosts 22,545 species, among them 20,052 species or 93% of the species are endemic to the region. The number of species actually known in the Oriental Region (22,545) is only half the number of species recorded in the Palaearctic Region (45,198) and about 15% of the total number of species known in the world. Considering the very high diversity in wet tropical biomes in the Oriental region, these figures give already a first indication that the Oriental fauna is poorly known. Probably only a regional comparison of well studied taxa like Culicidae will provide an idea of the differences between the Palaearctic (268 species) and the Oriental (975 species) regions. With 3.6 times more Oriental than Palaearctic species of Culicidae, and extrapolating this ratio to all Diptera would give a rough estimation of about 81,000 fly species occurring in the Oriental Region.

Appendix is based on data from the *BioSystematic Database of World Diptera* version 10 (Evenhuis *et al.* 2007). 117 families of Diptera are recorded so far in the Oriental region. This is 76% of the 152 Diptera families globally recognized at the beginning of 2006. It should be noted that the families that are missing are generally families with low species numbers. The future will show if these data have an implication for the understanding of the global distribution. It should be noted that the Nothybidae, a family of the superfamily Diopsioidea, with 8 species is the only endemic family occurring in the Orient. So, most families in the Oriental Region have a worldwide distribution.

Oosterbroek (1998) provided a key to the Oriental families with notes on their biology. In Table 7.2 is given a list of regional counts for the most speciose families in the world. It is tempting to consider some families as poorly represented or studied in the Orient by comparing the regions as

Table 7.2. Comparison of the number of species of the most speciose Diptera families in the biogeographical regions. AU = Australasian/Oceanian, OR = Oriental, PA = Palae- arctic, NE = Nearctic, NT = Neotropical. Data from the *BioSystematic Database of World Diptera* (Evenhuis *et al.* 2007).

Families	AU	OR	PA	AF	NE	NT	Total
Limoniidae	1,917	2,323	1,625	1,027	926	2,648	10,334
Tachinidae	864	835	3,051	1,032	1,439	2,729	9,629
Asilidae	579	1,017	1,673	1,686	1,073	1,485	7,413
Dolichopodidae	1,186	1,057	1,716	766	1,382	1,189	7,118
Chironomidae	530	816	3,579	569	1,111	790	6,951
Cecidomyiidae	321	569	3,275	215	1,247	561	6,051
Syrphidae	417	878	2,058	591	818	1,516	5,935
Ceratopogonidae	839	876	1,537	916	614	1,084	5,621
Muscidae	743	845	1,502	992	631	882	5,153
Bombyliidae	448	317	1,370	1,437	988	717	5,030
Tephritidae	829	1,051	891	995	372	785	4,625
Tabanidae	468	819	778	816	394	1,168	4,387
Tipulidae	394	911	1,303	375	620	779	4,324
Mycetophilidae	319	370	1,549	258	673	1,056	4,105
Phoridae	291	587	906	452	422	1,487	4,022
Drosophilidae	1,141	1,019	416	460	248	876	3,925
Culicidae	627	975	268	810	182	952	3,616
Sarcophagidae	183	249	1,013	429	451	872	3,071
Agromyzidae	287	325	1,274	279	763	383	3,013
Empididae	188	249	1,425	214	468	377	2,911

such. Such a statement would be incorrect because the area of the regions is different as well as the biomes they contain. We can suspect that, e.g., Asilidae are poorly studied in the Orient, but it is only when we look to the sampling and the description effort done in the region itself that we will get a scientific indication that the region is poorly studied. For example, a recent faunistic study on the robber flies around Angkor Wat reported 16 first records of genera for Cambodia (Tomasovic 2006). A compilation of the Diptera of Thailand by Papp *et al.* (2006) recorded 25 first country records of Diptera families and that was based on only two months of sampling. It is evident that the sampling effort is the main problem in the region as will be illustrated in the next paragraphs.

Diptera Diversity: Status, Challenges and Tools
(eds T. Pape, D. Bickel & R. Meier). © 2009 Koninklijke Brill NV.

Table 7.3. Number of species of *Elaphropeza* Macquart in the countries of the Oriental Region, including surface area and forest cover in km² and main type of biome. M = mangrove; STMF = subtropical mixed forest; TDFS = dry forest and savanna; TWF = tropical wet forest. Note high number of species in Singapore (marked in red).

Country	Spp.	Surface	Forest	
Indian subcontinent				
Pakistan	0	796,095	29,622	TDFS, TWF
India	4	3,165,596	568,853	TDFS, TWF
Sri Lanka	8	65,610	21,295	TWF
Bhutan	0	47,000	26,334	TWF
Nepal	6	147,181	74,648	TWF
Bangladesh	0	147,570	13,570	TWF, M.
Southeast Asia				
Myanmar	1	676,552	459,451	TWF, TDFS
Thailand	1	513,115	171,069	TDFS, TWF, STMF
Laos	0	236,800	129,663	STMF
Viet Nam	0	331,690	50,146	TWF
Cambodia	1	181,035	116,394	TDFS
Malaysia	10	329,750	46,633	TWF
Brunei	0	5,765	3,454	TWF
Indonesia	8	1,904,570	911,337	TWF,TDFS
Singapore	52	685	250	TWF
Philippines	6	300,000	62,768	TWF
Taiwan	10	35,961	21,024	TWF
China				
South China	38	?	?	STMF

6. Diversity and the Poor Sampling Effort: A Case Study of *Elaphropeza* (Empidoidea, Hybotidae)

Only recently has a renewed interest in the study of the Oriental Diptera fauna emerged. In contrast, the study of systematic botany has flourished in the region for decades.

So far, faunistic and taxonomic studies are based on small samples and short-term studies. Rarely are any long-term samplings carried out at the same locality. The collections that H. Sauter made on the former Formosa

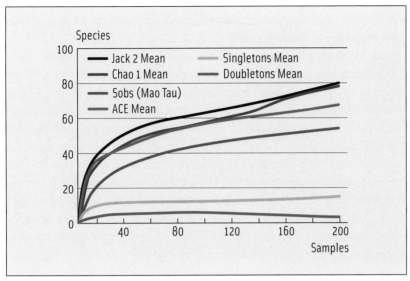

Figure 7.4. Cumulative species number of *Elaphropeza* Macquart (Hybotidae) recorded in Singapore (after Shamshev & Grootaert 2007). Sobs = species observed; Singletons = species known from a single specimen only; Doubletons = species known from two specimens only.

Island, now Taiwan, are an exception and provide us with a realistic insight in the fauna of Taiwan at the beginning of the 20th century (1907–1912). Sauter's collections are well preserved, but unfortunately there are no data about the biotopes sampled (Esaki 1941, Sachtleben 1941, Becker 1922).

A revision of *Elaphropeza* Macquart, a genus of small, leaf-dwelling, predatory flies belonging to the Hybotidae illustrates this problem (Shamshev & Grootaert 2007). Weekly collecting with Malaise traps in Singapore during one year at 8 stations in wet forest (6 traps) and mangrove (2 traps) supplemented with standardized hand collecting resulted in 544 samples. 210 samples contained *Elaphropeza* specimens belonging to 52 species. Only four species were already described, the rest were new to science. In fact the 52 species represent 43% of all described Oriental *Elaphropeza*. The species accumulation curve for Singapore (Fig. 7.4) shows no plateau, and the number of singletons and doubletons is around 33% and not declining. This clearly shows that even for a small area like Singapore, a one-year sampling effort is not enough to get a reliable estimate of the total *Elaphropeza* fauna present.

If we look on a regional basis at the distribution of *Elaphropeza* (Table 7.3), namely the numbers of species per country, the land area and the forest coverage, then the discrepancies are even more apparent. A huge country like India has only four described species. In addition, almost all species that have been described are still known from single records. Very few species are known from more than one country. All these elements point to a very high diversity of *Elaphropeza*, which remains poorly studied due to the low sampling effort. There are many similar examples in the Hybotidae (e.g., *Stilpon*: Shamshev & Grootaert 2004), Empididae (e.g., *Empis* (*Coptophlebia* Bezzi): Daugeron & Grootaert 2005) and Dolichopodidae, and it may be assumed that this lack of data can be extrapolated to all the species-rich Diptera families in the region.

7. Hotspots

In order to get an idea of diversity hotspots, we have to refer to other organisms than flies. Kier *et al.* (2005) presented a first global map of vascular plant species richness by ecoregion. The highest plant species richness in the world was found in the Borneo lowlands (10,000 species) followed by ecoregions located in Central and South America (8,000 species). The highest diversity was found in the tropical and subtropical moist broadleaf forest biome. Among biomes, adequate data are especially lacking for mangroves, flooded grasslands and flooded savannas. Indochina, Borneo (except northeast), sub-Himalayan regions and South India have poor quality data available, while Central India, Malaysia, Sumatra and North Borneo have good to moderate quality data available (Kier *et al.* 2005).

Care should be taken by extrapolation, because species-rich ecoregions are better studied than those poor in vascular plants (Kier *et al.* 2005). A data set for global representation analyses exists for vertebrates, with richness and endemism data now available for 30,000 species of birds, mammals, reptiles and amphibians (WWF: Wildfinder: global species data base, 2007; www.worldwildlife.org). For each respective group the hotspots are slightly different. The total richness for amphibians, reptiles, birds and mammals is highest in South China and North Indochina. The total endemism for organisms is very high in the Western Ghats, Yunnan, Laos, North Burma, North Thailand, North Viet Nam, Borneo and the southern Philippines. It is very tempting to concentrate research on hotspots for Diptera in the same hotspots as for the vascular plants and vertebrate groups.

8. Regional Diversity and Endemics

The Oriental Region can be subdivided into three major subregions: the Indian subcontinent, South-China and Southeast Asia. In taxonomic and systematic studies we are interested to see areas of endemism that are generally predictive for delineating taxa and for understanding distribution patterns. Unfortunately, to illustrate the endemic areas we have to refer to other insect groups since no dipterist ever made a complete zoogeographical analysis of the region. Beuk (2002) groups the cicadid fauna into 10 endemic areas and it is likely, but not certain that a similar pattern will be found in most other insect groups. Liang (1998) made a cladistic biogeographical analysis of the Cercopoidea and Fulgoroidea (Homoptera) of China and the adjacent regions, and he recognizes 16 zones that are discussed below in the paragraph on South China. Since these authors do not cover the same area and their aims are different, a combination of their proposals would provide a more detailed biogeographical pattern for the Orient.

The species richness of most taxa follows a latitudinal gradient with higher richness towards the equator (Rosenzweig 1995). A recent study on hawkmoths (Sphingidae) shows a reverse pattern with highest richness in northeast Thailand's Chiang Mai province and surrounding areas and with a decline towards the south (Beck *et al.* 2007). A peninsula effect, in relation to area size, is probably an explanation of the lower species richness in the south. The authors do not exclude the effect of strong congruencies in estimated species richness and sampling effort and suggest that unusual latitudinal patterns could be induced by ecological effects. However, more factors probably contribute to the higher species richness in northeast Thailand such as the altitudinal relief and higher endemism in montane regions. In empidoid flies (Empididae, Hybotidae, Dolichopodidae) there is a distinct intermingling of Palaearctic and Oriental 'genera' in low montane areas in the northeast Thai province of Loei (Grootaert & Verapong 2003) thus increasing the local diversity of this group. Beck *et al.* (2007) found a similar effect in north Viet Nam in hawkmoths. Never-

theless various studies on insects confirm that the northern area of Southeast Asia and the Chinese Yunnan province are among the most diverse
regions in the world (mammals: Olson *et al.* 2001; hawkmoths: Beck *et al.*
2007).

8.1 India

The Indian subcontinent is actually the 'great unknown' as to the origin
and composition of its fauna. It is a major region that has Gondwana elements, but this is difficult to demonstrate without more research. There
are assumptions that its fauna disappeared completely when the Indian
plate submerged during its drift towards Eurasia (Cranston 2005), but
Cranston also suggests that maybe the fauna of the southern mountains
of the southern Ghats retained faunal elements that later colonised northern India.

Reinert (1970) presents an interesting hypothesis for the Gondwanan
distribution of the culicid subgenus *Aedes* (*Diceromyia* Theobald). Actually, species of this subgenus are confined chiefly to the savanna and forest
fringe areas of the Afrotropical and Oriental Regions. Eight species are restricted to the former area and eleven to the latter, with no species occurring in both regions. The hypothesis is presented that the ancestor of the
subgenus evolved from an *Aedimorphus*-like stock in the Indian area and
dispersed into the Afrotropical Region when the climate and vegetation
of the intervening area were more propitious during the Oligocene when
a land bridge (Arabia) was established between Africa and Eurasia. As
the climate changed in the connecting area, the favourable environmental
corridor ceased to exist and the populations of the two areas radiated and
evolved in their respective regions.

Briggs (2003) proposes the following sequence of events when India
moved northwards: 1) India-Madagascar rifted from East Africa 158–
160 Mya; 2) India-Madagascar rifted from Antarctica about 130 Mya; 3)
India-Seychelles split from Madagascar 84–96 Mya; 4) India split from
Seychelles 65 Mya; 5) India began collision with Eurasia 55–65 Mya and
6) final suturing took place about 45–55 Mya. Fossil records and contemporary faunas indicate that throughout the late Cretaceous, India maintained exchanges with adjacent lands. During suturing of the subcontinent to Eurasia, the uplifting of the Himalayas took place and so probably
the mountains cutting off Yunnan and south China from Southeast Asia.
That brings us back to the original hypothesis that the tropical Oriental

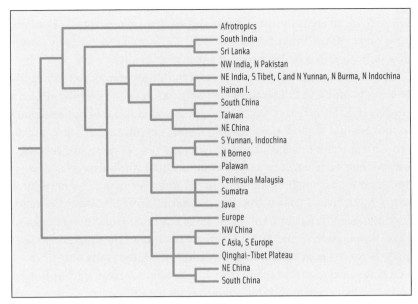

Figure 7.5. Composite biogeographical and phylogenetic cladogram of
the Homoptera Cercopoidea and Fulgoroidea
(after Liang 1998).

fauna originated from a Gondwanan stock that dispersed into Asia via India.

It is also hypothetised that India originally had a fauna adapted to a wet tropical climate that after the collision with Eurasia became gradually dryer due to the uplifting of the Himalayas and the Western Ghats.

The biodiversity hotspot of the Western Ghats-Sri Lanka is one of the world's most important hotspots. Bossuyt *et al.* (2004) showed by using molecular phylogenies that the Western Ghats and Sri Lanka have both a unique, related fauna, but distinct from each other.

Finally we refer to Mani (1974) for general data on the ecology and biogeography in India.

8.2 South China

Zhang (1998) gives a revision of the zoogeographical regions, which could be used as a reference for students of the Chinese fauna. There are no zoogeographical studies available for Diptera covering the whole of south China. Again, we have to refer to other insect groups to get an idea about the relation of Oriental China and the rest of the Orient, being well aware that the areas of distribution and endemism are not the same for all groups, and will not be the same for the various fly families simply depending on their biology and ecological preferenda.

Liang (1998) examined the relation of the biogeography and phylogeny of Cercopoidea and Fulgoroidea (Homoptera) in China and the adjacent regions (Fig. 7.5). He recognises 16 areas of endemism, but not all belong to the Oriental Region: northeastern China, northwestern China, southern China, Hainan Island, Taiwan, Qinghai-Tibet plateau, northwestern India + northern Burma + south western China (including southern Tibet, central and northern Yunnan and western Sichuan), southern Yunnan + Indochina, Malay peninsula, northern Borneo, Palawan, Sumatra, Java, southern India, Sri Lanka. These regions correspond more or less to those proposed by Beuk (2002), but there are important differences especially because Liang's aim was to explain the Chinese fauna only.

Liang concludes that the main Chinese fauna is related to European or Eurasian elements and that there was an invasion of old Gondwanan elements to the southwest and south China, Indochina and southeast Asia following the accretion of the Indian subcontinent. He claims that many tropical cercopoids and fulgoroids from India invaded southwest and south China, Indochina, Southeast Asia and the Greater Sunda Islands via Assam and Burma. He further explains that the continuing rise of the Himalayas due to the impact of the subduction caused desertification of central India, leaving south India and Sri Lanka as isolated tropical refugia resulting in discontinuous distributions and a sister relationship between south India+Sri Lanka and southern China, southern Yunnan; Indochina, Malaya and the Greater Sunda Islands seen today.

9. Southeast Asia

The third huge region in the Orient is Southeast Asia composed of Indochina (Laos, Thailand, Cambodia, Viet Nam), western Indonesia and the Philippines. Indochina is well delineated on its northern border from China by mountain chains. Again, we have to refer to other insect groups such as cicadas (Beuk 2002) to see that Southeast Asia is composed of a number of subregions that correspond to areas with separate geological evolution (see Fig. 7.6: Indochina, the Malay Peninsula with the South of Thailand, Sumatra, Java, lesser Sunda Islands, Borneo, Palawan, the Philippines, Taiwan with the Ryukyu Islands). At the moment we have no idea

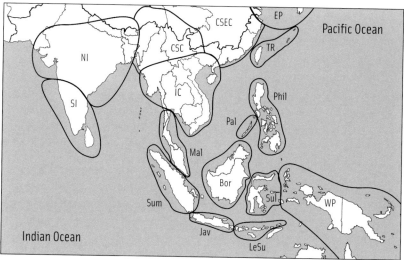

Fig. 7.6. Regions of endemism of cicadids after Beuk (2002). Abbreviations: Bor: Borneo; CSV: Central Southern China; CSEC: Central Southeastern China; EP: East Palaearctic; IC: Indochina; Jav: Java; LeSu: Lesser Sunda Islands; Mal: Malayan Peninsula; NI: Northern India; Pal: Palawan; Phil: Philippines; SI: Southern India; Sul: Sulawesi; Sum: Sumatra; TR: Taiwan and Ryukyu Archipelago; WP: West Pacific
(after Beuk 2002).

if the many areas of endemism have their counterparts in the distribution of flies.

Cubitt & Stewart-Cox (1995) give interesting illustrations of the biogeographical regions of Thailand. A compendium of the biogeography of Viet Nam is given by Sterling *et al.* (2006). A faunistic study of the Diptera and more precisely the phenology of particular species in the four biogeographical regions of Viet Nam, which are arranged along a latitudinal gradient, would give much information of the adaptation of species to different ecological zones.

10. Biogeographical Patterns

The presence of Palaearctic elements on mountains in the Oriental Region is still a puzzle. Mountaintops on the islands throughout Indonesia have been colonised by rather eurytopic plant species that eventually radiated through vicariant speciation (Oosterzee 1997). We have no evidence that this phenomenon took place in insects as well.

Hilara Meigen (Empididae), a genus of dance flies famous for the horizontal swarms in which males present a prey as nuptial gift to the females, has a worldwide distribution. It radiated in the northern as well as in the southern hemisphere. It is present in Mediterranean Australia and in areas with a more temperate climate as, e.g., above 500 m in Southeast Asia (Grootaert & Verapong 2001). It is not present in the Oriental tropical lowland. It is likely to have a Pangean origin with a strong adaptation to a temperate climate. Apparently, this adaptation is so strong that no species have yet invaded wet, tropical, lowland habitats.

The genus *Platypalpus* Macquart (Hybotidae), which consists of 2–5 mm long predatory species that hunt small insects on horizontal substrates, is a second example of such a Pangean distribution. The genus is particularly speciose in the mountains of south China (Yang *et al.* 2006a), and a few species occur in the hills and mountains of northeast Thailand (Grootaert & Shamshev 2006). Species are sporadically reported from higher elevations throughout the Oriental Region (P. Grootaert, unpubl.). Did these species reach these mountains during colder spells, or are other mechanisms required to explain their presence?

The very speciose, cosmopolitan genus *Empis* Linnaeus (Empididae) also has a number of representatives in the Oriental Region. However, they all belong to a single species group: the *Empis* (*Coptophlebia*) *hyalea*-group (Daugeron & Grootaert 2005). It probably has a Gondwanan origin and has only a circumtropical distribution. Apart from this species group, none of the numerous subgenera and species-groups of *Empis* colonised the tropical Oriental lowland.

The genus *Lichtwardtia* Enderlein (Dolichopodidae) has a very similar distribution to *Empis* (*Coptophlebia*). It is quite speciose in the Afrotropical Region and is very common throughout the Oriental Region. Only three species are actually described from the Oriental Region, and most records are attributed to *Lichtwardtia ziczac* (Wiedemann). In fact numerous species are actually confused with it. In Angkor Wat (Cambodia) alone, five species occur together and are very abundant (P. Grootaert, unpubl.). Recently, the genus was merged with *Dolichopus* Latreille by Brooks (2005), but 'true' *Dolichopus* do not occur in Oriental tropical lowland, only the very distinctive *Lichtwardtia*-group is present. In merging these genera, the ecological and distributional information linked to these taxa are lost. *Dolichopus* is a genus adapted to temperate climate and it radiated in the Holarctic. Of course, the possibility remains that *Dolichopus* is polyphyletic, because the constituent groups cannot be recognised and classified on morphological grounds yet.

Care should be taken not to conclude too quickly that species have a wide, circumtropical distribution. *Neurigona angulata* de Meijere, a *Neurigona* Rondani with quadriserial acrostichals, is reported from the Seychelles, India, Sri Lanka, Borneo, Java, Singapore, Taiwan, the Bismarck Archipelago and the Solomon Islands (Dyte 1975, Yang *et al.* 2006b). However, the distribution of *N. angulata* is much more restricted as proved by a recent revision (P. Grootaert, unpubl.) and is actually limited to Java, Singapore and Borneo. All the other records concern similar but different species. There are probably many more such examples.

Nevertheless, there are probably a number of good examples of Gondwanan distribution. The example of the genus *Teuchophorus* Loew is obvious, though it is not present in the Neotropics. *Teuchophorus* radiated in the Oriental and Australasian Regions, with 53 and 37 described species respectively. Fifteen species have been recorded from the Palaearctic Region, four from the Nearctic Region, seven from the Afrotropical Region (including six in *Olegonegrobovia* Grichanov). The species from the Afrotropical Region all belong to the *conspicuus-notabilis* group. Grichanov (1995) described them in the genus *Olegonegrobovia* because they have a peculiar morphology and at that time the link with *Teuchophorus* was not clear. Some species bear a set of strong bristles on the hind margin of the wing; the mid femur is swollen at its base and bears there a set of three generally very long bristles. The mid tibia bears always a set of long flattened ventral bristles. In addition the male genitalia are quite characteristic for this species-group. A number of very similar species have been found in Thailand, Singapore and south China (Grootaert 2006), and it is quite obvious that they are closely related to the Afrotropical species. The similarities cannot be explained by homoplasy. So, the question is how did this species group get into the Oriental Region or into the Afrotropics? It is unknown from the Indian subcontinent, but that is probably due to undersampling. In addition the 'plesiomorphic' type of *Teuchophorus* is not present in the Afrotropics or has been overlooked up to now.

11. Circumtropical Species

Several Oriental Diptera are considered to have a circumtropical distribution. In some cases Bickel (1996) calls them tramp species. However, is this really so in all cases? Let me comment on a few examples:

Tachytrechus tessellatus Macquart has a very wide distribution and is reported from Africa over India to Taiwan (Yang *et al.* 2006b). It is a large Dolichopodinae that is active in open, sunny, wet habitats, usually with only temporary, shallow waterbodies such as shallow pools after rain, upwelling water downhill, in open rainwater drains and on the sandy or muddy banks of small streams. All these habitats are generally disturbed (anthropogenic) habitats and often there is an algal bloom in the water bodies. It is likely that a species adapted to these temporary habitats possesses a good dispersal capacity and eventually has a circumtropical distribution.

The *Thinophilus indigenus* Becker complex is found in the same conditions as *T. tessellatus*, though the situation is more complex. Becker described *Thinophilus indigenus* from Egypt and later reported it from Egypt and Taiwan. Grichanov (1997) even reports it from Afrotropical countries. Grootaert & Meuffels (2001) described the closely related *Th. setiventris* Grootaert & Meuffels from Thailand, and it was later found in Singapore and Cambodia. The structure of the male genitalia looks almost identical in both species. However, the number of dull black dots of the mesonotum and on the occiput divides the group into three species. The pattern of the black spots might be a specific signal for the recognition of the species when males hover over a possible partner or rival. Molecular studies could help solve that problem, although a study on the phylogeny of Southeast Asia Dolichopodidae (Lim *et al.* 2006) show that the genetic

Table 7.4. Genetic distances between populations of the mangrove-associated *Teuchophorus simplicissimus* Meuffels & Grootaert and the rainforest-dwelling *T. stenostigma* Meuffels & Grootaert (Dolichopodidae) on Pulau Tioman (Malaysia) and Singapore as evidenced by the genes Cytb, COIa and COIb; after Lim *et al.* (2006). [NA = not analysed.]

	Cytb (%)	COIa (%)	COIb (%)
T. simplicissimus	3.5-3.9	NA	6.5
T. stenostigma	8.4	9.3	8.2

distances between populations of *Teuchophorus simplicissimus* Meuffels & Grootaert are small on the island Pulau Tioman (Malaysia), but are more than 10% between Pulau Tioman and Singapore (Table 7.4). The genetic distance between *Teuchophorus stenostigma* Meuffels & Grootaert in Singapore and Pulau Tioman is even larger. It might be explained by the fact that these species live in more or less undisturbed rain forest and so have a more limited dispersal capacity, leading to less genetic exchange. Contact between Tioman Island and Peninsular Malaysia date from the last ice ages when sea level dropped considerably. Dispersal is probably easier for *T. simplicissimus* that lives in the inner part of mangroves around pools with low but variable salinity. Dispersal along coasts by hopping from one mangrove patch to another could be easier and therefore explain the smaller genetic distance. However, if we follow the barcode paradigm hypothesis (Hebert *et al.* 2003), even the population on Tioman Island should be considered as a different species from the one in Singapore. Apart from the fact that the males are a bit larger in Tioman than in Singapore, we found no morphological indications that could separate them.

12. Disjunctive Distributions and Glaciations

Ecological adaptations are generally little flexible. As seen above, species of the dance fly genus *Hilara* are adapted to temperate climate conditions and do not invade areas with tropical climate. However, there are some disjunct distributions of species or clades that are present in mountains on the one hand and in lowlands or mangroves on the other hand. There are many examples in various insects groups and also in flies such as in the *Empis* (*Coptophlebia*) *hyalea*-group, *Platypalpus* Macquart and several dolichopodids. For example, *Neurigona pectinata* Becker, a very distinct species with an anteroventral comb of fine erect hairs on mid tibia and tarsus, is found in the hills of northeast Thailand at an altitude of 500m. It is very common in the mangroves of Singapore, but it was never observed in the nearby rainforests. *Platypalpus* spp. are relatively common in spring in the mid altitudinal mixed forests of northeast Thailand. They are not present in true tropical forests, and only a single, very rare species was found in the lowland forest in Singapore. These disjunctive distributions may be explained as a consequence of the Pleistocene glaciations. The climate was arid and the humid areas where insects could thrive were restricted to mountaintops, areas near the coast and saline coastal habi-

DIPTERA DIVERSITY: STATUS, CHALLENGES AND TOOLS
(EDS T. PAPE, D. BICKEL & R. MEIER). © 2009 KONINKLIJKE BRILL NV.

tats suitable for mangroves. These areas served as refugia or reservoirs for a number of species, explaining their current disjunct distribution. The following data on mangrove species further illustrate the phenomenon.

13. Mangroves, an Underestimated Source of Biodiversity

Mangroves are forests bordering the sea. They have characteristic trees with aerial roots and grow on a muddy soil. The forests can be quite wide with a gradient of salinity from pure seawater to areas with low, but variable salinity depending on spring tides and rainfall. The classical point of view that marine habitats have not been or are very poorly colonised by insects is quite wrong for mangroves. This idea has probably been nourished by the difficult access and dirty working conditions in mangroves. For Dolichopodidae we recorded after a year of sampling with Malaise traps in the mangroves of Singapore no less than 95 species of Dolichopodidae of which 50 species were exclusively found here, in addition nine species were bound to marine rocky shores or sandy beaches, and the remaining 36 species were invaders from freshwater or terrestrial habitats. In Table 7.5 it can be seen that almost all subfamilies have species that occur in mangroves and that are exclusively found in this habitat. The genera *Ngirhaphium* Evenhuis & Grootaert, *Nanothinophilus* Grootaert & Meuffels and *Phacaspis* Grootaert & Meuffels are also exclusively found in mangroves. Worth noting is that *Phacaspis* has been described on the basis of two species from the island of Motupore (near Port Moresby) on the south coast of Papua New Guinea. They are both recorded in mangroves of Singapore and Thailand (Grootaert & Meuffels 2001) and are in fact among the rare dolichopodid species recorded so far with a confirmed Oriental and Australasian distribution.

Of the 52 species of the small hybotid *Elaphropeza* (Shamshev & Grootaert 2007) occurring in Singapore, 30 species are recorded from mangroves. Fourteen species are even exclusively found here. Ubiquitous or eurytope species that are dominant in terrestrial habitats occur only in very small numbers in mangrove. Apparently adaptation to the marine environment is specific so that even ubiquitous species that can tolerate anthropogenic habitats are rare in marine habitats.

These two examples illustrate the possibility of mangroves to act as refugia for species, and when certain species adapt to variable saline conditions, mangroves might act as corridors for dispersal. Available samples

Table 7.5. Dolichopodid genera and number of species in mangrove in Singapore

Subfamilies	Genera		Species	Only man-grove	Marine
Aphrosylinae	3	*Thinolestris*	1		1
		Cymatopus	2		2
		Thambemyia	1		1
Diaphorinae	4	*Asyndetus*	1		1
		Chrysotus	2		
		Diaphorus	3		
		Trigonocera	1		
Dolichopodinae	6	*Hercostomus*	10	10	
		Lichtwardtia	2	1	
		Paraclius	8	8	
		prope *Argyrochlamys*	1		1
		Setihercostomus	1	1	
		Tachytrechus	1		
Hydrophorinae	2	*Nanothinophilus*	1	1	
		Thinophilus	17	14	
Incertae sedis	1	*Phacapis*	3	3	
Medeterinae	2	*Medetera*	3	1	
		Paramedetera	1		
Neurigoninae	1	*Neurigona*	3	2	
Parathalasiinae	2	*Eothalassius*	1		1
		Microphorella	1		1
Peloropeodinae	1	*Acropsilus*	3	1	
Rhaphiinae	1	*Ngirhaphium*	2	2	
Sciapodinae	4	*Amblypsilopus*	2		
		Chrysosoma	3	1	
		Mesorhaga	1	1	
		Plagiozopelma	1		
		Sciapodinae	5		
Sympycninae	4	*Chaetogonopteron*	7		
		Hercostomoides	1		
		Phrudoneura	1		
		Teuchophorus	5	4	1
Total	**31**		**95**	**50**	**9**

show similar high diversity in other groups such as Culicidae, Asilidae, Stratiomyidae and Phoridae.

14. General Conclusions and Challenges

The Oriental Region probably has one of the richest Diptera faunas in the world and is estimated to host at least 85,000 species. Especially the belt south of the Himalayas over south China (Yunnan, Guangxi and Guangdong province), north Burma, north Thailand, Laos, north Cambodia and north Viet Nam are very species rich areas. Moreover, these areas host many ancestral groups important to understand the systematics of Diptera on a world scale. The fauna of the Indian subcontinent is the big unknown and needs study before its presumed influence on the eastern part of the Oriental can be well understood. Apart from the Culicidae and Tephritidae, all other families are in urgent need of alpha-taxonomical research. It is demonstrated that the entire region is under-sampled. Rain forests, the most threatened habitats on Earth, are undoubtedly the most speciose habitats in need of urgent study. Mangroves have proven to be an underestimated source of biodiversity with huge diversity and strange, disjunctive distribution patterns of species. Mangroves are as threatened as rain forests.

Acknowledgements

I should like to thank Rudolf Meier and his team, who gave me insights in the genetic problems and hosted me during my one-year sabbatical stay in Singapore. Further, I acknowledge the help of my colleagues Christophe Daugeron, Maurice Leponce, Igor Shamshev and Ding Yang for their co-operation in empidoid research. Irina Brake and Chris Thompson provided the original datasets of the *BioSystematic Database of World Diptera*. Finally, I thank Dan Bickel, Adrian Plant and an anonymous referee for their many useful comments on the manuscript, and Thomas Pape for an in depth review of the manuscript.

References

Beck, J., Kitching, I.J. & Linsenmair, K.E. (2006) Wallace's line revisited: has vicariance or dispersal shaped the distribution of Malesian hawkmoths (Lepidoptera: Sphingidae)? *Biological Journal of the Linnaean Society* 89: 455–468.

Beck, J., Kitching, I.J. & Haxaire, J. (2007) The latitudinal distribution of sphingid species richness in continental southeast Asia: what causes the biodiversity 'hot spot' in northern Thailand? *The Raffles Bulletin of Zoology* 55: 179–185.

Becker, T. (1922) Dipterologische Studien, Dolichopodidae der Indo-Australischen Region. *Capita Zoologica* 1: 1–247.

Beuk, P.L.T. (2002) *Cicadas spreading by island or by spreading the wings? Historic biogeography of dundubiine cicadas of the Southeast Asian continent and archipelagos.* Academisch Proefschrift, Universiteit Amsterdam, 323 pp.

Bickel, D.J. (1996) Restricted and widespread taxa in the Pacific: Biogeographic processes in the fly family Dolichopodidae (Diptera). Pages 331–346 *in*: Keast, A. & Miller, S.E. (eds), *The origin and evolution of Pacific Islands biotas, New Guinea to eastern Polynesia: patterns and processes.* SPB Academic Publishing, Amsterdam.

Bickel, D. (2009) Why *Hilara* is not amusing: the problem of open-ended taxa and the limits of taxonomic knowledge. Pages 279-301 *in*: Pape, T., Bickel, D. & Meier, R. (eds), *Diptera Diversity: Status, Challenges and Tools.* Brill, Leiden.

Bossuyt, F., Meegaskumbura, M., Beernaert, N., Gower, D.J., Pethiyagoda, R., Roelants, K., Mannaert, A., Wilkinson, M., Bahir, M.M., Manamendra-Arachchi, K., Ng, P.K.L., Schneider, C.J., Oomen, O.V. & Milinkovitch, M.C. (2004) Local endemism within the Western Ghats-Sri-Lanka Biodiversity Hotspot. *Science* 306: 479–481.

Briggs, J.C. (1989) The historic biogeography of India: isolation or contact. *Systematic Zoology* 38: 322–332

Briggs, J.C. (2003) The biogeographic and tectonic history of India. *Journal of Biogeography* 30: 381–388.

Brooks, S.E. (2005) Systematics and phylogeny of Dolichopodinae (Diptera: Dolichopodidae). *Zootaxa* 857: 1–158.

Cranston, P. (2005) Biogeographic patterns in the Evolution of Diptera. Pages 274–311 *in*: Yeates, D.K. & Wiegmann, B.M. (eds), *The Evolutionary Biology of Flies.* Columbia University Press, New York.

Cubitt, G. & Stewart-Cox, B. (1995) *Wild Thailand.* Asia Books, Bangkok, 208 pp.

Daugeron, C. & Grootaert, P. (2005) Phylogenetic systematics of the *Empis (Coptophlebia) hyalea*-group (Insecta: Diptera: Empididae). *Zoological Journal of the Linnean Society* 145: 339–391.

Dyte, C.E. (1975) Family Dolichopodidae. Pages 212–258 *in*: Delfinado, D. & Hardy, D.E. (eds), *A catalog of the Diptera of the Oriental Region.* Vol. II. Suborder Brachycera through division Aschiza, Suborder Cyclorrhapha. University Press of Hawaii, Honolulu.

Esaki, T. (1941) Hans Sauter. *Arbeiten über morphologische und taxonomische Entomologie aus Berlin-Dahlem* 8: 81–86.

Evenhuis, N.L. (1994) *Catalogue of the fossil flies of the world (Insecta: Diptera)*. Backhuys Publishers, Leiden, 600 pp.

Evenhuis, N.L., Pape, T., Pont, A.C. & Thompson, F.C. (eds) (2007) *BioSystematic Database of World Diptera*, Version 10. Available at http://www.diptera.org/biosys. htm, accessed 4 March 2007.

Frith, C.B. & Frith, D.W. (2006) Seasonality of insect abundance in an Australian upland tropical rainforest. *Austral Ecology* 10: 237–248.

Grichanov, I.Y. (1995) *Olegonegrobovia* (Diptera: Dolichopodidae), new genus from Uganda. *International Journal of Dipterological Research* 6: 125–128.

Grichanov, I.Y. (1997) Notes on Afrotropical and Palaearctic species of the genus *Thinophilus* Wahlberg (Diptera: Dolichopodidae) with description of new species. *International Journal of Dipterological Research* 8: 135–147.

Grimaldi, D. (1999) The co-radiations of pollinating insects and angiosperms in the Cretaceous. *Annals of the Missouri Botanical Garden* 86: 373–406.

Grimaldi, D. & Cumming, J. (1999) Brachyceran Diptera in Cretaceous ambers and Mesozoic diversification of the Eremoneura. *Bulletin of the American Museum of Natural History* 239: 1-124.

Grootaert, P. (2006) *Dolichopodidae in mangroves of Southeast Asia: diversity, community structure, zonation and phenology: a case study in Singapore*. Pages: 91–92 *in*: Suwa, M. (ed.), Abstracts Volume, 6ᵗʰ International Congress of Dipterology, Fukuoka, 23–28 September 2006.

Grootaert, P. & Meuffels, H. (2001) A note on the marine dolichopodid flies from Thailand (Insecta, Diptera, Dolichopodidae). *Raffles Bulletin of Zoology* 49: 339–353.

Grootaert, P. & Shamshev, I. (2006) The genus *Platypalpus* Macquart (Diptera: Hybotidae) from Northeast Thailand with comments on the species groups in the Oriental region. *Journal of Natural History* 39: 4031–4065.

Grootaert, P. & Verapong Kiatsoonthorn (2001) First records of the dance fly genus *Hilara* in Thailand with the description of five new species. *Natural History Bulletin of the Siam Society* 49: 17–27.

Grootaert, P. & Verapong Kiatsoonthorn (2003) Insects of Na Haeo: a preliminary survey and seasonal dynamics of dolichopodid and empidid flies. Pages 121–135 *in*: La-aw Ampornpan & Shivcharn S. Dhillion (eds), *The Environment of Na Haeo, Thailand: Biodiversity, non-timber products, land use and conservation*. Craftsman Press Ltd, Bangkok.

Hall, R. (1998) The plate tectonics of Cenozoic SE Asia and the distribution of land and sea. Pages 99–131 *in*: Hall, R. & Holloway, J.D. (eds), *Biogeography and Geological Evolution of SE Asia*. Backhuys Publishers, Leiden.

Hebert, P.D.N., Ratnasingham, S. & deWaard, J.R. (2003) Barcoding animal life: cytochrome C oxidase subunit 1 divergences among closely related species. *Proceedings of the Royal Society of London B (Supplement)* 270: S96–S99.

Kier, G., Mutke, J., Dinerstein, E., Ricketts, H., Küper, W., Kreft, H. & Barthlott, W. (2005) Global patterns of plant diversity and floristic knowledge. *Journal of Biogeography* 32: 1107–1116.

Kitching, R.L., Bickel, D., Creagh, A.C., Hurley, K. & Symonds, C. (2004) The biodiversity of Diptera in Old World rain forests surveys: a comparative faunistic analysis. *Journal of Biogeography* 31: 1185–1200.

Liang, A.P. (1998) [Cladistic biogeography of Cercopoidea and Fulgoroidea (Insecta Homoptera) in China and adjacent regions]. *Acta Zootaxonomica Sinica (Supplement)* 23: 132–166. [In Chinese with English abstract and figure captions.]

Lim, S.G., Hwang, W.S., Grootaert, P. & Meier, R. (2006) Surviving the test of time: Testing species boundaries and the subfamily concepts in Dolichopodidae using South-east Asian species. Page 153 *in*: Suwa, M. (ed.), Abstracts Volume, 6th International Congress of Dipterology, Fukuoka, 23–28 September 2006.

Lydekker, R. (1896) *A geographic history of mammals*. Cambridge University Press, Cambridge, 400 pp.

Mani, M.S. (1974) *Ecology and biogeography in India*. W. Junk, The Hague, 773 pp.

Metcalfe, I. (1998) Palaeozoic and Mesozoic geological evolution of the SE Asian region: multidisciplinary constraints and implications for biogeography. Pages 25–41 *in*: Hall, R. & Holloway, J.D. (eds), *Biogeography and Geological Evolution of SE Asia*. Backhuys Publishers, Leiden.

Nagatomi, A. (1991) History of some families of Diptera, chiefly those of lower Brachycera (Insecta: Diptera). *Bulletin of the Biogeographical Society of Japan* 46: 1–22.

Nagatomi, A. & Yang, D. (1998) A review of extinct Mesozoic genera and families of Brachycera (Insecta, Diptera, Orthorrhapha). *Entomologist's Monthly Magazine* 134: 95–192.

Olson, D.M., Dinerstein, E., Wikramanayake, D., Burgess, N.D., Powell, G.V.N., Underwood, E.C., D'Amico, J.A., Itoua, I., Strand, H.E., Morrison, J.C., Loucks, C.J., Allnutt, T.F., Ricketts, T.H., Kura, Y., Lamoreux, J.F., Wettengel, W.W., Hedao, P. & Kassem, K.R. (2001) Terrestrial ecoregions of the world: a new map of life on Earth. *Bioscience* 51: 933–938.

Oosterbroek, P. (1998) *The families of Diptera of the Malay Archipelago*. Fauna Malesiana Handbooks Volume 1. Brill, Leiden, xii + 227 pp.

Oosterzee, P. van (1997) *Where Worlds Collide: the Wallace Line*. Reed International Books Australia Pty Ltd, Victoria, 234 pp.

Papp, L., Merz, B. & Földvári, M. (2006) Diptera of Thailand. A summary of the families and genera with references to the species representations. *Acta Zoologica Academiae Scientiarum Hungaricae* 52: 97–269.

Reinert, J.F. (1970) The zoogeography of *Aedes* (*Diceromyia*) Theobald (Diptera: Culicidae). *Journal of the entomological Society of South Africa* 33: 129–141.

Rosenzweig, M.L. (1995) *Species diversity in space and time*. Cambridge University Press, Cambridge (Massachusetts), xxi + 436 pages.

Sachtleben, H. (1941) Die Formosa-Sammlung des Deutschen Entomologischen Instituts. *Arbeiten über morphologische und taxonomische Entomologie aus Berlin-Dahlem* 8: 87–90.

Shamshev, I.V. & Grootaert, P. (2004) A review of the genus *Stilpon* Loew (Diptera, Empidoidea, Hybotidae) from the Oriental region. *The Raffles Bulletin of Zoology* 52: 315–346.

Shamshev, I.V. & Grootaert, P. (2007) Revision of the genus *Elaphropeza* Macquart (Diptera: Hybotidae) from the Oriental Region, with a special attention to the fauna of Singapore. *Zootaxa* 1488: 1–164.

Soós, Á. & Papp, L. (1991) *Catalogue of Palaearctic Diptera. Volume 7. Dolichopodidae — Platypezidae.* Akadémiai Kiadó, Budapest, 291 pp.

Sterling, E.J., Hurley, M.M. & Le Duc Minh. (2006) *Viet Nam a Natural History.* Yale University Press, New Haven (Connecticut) and London, 423 pp.

Tomasovic, G. (2006) Etude sur les genres d'Asilidae (Diptères) recensés de 2003 à 2005 sur le site d'Angkor (Cambodge). *Belgian Journal of Entomology* 8: 11–15.

Voris, H.K. (2000) Maps of Pleistocene sea levels in Southeast Asia: shorelines, river systems and time durations. *Journal of Biogeography* 27: 1153–1167.

Wallace, A.R. (1876) *The geographical distribution of animals; with a study of the relations of living and extinct faunas as elucidating the past changes of the Earth's surface.* Volume 1, MacMillan & Co., London; 503 pp.

Wiegmann, B.M., Yeates, D.K., Thorne, J.L. & Kishino, H. (2003) Time flies, a new molecular time-scale for brachyceran fly evolution without a clock. *Systematic Biology* 52: 745–756.

Wolda, H. (1968) Altitude, habitat and tropical insect diversity. *Biological Journal of the Linnean Society* 30: 313–323.

Wolda, H. (1988) Insect seasonality: why? *Annual Review of Ecology and Systematics* 19: 1–18.

Yang D., Merz, B. & Grootaert, P. (2006a) Description of three new species of *Platypalpus* Macquart from Guangdong (Diptera: Hybotidae, Tachydromiinae). *Revue Suisse de Zoologie* 113: 229–238.

Yang D., Zhu Y., M. Q. Wang & Zhang L. (2006b) *World catalog of Dolichopodidae (Insecta: Diptera).* China Agricultural University Press, Beijing, 704 pp.

Zhang Y.Z. (1998) The second revision of zoogeographical regions of China. *Acta zootaxonomica sinica (Supplement)* 23: 207–222.

CHAPTER EIGHT

DIVERSITY, RELATIONSHIPS AND BIOGEOGRAPHY OF AUSTRALIAN FLIES

DAVID K. YEATES[1], DANIEL BICKEL[2], DAVID K. MCALPINE[2] & DON H. COLLESS[1]

[1]*Australian National Insect Collection, Canberra, Australia;*
[2]*Australian Museum, Sydney, Australia*

INTRODUCTION

We live in a world largely filled with prokaryotes, fungi, flowering plants, nematodes, and a handful of ecological keystone insect groups. One of these keystone insect groups is the Diptera (true or two-winged flies, including mosquitoes, midges, etc.), and they probably comprise 12–15% of animal species diversity (Yeates *et al.* 2003, Thompson 2009). The flies span a wide range of anatomical and biological specialisations (Merritt *et al.* 2003) and are probably the most ecologically diverse of the four megadiverse insect orders (Kitching *et al.* 2005). The extant world Diptera fauna currently comprises approximately 153,000 described species in 159 families (Evenhuis *et al.* 2008); however, the total number including undescribed species is at least 300% higher. The living dipteran species have been classified into about 10,000 genera, 22–32 superfamilies, 8–10 infraorders and 2 suborders (Yeates & Wiegmann 1999). A phylogenetic classification of 8 suborders that avoids the use of the paraphyletic 'Nematocera' has been proposed recently (Amorim & Yeates 2006).

1. The Australian Continent

The continent of Australia covers approximately 7.5 million km^2 and spans about 34 degrees of latitude between 10° and 44° S. While most of Australia is in subtropical and temperate latitudes, over 70% of the continent, primarily in the central and western regions, is arid. The Australian

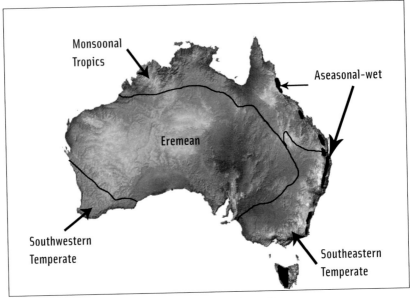

Figure 8.1. Biomes of Australia
(modified from Crisp et al. 2004).

land surface is relatively low altitude and low relief, with a modest range system rising to around 1,000 m along the east coast (Fig. 8.1). The highest mountain on the mainland, Mt Kosciuszko, at 2,220 metres altitude, is the focus of a very small Australian alpine biome in the southeastern corner of the continent. The east coast and a small section of southern Western Australia make up the mesic biomes, and in the east this contains small areas of rainforest now totalling only 30,000 km² (0.5% of the total land area) that represent remnants of the more widespread forests of the early and mid Tertiary. Temperate rainforests dominate large areas of western Tasmania in the south. The north and west of the continent is primarily in the monsoon tropical biome, with rainfall confined to a few summer months (Crisp *et al.* 2004). The Australian arid zone landforms only appeared in the last few million years; they are not biologically diverse but contain small mesic refugia in rangelands of the Pilbara region, MacDonnell Ranges and Flinders Ranges (Byrne *et al.* 2008). The semi-arid zone surrounding the arid interior contains more diversity and is probably a recent zone of speciation as aridification has progressed from the centre and environments have fragmented and changed. Since European settlement in the late 1700s, approximately 60% of Australia's forests have been cleared or significantly disturbed for agriculture and urban development (Raven & Yeates 2007).

2. The Australian Fauna

Australia is home to a species-rich and diverse component of the world fly fauna that is largely endemic at the species level (Colless & McAlpine 1991, Evenhuis 1989 and http://hbs.bishopmuseum.org/aocat/aocathome.html for checklist of described species; Bugledich 1999 for catalog of 'Nematocera'). Figs 8.2–8.15 comprise a variety of distinctive and unusual, or representative Australian species.

3. Total Estimated Fauna

Just over 6,400 Australian fly species have been described (Yeates *et al.* 2003, Austin *et al.* 2004) in 104 families. Estimating the total size of the fauna is difficult given the size of the continent. By a method of extrapolating the proportion of new species described in recent revisions, Yeates *et al.* (2003) estimated that about 80% of the fly fauna remained to be described, giving a total fauna of about 30,000 species. Most collecting and sampling has been undertaken in the eastern coasts and ranges of Australia where moist forests, especially rainforests, harbour a rich fauna. By contrast, the semi-arid zone has been more poorly surveyed, but recent studies suggest that some xeric-adapted families such as the Therevidae have undergone recent extensive radiation, and there are other families that have undergone extensive co-radiation with plant groups (e.g., Fergusoninidae) that are diverse outside rainforest (Scheffer *et al.* 2004). By comparison, the described Nearctic fauna is 21,454 species (Thompson 2009), and the European fauna has a total of 19,422 described species (Jong 2004, Pape 2004).

Alpha diversity of Diptera can be very high locally, for example 250–280 species of schizophoran Diptera on an altitudinal transect in rainforest at Mt Lewis (100–1,200 m altitude) in the wet tropics of North Queensland (Wilson *et al.* 2007). Intensive collecting at the Warrumbungle National Park in the western slopes of New South Wales over 15 years has revealed a fauna of 112 therevid species alone (Winterton *et al.* 2005).

Diptera Diversity: Status, Challenges and Tools
(eds T. Pape, D. Bickel & R. Meier). © 2009 Koninklijke Brill NV.

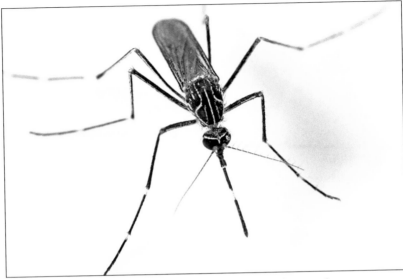

Figure 8.2. *Ochlerotatus notoscriptus* Skuse (Culicidae).
(© CSIRO Entomology)

Figure 8.3. Unknown Limoniidae.
(© Jiri Lochman, Lochman Transparencies)

4. Special Features of the Fauna

4.1 Families endemic to Australia

The two wholly endemic Australian families have relictual distributions from much wider distributions dating from Cretaceous times.

Valeseguyidae. This scatopsoid family comprises a single extant species, *Valeseguya rieki* Colless known only from the Otway Ranges, Victoria. However, *Valeseguya* Colless also occurs in Miocene Dominican Republic amber, and a related genus is known from mid-Cretaceous Burmese amber (Amorim & Grimaldi 2006).

Ironomyiidae. This lower cyclorrhaphan family was previously known from a single extant species, *Ironomyia nigromaculata* White, occurring from the Atherton Tablelands in the Queensland tropics to Tasmania. Two additional species are known from southeastern Queensland, New South Wales, and Tasmania (McAlpine 2008). As well, related fossil genera have been described from Cretaceous amber in Manitoba, Canada and Lebanon (Grimaldi & Cumming 1999).

4.2 Diptera families notably absent from Australia

The Australian Diptera fauna lacks the following major fly families: Ptychopteridae, Deuterophlebiidae (mountain midges), Vermileonidae (worm lions), Diopsidae (stalk eyed flies), Dryomyzidae, Opomyzidae, Scathophagidae (dung flies), Helcomyzidae and Pallopteridae.

4.3 Major families of Diptera in Australia

Current knowledge suggests that the largest fly families in Australia are (in order of species number) Limoniidae (part of Tipulidae *s.l.*) (Fig. 8.3), Dolichopodidae, Drosophilidae, Tephritidae (Fig. 8.12), Ceratopogonidae, Tachinidae, Muscidae, Asilidae, Chloropidae (Figs 8.10-8.11), and Bombyliidae. Field observations and material accumulated in collections suggests that the Mycetophilidae (*s.l.*), Sciaridae, Cecidomyiidae, Empididae, Phoridae, Lauxaniidae, Platystomatidae, and Chloropidae are likely to have many more species than currently described. Some of these families may have more than 1,000 species in Australia, and most are also diverse in other regions of the world. A recent estimate of the Australian tachinid fauna suggests there may be 3,500–4,000 species, based on data from surveys in central Queensland (O'Hara *et al.* 2004). The Australian fauna is also unusually rich in Pyrgotidae, although few species have been

Figure 8.4. *Odontomyia* sp. (Stratiomyiidae).
(© CSIRO Entomology)

Figure 8.5. Wingless female *Boreoides subulatus* Hardy, (Stratiomyiidae).
(© CSIRO Entomology)

described, and Coelopidae (McAlpine 1991). By comparison, some families that are species-rich in other continents, such as the Anisopodidae, Simuliidae and Syrphidae, have relatively depauperate Australian faunas.

As a result of recent revisionary work, some taxa are now shown to be highly diverse. For example, the dolichopodid subfamily Sciapodinae has 253 Australian species, up from 45 previously valid species (Bickel 1994). The Bombyliidae and Therevidae have come under intense scrutiny recently (e.g., Yeates 1991; Yeates & Lambkin 1998; Lambkin *et al.* 2003; Winterton *et al.* 1999, 2000, 2005), including extensive field work in remote areas of the continent. This has revealed a surprisingly diverse fauna in semi-arid Australia, with high alpha and beta diversity unexpected in many of these superficially uniform landscapes. These faunas include a number of large, recently evolved radiations of species in semi-arid Australia consisting of hundreds of species each, and extensive cryptic diversity at species level masked behind a few old species names. A good example of this is the bee fly 'species' *Anthrax angularis* Thompson, which was found to contain 20 cryptic species once scrutinised using modern morphological techniques (Yeates & Lambkin 1998). There may be as many as 700 species of Therevidae and over 1,000 species of Bombyliidae once the fauna is fully understood.

Our knowledge of species is relatively complete in a number of families, especially those that have economic, medical or evolutionary importance, such as Ceratopogonidae, Culicidae (Fig. 8.2), Tabanidae, Drosophilidae, Tephritidae, Agromyzidae, Calliphoridae, and Platystomatidae (Fig. 8.9).

By contrast, many Australian families are 'orphan taxa', in that they have been neglected or are without current taxonomic 'guardians.' Quite simply, keys do not exist and old descriptions are unusable, meaning that some of the most abundant taxa cannot be identified much beyond family level. For example, the Sciaridae can be extraordinarily abundant in Australian forests. The larvae feed variously on detritus, rotten wood, fungi, and are important in nutrient cycles. In the Australian fauna, some 65 species have been described, mostly between 1888–1890 by Frederick Skuse, who placed them in the then vaguely defined type-genus, *Sciara* Meigen. There has been little work since and there are no keys to the Australian fauna. By comparison, some 728 species are described for the Palaearctic region, keys are available, and there are several active workers, who estimate the Palaearctic fauna to be nearer 1,100 species. In the Aus-

Figure 8.6. *Pelecorhynchus maculipennis* Thompson (Pelecorhynchidae).
(© CSIRO Entomology)

Figure 8.7. *Miltinus musgravei* Mackerras (Mydidae).
(© CSIRO Entomology)

tralian fauna the following major families are orphaned: Mycetophilidae (*s.l.*), Sciaridae, Empididae (*s.l.*), Phoridae, Chloropidae, Muscidae, and Tachinidae.

5. Biogeography

An enormous literature exists on the geological, climatic, and biogeographical history of Australia (see Kroenke 1997, Hill 2001, McLoughlin 2001, numerous papers in Keast 1981, Raven & Yeates 2007). This can only be briefly summarized here. Australia was part of the southern Gondwanan supercontinent prior to the Cretaceous, when components started drifting apart. By the end of the Cretaceous, Australia, Antarctica, and New Zealand were loosely assembled at a higher latitude and had a warm to temperate humid climate. Up to this time, a Gondwanan biota, often in association with rainforest, is assumed to have existed in substantial parts of the region. As Australia separated from Antarctica and drifted northwards during the Tertiary, the continent became progressively drier and dominated by sclerophyll plant communities. Drying since the Miocene has reduced the original widespread rainforest to a series of 'island' refugia along the eastern ranges separated by gaps of sclerophyll vegetation. As the Australian plate drifted north during the Tertiary, opportunities for dispersal from Asia would have increased. There would have been continued and multiple invasions through time. The earliest invaders have spread into Australia and diversified among the widespread forests, possibly masking their northern origin. By contrast, recently arrived taxa on the northern Cape York Peninsula clearly show lowland Papuan affinities.

6. Affinities and History of the Australian Diptera

Mackerras (1950, modified in Main 1981) divided the Australian insect fauna into four major zoogeographic elements based on extralimital affinities, distributions, fossil history and dispersal capacity: Pangean (Archaic), Gondwanan (Southern), Asia Tertiary (Old Northern), and Modern (Young Northern). However, single families may contain components from multiple elements.

DIPTERA DIVERSITY: STATUS, CHALLENGES AND TOOLS
(EDS T. PAPE, D. BICKEL & R. MEIER). © 2009 KONINKLIJKE BRILL NV.

Figure 8.8. *Lucilia cuprina* Wiedemann (Calliphoridae).
(© CSIRO Entomology)

Figure 8.9. *Lamprogaster laeta* Macquart (Platystomatidae).
(© CSIRO Entomology)

6.1 Pangean (Archaic).

These include taxa that were well-differentiated and widespread before the breakup of Pangea (i.e., pre-Cretaceous). Amorim & Silva (2002) provided a useful review of such archaic elements on a world basis. They note that while the Diptera certainly have Jurassic representation, none are known as fossils from Gondwanan terranes. Nevertheless, many of the major lower Dipteran clades appear to have originated from that time. However, their degree of differentiation and correlation with modern groups is uncertain.

Apioceridae. The genus *Apiocera* Westwood has 67 Australian species, and is also known from South Africa, southern South America, and western North America. Yeates & Irwin (1996) suggest a Pangean origin for the family.

Apsilocephalidae. A small family comprising four monotypic genera, one from western North America, two from Tasmania, and a Cretaceous genus from Burmese amber (Nagatomi *et al.* 1991, Winterton & Irwin 2008).

Xylophagidae. This family has a relictual distribution with a few species/genera on each continent. The enigmatic genus *Exeretonevra* Macquart comprises four species restricted to highland areas of eastern Australia, from northern New South Wales to Tasmania (Palmer & Yeates 2000).

6.2 Gondwanan (Southern).

This element is characteristic of temperate regions of the southern landmasses which once formed Gondwana. The split-up of Gondwana and the isolation of Australia which occurred by the early Tertiary, ensured that little subsequent interchange occurred between these lands. In Australia, Gondwanan insects often display a southern temperate or Bassian distribution and are associated with cool and humid forests, although some have radiated into sclerophyll and cool semiarid habitats. Gondwanan biotas are characteristic of temperate regions of the southern landmasses which once formed Gondwana: South America, New Zealand, Australia, New Caledonia, India, Madagascar and southern Africa. Hennig's 1960 monograph on the New Zealand Diptera was a ground-breaking analysis of Pacific Gondwanan relationships, and Amorim & Silva (2002) provided a useful summary.

Although many Diptera show classical Gondwanan patterns, some of these are really relictual distributions, based on the extinction of Laurasian elements known from fossils. These now include the two endemic

Figures 8.10–8.11 *Batrachomyia* sp. (Chloropidae) larva parasitising a green-eyed tree frog (*Litoria genimaculata*) in North Queensland

Figure 8.10. Larva *in situ* under the frog's skin.
(© Dr Conrad Hoskin (Botany and Zoology, Australian National University))

Figure 8.11. Larva when removed from under the skin.
(© Dr Conrad Hoskin (Botany and Zoology, Australian National University))

Australian families, the Valeseguyidae and Ironomyiidae, discussed above, and the Sciadoceridae, currently known from Australia, New Zealand and South America (Hennig 1960, Grimaldi & Cumming 1999). Relatively recent trans-Tasman dispersal better explains the distributions of some plants and animals on the southern continents and islands than ancient Gondwanan vicariance (McGlone 2005), however, we have no well established examples in the Diptera yet.

Nevertheless, many Australian taxa of interest show Austral patterns (see Williams 2002 for additional examples). Taxa with the clearest Austral patterns mostly belong in the lower Diptera ('Nematocera') and lower Brachycera ('Brachycera-Orthorrhapha'), which probably reflect radiations in the mid to late Mesozoic.

Australimyzidae. This small family was recently revised, containing nine species in southern Australia, New Zealand and Macquarie Island (Brake & Mathis 2007).

Austroleptidae. This small family was recently removed from the Rhagionidae (Stuckenberg 2001) and comprises a single genus, *Austroleptis* Hardy, confined to temperate Australia and South America.

Blephariceridae. The Australian net-winged midge fauna is confined to eastern Australia and represented by the subfamilies Edwardsininae and Blepharicerinae (Zwick 1998). The genus *Edwardsina* Alexander is a relictual Gondwanan group restricted to Tasmania and temperate southeastern Australia and South America. On the other hand, the tribe Apistomyiini (Blepharicerinae) is absent from Tasmania and reaches its greatest diversity in subtropical and tropical areas of the east coast. It has additional species elsewhere in Australasia (including New Zealand and New Caledonia) and southeast Asia (Zwick 1998).

Bombyliidae: Lomatiinae. The bee fly family Bombyliidae is extremely diverse in Australia, and large genera such as *Comptosia* Macquart and *Aleucosia* Edwards have close relatives in South America and Africa (Yeates 1990). The recent discovery of Eocene European *Comptosia* fossils complicates the assignment of the group to a biogeographic category (Wedmann & Yeates 2008).

Brachystomatidae: *Ceratomerus* Philippi has 19 endemic Australian species restricted to rocky rainforest streams of eastern Australia from the Tablelands near Cairns to Tasmania. It has a classical Gondwanan distribution of Australia/Tasmania, New Zealand and South America, and the *C. campbelli* species-group of Australia appears to be sister group to

Figure 8.12. *Phytalmia mouldsi* McAlpine & Schneider (Tephritidae),
males fighting, Iron Range, Cape York.
(© Dr Gary Dodson (Ball State University, USA))

Figure 8.13. *Badisis ambulans* McAlpine (Micropezidae),
commensal in *Cephalotus* pitcher plants.
(© Jiri Lochmann)

the *C. paradoxus* group of Chile, while the Australian *C. ordinatus* group appears to be sister to a clade of several New Zealand and Ecuadorean (high altitude) species (Sinclair 2003).

Dolichopodidae: Sciapodinae. *Heteropsilopus* Bigot is known only from Australia and the Indian subcontinent. In Australia, the *cingulipes* group (16 spp.) displays a classical Bassian distribution in southern Australia and Tasmania, mostly in association with sclerophyll habitats. The *triligatus* group (13 spp.) is mostly restricted to montane habitats above 900 m in the Western Ghats on the Indian subcontinent, a region with many Gondwanan relict taxa. Despite having such a wide disjunct distribution, Indian and Australian species are clearly monophyletic, suggesting a Gondwanan distribution in place before India and Australia separated in the Lower Cretaceous. The genus *Parentia* Hardy has diversified in Australia, New Zealand, and New Caledonia, and one species occurs in Fiji. In Australia, 26 species occur mainly in the southern half of the continent, while *Parentia* is the dominant sciapodine genus in New Zealand with 27 species, New Caledonia has at least 17 species, and the single Fijian species has possibly dispersed from New Caledonia. *Parentia* has no relatives in the Neotropics and is a trans-Tasman Gondwanan genus. In New Zealand and New Caledonia, species of the genus usually occur in mesic forests, while in Australia they are found mostly in dry sclerophyll/heath associations (Bickel 1994, 2006b).

Keroplatidae. Species of *Arachnocampa* Edwards, commonly known as 'glow worms', have luminescent larvae that inhabit moist overhangs and caves. One species occurs in New Zealand and three in the moist forests of eastern Australia, from Tasmania to southeastern Queensland, with undescribed species from the Atherton Tablelands of Queensland (Matile 1990, Baker *et al.* 2008) and Victoria.

Perissommatidae. This remarkable bibionoid family has its compound eyes divided by a strip of cuticle into dorsal and ventral components (thus appearing four-eyed). It swarms in winter months and comprises four Australian and one southern South American species (Colless 1969).

Rhagionidae. *Pelecorhynchus* Macquart is a cool-adapted genus, restricted to montane swamps along the eastern massif between southern Queensland and Tasmania, and it is particularly rich in species in the high plateaus of eastern New South Wales, where 27 of the 31 described Australian species occur (Mackerras & Fuller 1942). This genus also occurs in Chile.

Figure 8.14. Male *Achias australis* Malloch (Platystomatidae),
far north Queensland.
(© *Paul Zabrosky*)

Figure 8.15. *Nycterimyia commoni* Paramonov (Nemestrinidae).
(© *Paul Zabrosky*)

Thaumaleidae. The genus *Austrothaumalea* Tonnoir is diversified in Australia (19 spp.), New Zealand (10 spp.), and southern South America (3 spp.), and is thought to date from Gondwanan times. They are weak fliers hovering near watercourses and therefore generally not strong dispersers. However, Sinclair (2008) described a species from New Caledonia which he regards as the result of overseas dispersal from Australia rather than relictual Gondwana distribution.

Therevidae. Phylogenetic work on the family has revealed that the Australian species belong to two clades, the *Anabarhynchus* Macquart, *Agapophytus* Guerin and *Taenogera* Kröber genus groups. These genera have links to groups in southern South America and other western Pacific continental land masses such as New Zealand and New Caledonia (Lambkin 2006).

6.3 Asia Tertiary (Old Northern) and Modern (Young Northern).

As the Australian plate drifted north during the Tertiary, dispersal from Oriental/Papuan sources would have increased through time. The earliest invaders would have been able to penetrate deep into Australia and speciate among the widespread forests, leading to a gradual masking of their northern origin. As forests contracted eastwards during post-Miocene drying, the descendants of these early arrivals might show a distribution pattern more like that of Gondwanan forest-associated taxa rather than Asia Tertiary elements.

Characteristically, Asia Tertiary taxa have strong links with extralimital palaeotropical groups, have radiated in northern tropical Australia, and/ or show a marked decrease in species richness southwards. These northern elements often have penetrated southward along the eastern Australian ranges, but mostly in association with tropical and subtropical rainforests. Also, in reverse, taxa that had major radiations in Australia (e.g., Fergusoninidae) may have dispersed into the Oriental/Papuan region.

Most taxa with northern connections are members of the younger Cyclorrhaphan families, reflecting major radiation in the early Tertiary. Nevertheless, many families have genera with both Gondwanan and northern patterns. Possible families include:

Calliphoridae. The Ameniinae comprise seven genera, of which six occur in Australia. It is probable that the ancestral Ameniinae migrated to Australia from the north (Colless 1998).

Dolichopodidae: Sciapodinae. Some genera in the Sciapodinae show clear Gondwanan patterns as discussed above. However, a rich fauna of Oriental-Papuan affinity (*Chrysosoma* Guerin-Meneville, *Amblypsilopus* Bigot, and *Plagiozopelma* Enderlein) dominates the northern tropics and has penetrated to varying degrees southward along the eastern Australian coast and ranges in association with tropical and subtropical rainforests, which are discontinuously distributed along the eastern coastal ranges. The southern limit of such taxa in eastern Australia is the southern limit of subtropical rainforest in New South Wales. Riparian rainforest corridors support many of these northern genera, but faunal richness drops sharply in areas where sclerophyll vegetation predominates (Bickel 1994).

Drosophilidae. This Australian drosophilid fauna is well-known and comprises some 230 species. The family is most diverse in warm, wet forests of northern and eastern Australia, and diminishes southward from tropical Queensland. The fauna is closely related to that of southeastern Asia and New Guinea, and genera undoubtedly reached Australia at different times from the North. *Scaptodrosophila* Duda is believed to have been an early invader as it is the largest Australian drosophilid genus and has radiated in southern as well as northern Australia (Bock 1982).

Empididae. Species of the Australian genus *Clinocera* Meigen are associated with rocky streams and appear to be related to east Asian species, since New Zealand and Chilean taxa are assigned to a different species-group. Sinclair (2000) suggested that the Australian species possibly originated from dispersal of ancestral populations southwards from New Guinea. Seven species of *Clinocera* occur in eastern Australia, with the greatest diversity again in the southeast.

Fergusoninidae. This acalyptrate family has its major radiation in Australia (27 described species so far), although species are also known from India, New Guinea, New Zealand and the Philippines. These flies have an intricate, symbiotic relationship with plant-parasitic nematodes. The fly larvae form leaf galls on *Eucalyptus* and other Myrtaceae, providing a habitat for nematodes of the genus *Fergusobia,* and nematodes travel between host plants in the bodies of female flies (Giblin-Davis *et al.* 2001).

Neminidae. This small family, based on the Australian tree trunk associate *Nemo* McAlpine, is known from Australia, Madagascar, South Africa and New Guinea (Freidberg 1994).

Neurochaetidae. This family, the upside down flies, is currently found in southern Africa, Madagascar, and the Oriental and Australasian regions, but it was also present in northern Europe where it is known from early Tertiary Baltic Amber (McAlpine 1993).

Pipunculidae. Pipuculids have major radiations of species in Australia (80 described species; likely ~200 species total). Linkages of this fauna are primarily with New Guinea, New Caledonia, New Zealand, Indonesia and more northern Oriental components (Skevington 1999, 2001, 2002; Skevington & Yeates 2001).

Rhinophoridae. The Axiniinae have unusual antennae and occur in Australia (15 spp.) and New Guinea (2 spp.). Although originally described as a distinct family (Colless 1994), they are now included in the Rhinophoridae (Pape & Arnaud 2001).

Sepsidae. The Australian species are grouped into three lineages; species derived from ancestors that reached Australia well before the Pleistocene, immigrants originating in or near New Guinea that entered Australia possibly during the Pleistocene, and an Oriental group that reached Australia during or after the Pleistocene (Colless 1980).

6.4 Uncertain affinity

Many taxa show no clear distributional patterns or phylogenetic affinities that would suggest either a Gondwanan or extralimital northern origin. For example, the following four families have predominantly temperate Australian distributions, and three also occur in New Zealand. However, they are all Acalyptratae, which suggests they originated in the late Cretaceous or early Tertiary, when Australia was already quite isolated. Also, the four families have undergone radiations or have sister taxa elsewhere in the world. Some may represent early Old Northern arrivals whose origin is masked.

Australimyzidae. The small acalyptrate family has a trans-Tasman distribution and comprises nine extant species in the genus *Australimyza* Harrison, five from New Zealand, three from Australia, and one from Macquarie Island (Brake & Mathis 2007). Although this may appear to be a Gondwanan distribution, the family is coastal and possibly readily dispersed. Its apparent sister taxon is the Neotropical rainforest family Inbiomyiidae.

Heleomyzidae. *Tapeigaster* Macquart comprises 11 species found in temperate Australia. McAlpine & Kent (1982) suggested that either the genus evolved in isolation on the Australian continent during much of the

Tertiary, or it is a relict group surviving from the extinction of its nearest relatives.

Lauxaniidae. *Poecilohetaerus* Hendel is essentially confined to a broad arc of southeastern Australia, southwestern Western Australia (with a gap marked by the Nullarbor Plain), the Cairns district Wet Tropics, and Tasmania. Most species are restricted to southern rainforest and wet sclerophyll forest. The genus also occurs in New Zealand (Schneider 1991).

Teratomyzidae. The fern flies comprise a little studied family that has mainly a south-temperate distribution (South America, Australia & New Zealand). However, one of the seven genera, *Teratomyza* Malloch (*s.l.*), has a wider distribution, extending from New Zealand to eastern Asia (McAlpine & deKeyzer 1994). This distribution suggests a south-temperate origin for the family, with later extension to the northern hemisphere via Malesia.

7. Ecological Biogeography of the Australian Fauna

The patchy distribution of rainforest in eastern Australia has been noted previously. However, these refugia or 'islands' of rainforest are not entirely isolated, to be reached only by chance dispersal across 'oceans' of dry sclerophyll forest or grassland. Suitable habitat for many rainforest Diptera occurs as mixed, wet sclerophyll-rainforest vegetation along creeks. These form secondary refugia and even pathways to the upland rainforests. Even coastal regions dominated by sclerophyll vegetation have wet gully forests. Many groups of Oriental/Papuan affinity in eastern Australia decrease in diversity southwards, but reach into temperate southern New South Wales in association with subtropical rainforest (including wet sclerophyll-rainforest mixtures). By comparison, southern Western Australia is depauperate compared to similar latitudes in eastern Australia. In marked contrast to the widespread genera found along the forest 'islands' of the eastern ranges, there are few shared taxa between southwestern and northwestern Australia, and the two regions are also distinctly different at higher taxonomic levels. This reflects two factors, the extensive, low-lying, arid tract that acts as a barrier between the Southwest and the monsoonal North. Also, unlike the eastern coast, there is no rainforest in the Southwest which would possibly support northern elements.

The phenology of many taxa may reflect adaptations to the Australian climate. Mackerras (1950) observed that cool-temperate Australian Diptera (often with Gondwanan affinities) emerge in winter to early spring in the northern part of their range, but later in montane or southern localities. A number of cool-adapted Diptera, e.g., the Perissommatidae (Ferguson 2007) and the hilarine genus *Eugowra* Bickel (Bickel 2006a), are known to swarm on sunny days in winter months. Such phenology may reflect an ancestral adaptation to life at cool higher latitudes. With the progressive drift of the Australian continent towards the tropics in the Tertiary, the shift of adult reproductive and dispersal stages into winter and early spring (especially in the north of the range), has allowed cool-adapted taxa to survive aridity and high summer temperature as cryptic immature stages.

In addition to the lowland New Guinea fauna in tropical Australia, there is a distinctive monsoonal fauna across much of the north, extending from Cape York Peninsula to Arnhem Land and the Kimberley Ranges. This comprises widespread species or sister species between Cape York Peninsula and Arnhem Land. Although most species are associated with vine forests along watercourses, they also occur in monsoonal woodlands during the wet season from December to April.

The relationship of the montane Papuan fauna with that of Australia is of particular interest. Hill (2001) has noted the vegetative and floristic similarity of montane New Guinea rainforest and the early Tertiary Tasmanian flora. The Australian *Nothorhaphium* Bickel (Dolichopodidae) has a southern temperate Australian distribution, and the widespread species *N. aemulans* (Becker) has its sister species, *N. oro* Bickel, in montane New Guinea above 2,000 m (Bickel 1999). How can this disjunction be explained? Gressitt (1982) noted that the highlands of New Guinea contain old Australian elements of Gondwanan affinity, often with a southern distribution and associated with cool and humid conditions. A similar pattern is displayed by the classic Gondwanan tree genus *Nothofagus*, which occurs in montane New Guinea but is disjunct by several thousand kilometers from its Australian congeners. Hope (1996) noted that based on pollen, *Nothofagus* has been in New Guinea since the late Miocene, and that the cool moist climate these trees require suggests a mid-Tertiary lowland biotic connection between Australia and proto-New Guinea. During this time the Australian Plate was further south and the overall climate was generally cooler. Subsequently, as the Australian Plate pushed northward into the tropics, the New Guinea mountains rose, and the overall climate became warmer, *Nothofagus* and other Gondwanan biota tracked the cool moist conditions higher into the mountains, and subsequently became disjunct from their temperate Australian congeners.

8. Ecological and Economic Importance

Flies are ubiquitous and often abundant in Australian terrestrial ecosystems and perform such important ecological functions as nutrient recycling, predation, pollination, and parasitism of other insects. Some are significant nuisances, such as the bush fly (*Musca vetustissima* Walker), and various species of blood sucking mosquitoes, sandflies and blackflies (Culicidae [Fig. 8.2], Ceratopogonidae: mostly *Culicoides* Latreille; and Simuliidae: mostly *Austrosimulium* Tonnoir respectively). Flies outrank other orders in terms of medical and veterinary significance, being responsible for the transmission of a wide variety of pathogens in man and animals. These diseases are mostly absent from Australia, with exceptions such as dengue (transmitted by *Aedes aegypti* Linnaeus) and various encephalitides (transmitted by several mosquito species). Human malaria has been eradicated from Australia but was transmitted by species of the *Anopheles farauti* Laveran complex. Phlebotomine sandflies (Psychodidae) are known from Australia but do not appear to transmit diseases of man or domestic animals here, as they do in other parts of the world. The sheep blowfly (*Lucilia cuprina* Wiedemann: Calliphoridae, Fig. 8.8) causes extensive economic losses in sheep when the larvae infest the host tissues causing myiasis. Probably the most important pests of Australian horticulture belong to the Tephritidae, most notably the Queensland fruit fly (*Bactrocera tryoni* Froggatt) in the east, and the introduced Mediteranean fruit fly (*Ceratitis capitata* Wiedemann) in western Australia. Mosquitoes were used as vectors of myxoma virus, which dramatically reduced Australia's rabbit plague (Fenner & Ratcliffe 1965), and some leaf mining Agromyzidae have been introduced for biological control of weeds such as lantana. Of quarantine concern are pest flies that are currently absent from Australia such as the Old World screw worm (*Chrysomyia bezziana* Villeneuve, Calliphoridae), and various invasive fruit flies in South East Asia of the genus *Bactrocera* Macquart, such as the oriental fruit fly *B. dorsalis* Hendel. A close relative, *B. papayae* Drew & Hancock, was recently eradicated from north Queensland after invading from the north (Cantrell *et al.* 2002).

9. Morphology and Behaviour

Some Australian fly taxa have unusual behaviour, morphology or life histories that are worthy of note. Larval *Arachnocampa* Edwards (Keroplatidae) are known as 'glow worms', possessing light-producing organs associated with the Malpighian tubules (Baker *et. al.* 2008). The larvae can be found in caves and rainforests along the east coast, where they construct mucus tubes from which they suspend snares with droplets of oxalic acid that capture and kill prey attracted by the light. *Planarivora* Hickman (Keroplatidae) larvae parasitise land planarians (Hickman 1965). Male flies of the genus *Mycomya* Rondani (Mycetophilidae) and some Sciaridae and Cecidomyiidae pollinate ground orchids. Evidence from *Mycomya* spp. suggests the pollination strategy is probably sexual deception, much better known in wasp-orchid systems (Hamilton *et al.* 2002). Larvae of the endemic, wingless micropezid fly *Badisis* McAlpine (Fig. 8.13) live in pitchers of the endemic western Australian pitcher plant *Cephalotus* (Yeates 1992). They can survive the plant digestive enzymes and feed on the insect prey of the pitcher, much as other insects do in association with pitcher-forming carnivorous plants in Queensland and other parts of the world. Males of some species of *Achias* Fabricius (Platystomatidae) (Fig. 8.14) have eyes on stalks similar to Diopsidae (McAlpine 1994). Males of the Australasian fruit fly genus *Phytalmia* Gerstaecker (Tephritidae) (Fig 8.12) defend oviposition sites on fallen logs in rainforest from conspecific males using bizarre antler-like head projections (Dodson 1997). Adult upside down flies of the genus *Neurochaeta* McAlpine (Neurochaetidae) rest with their head facing downwards on native arum lily stems in rainforest. The adult morphology is highly modified, and they can run as quickly backwards as forwards (McAlpine 1993). Larvae of *Batrachomyia* Krefft (Chloropidae) live beneath the skin of frogs, feeding on blood (Fig. 8.10–8.11). Hill-topping is also a common phenomenon in Australia, with males of species in the Bombyliidae, Pipunculidae, and Tachinidae commonly found aggregating on prominent hilltops (e.g., Yeates & Dodson 1991, Skevington 2001).

10. Future Directions

The Australian fly fauna is poorly known taxonomically, and there remains much to be understood about their different life history stages and ecological roles and impacts. The taxonomically orphaned families of flies such as Tachinidae, Sciaridae and Mycetophilidae are in greatest need of systematic research. These families require dedicated research focus — they are too large and complex for 'dabbling' to have much impact on their taxonomy. Increased funding for the Australian Biological Resources Study (ABRS) participatory program is the best mechanism to resource this and other taxonomic research needs (Raven & Yeates 2007). Australia's arid zone and monsoon tropical fly fauna has probably been the least surveyed and studied, and this is clearly a research need for the future.

The relationships of the Australian fauna to the world fauna are being investigated currently in the NSF-funded Assembling the Tree of Life grant on Diptera (http://www.inhs.uiuc.edu/cee/FLYTREE/), and the evolutionary relationships of flies have been reviewed recently (Yeates *et al.* 2007). Over recent decades NSF Partnerships for Enhancing Expertise in Taxonomy grants on Therevidae and Tabanidae have provided considerable resources to increase the taxonomic understanding of Australia's fly fauna. Australia's Commonwealth Environmental Research Facility (CERF) Taxonomic Research and Information Network (TRIN) provides a model for the research innovation and resources required to address the taxonomy of species-rich families of Diptera (http://www.taxonomy.org.au).

An important future challenge is to make identification of the Australian fly fauna more accessible for non-taxonomists. Australian researchers lead the world in providing user-friendly, web-based resources and computer-based interactive keys. A web-based Anatomical Atlas of Flies has been developed, allowing students and others to more easily learn fly morphology (http://www.ento.csiro.au/biology/fly/fly.html). Chironomidae are important indicators of the health of aquatic ecosystems, and an identification guide to the larvae is now available on the web (Cranston 2000). Web or CD-based interactive keys are now available for the Australasian genera of Therevidae (Winterton *et al.* 2005) and all Australian fly families (Hamilton *et al.* 2006).

REFERENCES

Amorim, D.S. & Grimaldi, D.A. (2006) Valeseguyidae, a new family of Diptera in the Scatopsoidea, with a new genus in Cretaceous amber from Myanmar. *Systematic Entomology* 31: 508–516.

Amorim, D.S. & Silva, V.C. (2002) Diptera evolution in the Pangea. *Annales de la Société entomologique de France (n.s.)* 38: 177–200.

Amorim, D.S. & Yeates, D.K. (2006) Pesky gnats: getting rid of Nematocera in Diptera Classification. *Studia Dipterologica* 13: 3–9.

Austin A.D., Yeates, D.K., Cassis, G., Fletcher, M.J., LaSalle, J., Lawrence, J.F., McQuillan, P.B., Mound, L.A., Bickel, D.J., Gullan, P.J., Hales, D.F. & Taylor, G.S. (2004) Insects 'Down Under' — diversity, endemism and evolution of the Australian insect fauna: examples from select orders. *Australian Journal of Entomology* 43: 216–234.

Baker, C.H., Graham, G., Scott, K., Yeates, D.K. & Merritt, D.J. (2008) Geography, not habitat structures Australian glow-worm Diversity: Habitats, species limits, relationships and distribution of *Arachnocampa* (Diptera, Keroplatidae). *Molecular Phylogenetics and Evolution* 48: 506–514.

Bickel, D.J. (1994) The Australian Sciapodinae (Diptera: Dolichpodidae), with a review of the Oriental and Australasian faunas, and a world conspectus of the subfamily. *Records of the Australian Museum, Supplement* 21, 394 pp.

Bickel, D.J. (1999) Australian Sympycninae II: *Syntormon* Loew and *Nothorhaphium* gen. nov., with a treatment of the Western Pacific fauna, and notes on the subfamily Rhaphiinae and *Dactylonotus* Parent (Diptera: Dolichopodidae). *Invertebrate Taxonomy* 13: 179–206.

Bickel, D.J. (2006a) *Eugowra,* a new fly genus from Australia (Diptera: Empididae). *Records of the Australian Museum* 58: 119–124.

Bickel, D.J. (2006b) *Parentia* (Diptera: Dolichopodidae) from Fiji: a biogeographic link with New Caledonia and New Zealand. *In:* Evenhuis, N.L. & Bickel, D.J. (eds), Fiji Arthropods V. *Bishop Museum Occasional Papers* 89: 45–50. Also available at http://hbs.bishopmuseum.org/fiji/fiji-arthropods/pdf/FAV-03.pdf

Bock, I.R. (1982) Drosophilidae of Australia. V. Remaining genera and synopsis (Insecta: Diptera). *Australian Journal of Zoology. Supplementary Series* 89, 164 pp..

Brake, I. & Mathis, W.N. (2007) Revision of the genus *Australimyza* Harrison (Diptera: Australimyzidae). *Systematic Entomology* 32: 252–275.

Bugledich, E-M. A. (1999) Diptera: Nematocera. *In:* Wells, A. & Houston, W.W.K. (eds), *Zoological Catalogue of Australia*, Volume 30.1. CSIRO Publishing, Melbourne, xiii + 627 pp.

Byrne, M., Yeates, D.K., Joseph, L., Kearney, M., Bowler, J., Williams, M.A., Cooper, S., Donnellan, S.C., Keogh, J.S., Leys, R., Melville, J., Murphy, D.J., Porch, N. & Wyrwoll, K.-H. (2008) Birth of a biome: insights into the assembly and maintenance of the Australian arid zone biota. *Molecular Ecology* 17: 4398-4417.

Cantrell, B., Chadwick, B. & Cahill, A. (2002) *Fruit Fly Fighters: Eradication of the Papaya Fruit Fly.* SCARM Report 81. SCARM/CSIRO Publishing, Melbourne.

Colless, D. H. (1969) The genus *Perissomma* (Diptera: Perissommatidae) with new species from Australia and Chile. *Australian Journal of Zoology* 17: 719–728.

Colless, D.H. (1980) Biogeography of Australian Sepsidae (Diptera). *Australian Journal of Zoology* 28: 65–78.

Colless, D.H. (1994) A new family of muscoid Diptera from Australasia, with sixteen new species in four new genera (Diptera: Axiniidae). *Invertebrate Taxonomy* 8: 471–534.

Colless, D.H. (1998) Morphometrics in the genus *Amenia* and revisionary notes on the Australian Ameniinae (Diptera: Calliphoridae), with the description of eight new   species. *Records of the Australian Museum* 50: 85–123.

Colless, D.H. & McAlpine, D.K. (1991) Diptera, pp. 717–786 *in*: CSIRO (ed.), *The Insects of Australia*, 2nd Edition. Melbourne University Press, Carlton.

Cranston, P.S. (2000) Electronic identification guide to the Australian Chironomidae. Available at http://www.science.uts.edu.au/sasb/chiropage/, accessed 10 October 2007.

Crisp, M, L. Cook & Steane, D. (2004) Radiation of the Australian flora: what can comparisons of molecular phylogenies arcoss multiple taxa tell us about the evolution of diversity in present day communities? *Philosophical Transactions of the Royal Society of London* B 359: 1551–1571.

Dodson, G. (1997) Resource defense mating system in antlered flies, *Phytalmia* spp. (Diptera: Tephritidae). *Annals of the Entomological Society of America* 90: 496–504.

Evenhuis, N.L. (ed.) (1989) *Catalog of the Diptera of the Australasian and Oceanian Regions.* Bishop Museum Press, Honolulu, 1155 pp. [Also updated website: http://hbs.bishopmuseum.org/aocat/aocathome.html]

Evenhuis, N.L., Pape, T., Pont, A.C. & Thompson, F.C. (eds) (2008) *BioSystematic Database of World Diptera.* Version 10.5. Available at http://www.diptera.org/names/, accessed 9 September 2008.

Fenner, F. & Ratcliffe, F.N. (1965) *Myxomatosis.* Cambridge University Press: Cambridge, 379 pp.

Ferguson, D. (2007) Field observations of *Perissomma mcalpinei* Colless (Diptera: Perissommatidae). *Australian Entomologist* 34: 93–96.

Freidberg, A. (1994) *Nemula*, a new genus of Neminidae (Diptera) from Madagascar. *Proceedings of the Entomological Society of Washington* 96: 471–482.

Giblin-Davis, R.M., Makinson, J.R., Center, B.J., Davies, K.A., Purcell, M.F., Taylor, G.S., Scheffer, S., Goolsby, J. & Center, T.D. (2001) *Fergusobia/Fergusonina*-induced shoot bud gall development on *Melaleuca quinquenervia*. *Journal of Nematology* 33: 239–247.

Gressitt, J.L. (1982) Zoogeographical summary. Pages 897–918 *in*: Gressitt, J.L. (ed.), *Biogeography and ecology in New Guinea.* 2 vols, W. Junk, The Hague, 1437 pp.

Grimaldi, D.A. & Cumming, J.M. (1999) Brachyceran Diptera in Cretaceous ambers and Mesozoic diversification of the Eremoneura. *Bulletin of the American Museum of Natural History* 239, 123 pp.

Hamilton, J.R., Yeates, D.K., Hastings, A., Colless, D.H., McAlpine, D.K., Bickel, D., Cranston, P.S., Schneider, M.A., Daniels, G. & Marshall, S. (2006) *On The Fly: The Interactive Atlas and Key to Australian Fly Families.* (CD ROM). Australian Biological Resources Study/Centre for Biological Information Technology (CBIT).

Hamilton, J., Peakall, R. & Yeates, D.K. (2002) Sex, Flies and Videotape: Pollination and Sexual Deception in the Australian Terrestrial Orchid *Pterosylis.* Page 96 *in*: Abstracts Volume, 5th International Congress of Dipterology, Brisbane, 29th Sept–4th Oct.

Hennig, W. (1960) Die DipterenFauna von Neuseeland als systematisches und tiergeographisches Problem. *Beiträge zur Entomologie* 10: 221–329. [English translation: Hennig, W. (1966) New Zealand as a problem in systematics and zoogeography. *Pacific Insects Monograph* 9: 1–81.]

Hickman, V.V. (1965) On *Planarivora insignis* gen. et sp. n. (Diptera: Mycetophilidae), whose larval stages are parasitic in land planarians. *Papers and Proceedings of the Royal Society of Tasmania* 99: 1–8.

Hill, R.S. (2001) Biogeography, evolution and palaeoecology of *Nothofagus* (Nothofagaceae): the contribution of the fossil record. *Australian Journal of Botany* 49: 321–332.

Hope, G.S. (1996). History of *Nothofagus* in New Guinea and New Caledonia. Pages 257–270 *in*: Veblen, T.T., Hill, R.S. & Read, J. (eds), *The ecology and biogeography of* Nothofagus *forests.* Yale University Press, New Haven.

Jong, H. de (ed.) (2004) *Diptera: Brachycera, Fauna Europaea*, Version 1.1. Available at http://www.faunaeur.org, accessed 10 October 2007.

Keast, A. (ed.) (1981) *Ecological Biogeography of Australia*, 3 vols. W. Junk, The Hague, 2142 pp.

Kitching, R.L., Bickel, D.J. & S. Boulter (2005) Guild analyses of dipteran assemblages: a rationale and investigation of seasonality and stratification in selected rainforest faunas. Pages 388–415 *in*: Yeates, D.K. & Wiegmann, B.M. (eds), *The Evolutionary Biology of Flies.* Columbia University Press, New York.

Kroenke, L. W. (1997) Plate tectonic development of the western and southwestern Pacific: Mesozoic to the present. Pages 19–34 *in*: Keast, A. & Miller, S. (eds), *The origin and evolution of Pacific Island biotas, New Guinea to eastern Polynesia: Patterns and processes.* SPB Academic Publishing bv, Amsterdam.

Lambkin, C.L. (2006) Australasian stiletto flies (Diptera: Therevidae): Divergence time estimates indicate Gondwanan separation, evolutionary radiations with aridification and recent dispersal. Page 149 *in*: M. Suwa (Ed.) , *Abstracts volume, 6th International Congress of Dipterology,* Fukuoka, Japan.

Lambkin, C.L., Yeates, D.K. & Greathead, D.J. (2003) An evolutionary radiation of bee flies in semi-arid Australia: Systematics of the Exoprosopini (Diptera: Bombyliidae). *Invertebrate Systematics* 17: 735–891.

Mackerras, I.M. (1950) The zoogeography of Diptera. *The Australian Journal of Science* 12: 157–161.

Mackerras, I.M. & Fuller, M.E. (1942) The genus *Pelecorhynchus* (Diptera: Tabanidae). *Proceedings of the Linnean Society of New South Wales* 67: 9–76.

Main, B.Y. (1981) A comparative account of the biogeography of terrestrial invertebrates in Australia: some generalizations. Pages 1055–1077 *in*: Keast, A. (ed.), *Ecological biogeography of Australia*. W. Junk, The Hague.

Matile, L. (1990) Recherches sur la systématique et l'évolution des Keroplatidae (Diptera, Mycetophiloidea). *Memoirs du Museum d'Histoire Naturelle de Paris. Serie A (Zoologie)* 148: 1–682.

McAlpine, D.K. (1991) Review of the Australian kelp flies (Diptera: Coelopidae). *Systematic Entomology* 16: 29–84.

McAlpine, D.K. (1993) Review of the Upside-down flies (Diptera: Neurochaetidae) of Madagascar and Africa, and evolution of neurochaetid host plant associations. *Records of the Australian Museum* 45: 221–239.

McAlpine, D.K. (1994) Review of the species of *Achias* (Diptera: Platystomatidae). *Invertebrate Taxonomy* 8: 117–279.

McAlpine, D.K. (1998) A review of the Australian stilt flies (Diptera: Micropezidae) with a phylogenetic analysis of the family. *Invertebrate Taxonomy* 12: 55–134.

McAlpine, D.K. (2008) New extant species of Ironic flies (Diptera: Ironomyiidae) with notes on Ironomyiid morphology and relationships. *Proceedings of the Linnean Society of New South Wales* 129: 17–38.

McAlpine, D.K. & deKeyzer, R.G. (1994) Generic classification of the fern flies (Diptera: Teratomyzidae), with a larval description. *Systematics Entomology* 19: 305–326.

McAlpine, D.K. & Kent, D.S. (1982) Systematics of *Tapeigaster* (Diptera: Heleomyzidae) with notes on biology and larval morphology. *Proceedings of the Linnean Society of New South Wales* 106: 33–58.

McGlone, M.S. (2005) Goodbye Gondwana. *Journal of Biogeography* 32: 739–740.

McLoughlin, S. (2001) The breakup of Gondwana and its impact on pre-Cenozoic floristic provincialism. *Australian Journal of Botany* 49: 271–300.

Merritt, R.W., Courtney, G.W. & Keiper, J.B. (2003) Diptera (Flies, Mosquitoes, Midges, Gnats). Pages 324–340 *in*: Resh, V.H. & Cardé, R.T. (eds), *Encyclopedia of Insects*. Academic Press, San Diego.

Nagatomi, A., Saigusa, T., Nagatomi, H., Lyneborg, L. (1991) Apsilocephalidae, a new family of the orthorrhaphous Brachycera (Insecta, Diptera). *Zoological Science* 8: 579–591.

O'Hara, J.E., Skevington, J.H. & Hansen, D.E. (2004) A reappraisal of tachinid diversity in Carnarvon N.P., Australia, and estimation of the size of the Australian Tachinidae fauna. *The Tachinid Times* 17: 8–10.

Palmer, C.M. & Yeates, D.K. (2000) The phylogenetic importance of immature stages: Solving the riddle of *Exeretonevra* Macquart (Diptera: Xylophagidae). *Annals of the Entomological Society of America* 93: 15–27.

Pape, T. (ed.) (2004) *Diptera: Brachycera, Fauna Europaea*, Version 1.1. Available at http://www.faunaeur.org, accessed 10 October 2007.

Pape, T. & Arnaud, P.H., Jr. (2001) *Bezzimyia* — a genus of New World Rhinophoridae (Insecta, Diptera). *Zoologica Scripta* 30: 257-297.

Raven, P.H. & Yeates, D.K. (2007) Australian Biodiversity: Threats for the Present, Opportunities for the Future. *Australian Journal of Entomology* 46: 177–187.

Schneider, M.A. (1991) Revision of the Australasian genus *Poecilohetaerus* Hendel (Diptera: Lauxaniidae). *Journal of the Australian Entomological Society* 30: 143–168.

Scheffer, S.J., Giblin-Davis, R.M., Taylor, G.S., Davies, K.A., Purcell, M., Lewis M.L., Goolsby, J., Center T.D. (2004) Phylogenetic relationships, species limits, and host specificity of gall-forming *Fergusonina* flies (Diptera : Fergusoninidae) feeding on *Melaleuca* (Myrtaceae). *Annals of the Entomological Society of America* 97: 1216–1221.

Sinclair, B.J. (2000) Revision of the genus *Clinocera* Meigen from Australia and New Zealand (Diptera: Empididae: Clinocerinae). *Invertebrate Taxonomy* 14: 347–361.

Sinclair, B.J. (2003) Taxonomy, phylogeny and zoogeography of the subfamily Ceratomerinae of Australia (Diptera: Empidoidea). *Records of the Australian Museum* 55: 1–44.

Sinclair, B.J. (2008) A new species of *Austrothaumalea* from New Caledonia (Diptera: Thaumaleidae). *Zoologia Neocaledonica* 6, *Mémoires du Muséum national d'Histoire Serie A, Zoologie* 197: 277–281.

Skevington, J.H. (1999) *Cephalosphaera* Enderlein, a genus of Pipunculidae (Diptera) new for Australia, with descriptions of four new species. *Australian Journal of Entomology* 38: 247–256.

Skevington, J.H. (2001) Revision of the Australian *Clistabdominalis* (Diptera: Pipunculidae). *Invertebrate Taxonomy* 15: 695–761.

Skevington, J.H. (2002) Phylogenetic revision of Australian members of the *Allomethus* genus group (Diptera: Pipunculidae). *Insect Systematics and Evolution* 33: 133–161.

Skevington, J.H. & Yeates, D.K. (2001) Phylogenetic classification of Eudorylini (Diptera: Pipunculidae). *Systematic Entomology* 26: 421–452.

Stuckenberg, B.R. (2001) Pruning the tree: a critical review of classifications of the Homeodactyla (Diptera, Brachycera), with new perspectives and an alternate classification. *Studia dipterologica* 8: 3–41.

Thompson, F.C. (2009) Nearctic Diptera: Twenty years later. Pages 3-46 in: Pape, T., Bickel, D. & Rudolf, M. (eds), *Diptera Diversity: Status, Challenges and Tools*. Brill, Leiden.

Wedmann, S. & Yeates, D.K. (2008) Eocene records of bee flies (Insecta, Diptera, Bombyliidae, *Comptosia*), their paleobiogeographic implications and remarks on the evolutionary history of bombyliids. *Palaeontology* 51: 231–240.

Williams, G. (2002) A taxonomic and biogeographic review of the invertebrates of the Central Eastern Rainforest Reserves of Australia (CERRA) World Heritage Area, and adjacent regions. *Technical Reports of the Australian Museum* 16, 208 pp.

Wilson, R., Williams, S.E., Trueman, J.W.H. & Yeates, D.K. (2007) Communities of Diptera in the Wet Tropics of North Queensland are vulnerable to Climate Change. *Biodiversity and Conservation* 16: 3163–3177.

Winterton, S.L. & Irwin, M.E. (2008) *Kaurimyia* gen. nov.: discovery of Apsilocephalidae (Diptera: Therevoid clade) in New Zealand. *Zootaxa* 1779: 38–44.

Winterton, S.L., Irwin, M.E. & Yeates, D.K. (1999) Systematic revision of the *Taenogera* Krober genus-group (Diptera: Therevidae), with descriptions of two new genera. *Australian Journal of Entomology* 38: 274–290.

Winterton, S.L., Skevington, J.H. & Lambkin, C.L. (2005) *Stilletto Flies of Australasia*. Available at http://www.cdfa.ca.gov/phpps/ppd/Entomology/Lucid/Therevidae/Austherevid/key/Austherevid/Media/Html/opening_page.html, accessed July 2007.

Winterton, S.L., Yang, L.L., Wiegmann, B.M. & Yeates, D.K. (2000) Phylogenetic revision of the Agapophytinae subf. n. (Diptera: Therevidae) based on molecular and morphological evidence. *Systematic Entomology* 26: 173–211.

Yeates, D.K. (1990) Phylogenetic relationships of the Australian Lomatiinae (Diptera: Bombyliidae). *Systematic Entomology* 15: 491–509.

Yeates, D.K. (1991) Revision of the bee fly genus *Comptosia* (Diptera: Bombyliidae). *Invertebrate Taxonomy* 5: 1023–1178.

Yeates, D.K. (1992) Immature stages of the apterous fly *Badisis ambulans* McAlpine (Diptera: Micropezidae). *Journal of Natural History* 26: 417–424.

Yeates, D.K. & Dodson, G.N. (1990) The mating system of a bee fly (Diptera: Bombyliidae): 1. Nonresource based hilltop territoriality and a resource-based alternative. *Insect Behavior* 3: 603–617.

Yeates, D.K. & Irwin, M.E. (1996) Apioceridae (Insecta: Diptera): Cladistics and biogeography. *Zoological Journal of the Linnean Society* 116: 247–301.

Yeates, D.K. & Lambkin, C.L. (1998) Cryptic species diversity and character congruence: review of the tribe Anthracini (Diptera: Bombyliidae) in Australia. *Invertebrate Taxonomy* 12: 977–1078.

Yeates, D.K. & Wiegmann, B.M. (1999) Congruence and controversy: toward a higher-level classification of the Diptera. *Annual Review of Entomology* 44: 397–428.

Yeates, D.K., Wiegmann, B.M., Courtney, G.W., Meier, R., Lambkin, C.L. & Pape, T. (2007) Phylogeny and systematics of Diptera: two decades of progress and prospects. *Zootaxa* 1668: 565–590.

Yeates, D.K., Harvey, M. & Austin, A.D. (2003) New estimates for terrestrial arthropod species-richness in Australia. *Records of the South Australian Museum Monograph Series* 7: 231–241.

Zwick, P. (1998) Australian net-winged midges of the tribe Apistomyiini (Diptera: Blephariceridae). *Australian Journal of Entomology* 37: 289–311.

CHAPTER NINE

BIOGEOGRAPHY OF DIPTERA IN THE SOUTHWEST PACIFIC

DANIEL BICKEL

Australian Museum, Sydney, Australia

INTRODUCTION

The Southwest Pacific is normally considered part of the Australian zoogeographic region, but is here treated separately. It includes New Guinea (Indonesian Papua, West Papua, and Maluku, as well as Papua New Guinea), the Bismark Archipelago, the Solomon Islands, Vanuatu, Fiji, New Caledonia, and New Zealand (Fig. 9.1). It comprises a complex of land masses and climatic zones, ranging from mini-continents like New Guinea to tiny oceanic atolls, from equatorial tropics to the southern temperate/alpine zones. However, a common feature is that all these land masses are surrounded by ocean, and are islands, however large. Therefore, Diptera distributions in this region must be viewed in light of island biogeography, or at least 'island isolation.'

The central question in Pacific Basin biogeography is the disjunct distribution of terrestrial taxa, where closely related species are separated by inhospitable saltwater barriers. Of the explanations advanced to explain such distributions, two are widely utilized: *a) Dispersal.* The taxon spread, either actively or passively, across inhospitable barriers from a source area. Source areas for Southwest Pacific biotas are surrounding continental landmasses (the Orient, Australia and New Guinea region), and secondarily, large islands. It should be noted that anthropogenic introduction must be regarded as a special type of dispersal, and natural *vs.* accidental human dispersal are not always distinguishable. [For example, the dolichopodid *Austrosciapus connexus* (Walker) is abundant in both natural and disturbed habitats along the eastern Australian coast. It also occurs in Hawaii and French Polynesia suggesting accidental in-

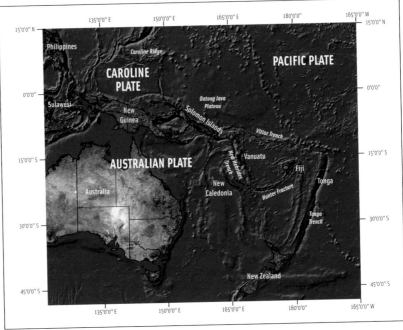

Figure 9.1. Map of Southwest Pacific, with major island groups and tectonic features.

troduction. However, it also occurs on Norfolk Island, which could be the result of accidental introduction or unaided dispersal, the latter being documented for many Australian moths (Bickel 1996).] *b) Vicariance*. The taxon always occupied the landmasses on which it occurs, and ancient geographies have been changed by geological events, with barriers subsequently breaking up once contiguous distributions. Variants of the vicariance model that are relevant here include panbiogeography (e.g., Craw *et al.* 1999) and the expanding earth hypothesis (e.g., McCarthy 2005).

Dispersal and vicariance, the two major patterns of origin are not mutually exclusive, and both are apparent in various taxa and not always clearly separable. Since biogeographic processes are largely historical and unobserved, one must deduce the most probable histories from available evidence. Young, remote volcanic islands probably received their biota by dispersal. However, Heads (2006) suggested that the young age of current Melanesian and Polynesian islands does not exclude the possibility that the biota may have evolved *in situ*, as meta-populations on island arcs. Biotic components of these island arcs could have moved progressively to younger islands as the older islands were submerged.

The interpretation of processes that lead to distribution patterns is not always clear. The biotas of old isolated 'continental' islands such as New Zealand and New Caledonia could have achieved their biotas either by vicariance and/or dispersal. Indeed, the same patterns have been interpreted by biogeographers to support quite different hypotheses of historical processes. And where readily dispersed taxa undergo rapid speciation accompanied by extensive homoplasy within tectonically active regions, the search for a unique set of area cladograms based on dichotomous branching becomes an impossible task (also see Polhemus 1996).

Apart from distinguishing between dispersal and vicariance, the geographic affinity of insular taxa may not be clear. Many genera or generic groups are widely distributed or cosmopolitan. Taxonomic literature for tropical Diptera is fragmentary, and few groups have been analyzed to the extent that biogeographic provenance can be associated with unique synapomorphies (if indeed that is always possible). The tropical islands of the western Pacific are likely to have many pantropical genera, and it is desirable that ancestral provenance be unequivocally demonstrated. For many terrestrial arthropod genera this is impossible.

Proper assessment ideally requires knowledge of basic taxonomy (for described Australasian and Oceanian species, see Evenhuis [1989, incl. web updates]). Most Diptera families have received only superficial attention, and this is particularly true for Melanesia, where not only are most species undescribed, but most collections have been made at a few coastal or selected sites, and vast landscapes remain unsampled. Many highly diverse families comprising small-sized species are poorly known throughout the Southwest Pacific, in particular the Cecidomyiidae, Sciaridae, Chironomidae, and Phoridae. One such family provides an example. Quate & Quate (1967) treated 228 species of Psychodidae from the Papuan subregion, describing 191 new species. They found that 129 species or 57% of the total Papuan psychodid fauna are known only from a single site. Such high local endemism suggests the entire Papuan subregion is under-collected. They estimated that the 228 species were but a small part of the total, possibly not more than 25% of the actual Papuan Psychodidae. Considering the undescribed diversity in museum collections, often from limited sampling, it is doubtful even 20% of the entire New Guinea Diptera fauna is described (and it must be noted that the Quate & Quate monograph remains one of the most comprehensive treatments for any fly family in the Papuan subregion).

DIPTERA DIVERSITY: STATUS, CHALLENGES AND TOOLS
(EDS T. PAPE, D. BICKEL & R. MEIER). © 2009 KONINKLIJKE BRILL NV.

Some island groups are better known than others, and certainly New Zealand is among the best-studied, but even there, major gaps exist in taxonomic description (MacFarlane & Andrew 2001). New Caledonia has had considerable collecting activity in recent years, but major revisionary work is needed. Fiji has been the subject of recent intensive collecting with Malaise traps (Evenhuis & Bickel 2005) and the results are sometimes surprising, revealing great hidden diversity. For example, prior to recent surveys, 14 species of Sciapodinae (Dolichopodidae) were known from Fiji, and now 95 species are known, 79 new to science (D.J. Bickel, unpubl.).

This review will focus on the distribution patterns of representative taxa, in so far as are they are sufficiently known, and how they reflect major biogeographic processes of the Southwest Pacific (for a general review of Diptera biogeography, see Cranston 2005). Of course there is a bias towards taxa that provide a biogeographic 'signal.' Although many taxa are speciose throughout the region, they lack distinctive synapomorphies and cannot be used to show clear biogeographic patterns. So, based on selected taxa, I will first review wide general patterns, and then focus on individual archipelagoes, especially noting their geological history and how this has affected their Diptera faunas.

1. General Regional Considerations

The overall biogeography of the Southwest Pacific has been reviewed by Keast (1996), Miller (1996), Gressitt (1961), and various authors in Keast & Miller (1996). For a general geological setting, see Hall (2001), Kroenke (1996), and Neall & Trewick (2008), and for vegetation and geography, see Mueller-Dombois & Fosberg (1998).

For many Pacific distributions, long-distance inter-island dispersal has been considered almost axiomatic. The transoceanic dispersal capability of Diptera is well established, as they are part of the small-sized arthropod fauna which is carried by wind (e.g., records in Holzapfel & Harrell 1968, Peck 1994a, and Farrow 1984) or on surface flotsam (e.g., Peck 1994b). Widely distributed tramp species exhibit the capacity to colonize across large water gaps. Such species often show broad ecological tolerance and are adapted to survive in disturbed lowland areas, thereby increasing their chance of successful dispersal by wind, rafting, or even human transport. Many generalist lowland species of insects are thereby preadapted to becoming tramps (Wilson 1959). For example, *Chrysosoma leucopogon* Wiedemann (Dolichopodidae) ranges from the East African coast and Madagascar to French Polynesia, occurring on isolated oceanic islands as well as continental landmasses. It is often abundant in coastal areas, increasing its chances of waif dispersal (Bickel 1996).

However, numerous anomalous taxa on Pacific islands either lack obvious dispersal traits or represent ancient lineages, which has led many authors to propose continental origins for both the flora and fauna of the Pacific islands. In recent years, molecular systematics has led to a revival of dispersalist hypotheses (e.g., Waters & Craw 2006, de Queiroz 2004). Even the presence of plant taxa such as the conifer genus *Agathis*, Proteaceae, and cycads, all long considered examples of a Gondwanan connection of many Pacific islands, has been challenged by some authors to be the result of relatively recent long-distance dispersal events (see discussion in Keppel *et al.* 2008). As well, the long-held view that New Caledonia and New Zealand are Gondwanan fragments with ancient and relictual biotas has been challenged by geological evidence suggesting that these two land masses were almost completely submerged in the early Tertiary (although under different tectonic regimes). This, combined with molecular phylogenies of selected taxa has led some workers to propose that the biota of these two 'continental' island groups are largely the result of diversification following post-Oligocene dispersal (Grandcolas, *et al.* 2008, Goldberg *et al.* 2008).

Craig *et al.* (2001) suggested a combination of vicariance and dispersal to account for the distribution of the black fly subgenus *Simulium* (*Inseliellum* Rubtsov) in the western Pacific, but also emphasized the importance of suitable aquatic habitats in determining colonization and subsequent radiation.

The general decrease in taxonomic diversity from the Indo-Malayan/ Papuan regions eastwards is a well-documented phenomenon, which is seen as a result of 'sweepstakes' dispersal. The zoogeography of the family Tabanidae in the western Pacific shows a progressive filtering of the horsefly fauna eastward through the Melanesian Archipelago. Mackerras (1961) claimed the western Pacific was colonized first by a stream of Afrotropical–Oriental taxa, and cites a direct connection with New Caledonia *Philoliche* Wiedemann species having its nearest relatives in Mauritius, and a second later stream of Oriental–Australian origin progressive dispersing eastwards into the Pacific.

Diptera Diversity: Status, Challenges and Tools
(eds T. Pape, D. Bickel & R. Meier). © 2009 Koninklijke Brill NV.

While some genera are widespread, others are clearly restricted: not all have successfully penetrated the Pacific Basin. For example, the subfamily Phlebotominae (Psychodidae) is widespread and diversified in both New Guinea and Australia, but it is absent from the rest of the Pacific, including New Zealand (Lewis & Dyce 1989). Overall, tropical Oriental-Papuan genera have dispersed far more widely than southern temperate taxa, which are often restricted in distribution. Possibly the combination of tropical habitats and 'stepping-stone' archipelagoes favor dispersal and radiation. As well, the Oriental-Papuan genera are warm-adapted and most Pacific islands are geographically within the tropics. Thus, New Caledonia received tropical elements from the nearby Melanesian Archipelago, but its fauna does not appear to disperse well. New Zealand is even more isolated and has few if any tropical taxa, possibly because there are no 'stepping-stone' islands linking it to the north.

Belkin's (1962) monumental study of the South Pacific mosquito fauna (Culicidae), although somewhat dated, is the most comprehensive treatment for any Diptera family in the region. In addition to careful descriptions and figures of all life stages, it includes detailed maps and distribution records. Although Belkin adopted a dispersalist position (but see below under Vanuatu-Fiji) he is careful to point out distributional anomalies. He noted major links with the Papuan-Australian and the Oriental/Indo-Malayan faunas, but also minor links with the Afrotropics and southern South America. Most major archipelagoes appear to have a high degree of endemism.

Polhemus (2007) suggested that there may have been a roughly continuous island arc system that developed in the Oligocene (45 Mya) and extended from the Bismarck Archipelago all the way to Fiji. This would have allowed, at least in theory, longitudinal dispersal from the Papuan region all the way to Viti Levu, although varying topography, island size and distance would have acted as filters to dispersal.

The breakup of the Gondwanan supercontinent and the resultant drift and isolation of the component landmasses have produced vicariant distribution patterns. In the Southwest Pacific, Gondwanan biotas are characteristic of the southern temperate regions that were once part of the supercontinent: New Zealand, Australia, and New Caledonia. For a taxon to be Gondwanan, it needs to have direct phylogenetic links to a decidedly Gondwanan taxon from Australia, New Caledonia, or New Zealand, and not indirectly via the Melanesian Archipelago. For example, although an ancestrally Gondwanan genus may occur in New Guinea and have dispersed to Fiji via the Melanesia Archipelago, it might not be considered from a Gondwanan source.

Dyce (2001) commented on the origin of the *Culicoides* Latreille (Ceratopogonidae) fauna of Australasia, which appears to be derived from two basic Gondwanan stocks. One group is associated with Australia and New Caledonia, with the Australian component radiating into the Papuan region, and the New Caledonian component contributing to some current Pacific island groups. The second Gondwana group originated on the Indian subcontinent and subsequently radiated through Southeast Asia and into New Guinea and Australia.

Parentia Hardy (Dolichopodidae) is an example of a Gondwanan genus in the Southwest Pacific. *Parentia* has no close relatives in the cool temperate southern Neotropical faunas, and it is exclusively a trans-Tasman element that has diversified primarily in Australia, New Zealand, and New Caledonia. In Australia, 26 species occur mainly in dry sclerophyll/heath associations in the southern half of the continent, while New Zealand has 27 species and New Caledonia 17 species, which occur in mesic forests and coastal habitats. There is a sister-group relationship between the derived Australian *dispar* and New Zealand *malitiosa* groups. Although trans-Tasman dispersal of insects from Australia is well-known, the extensive diversification of New Zealand *Parentia*, which includes the most plesiomorphic members of the genus, and the synapomorphies that unite the derived Australian *dispar* and New Zealand *malitiosa* groups suggest a vicariant pattern. The New Caledonia fauna appears to be closest to that of New Zealand (Bickel 1996, 2002). Recently, a single *Parentia* species was discovered on coastal Viti Levu, Fiji, and is therefore a Gondwanan element in Fiji, closely tied to coastal species groups from New Caledonia and New Zealand. Since there is no evidence for land connections or 'stepping-stone' archipelagoes further south, this species' ancestor probably dispersed to Fiji from New Caledonia (Bickel 2006a).

2. Specific Regions

2.1 New Guinea

This is the largest and highest, tropical island, and it is the center of the Papuan subregion, which also includes the Moluccas, the Bismarck and Admiralty archipelagoes, and Bougainville and Buka of the northern

Solomon Islands. New Guinea has a rich endemic biota that has radiated in response to the island's high habitat diversity, and in addition to major endemic diversification, the biota shows both Australian and Oriental affinities (Gressitt 1982). This high degree of regional endemism is possibly due to New Guinea's origin as a region of island-arc accretion along the leading, stable edge or northern cratonic margin of the Australian Plate. After the breakup of Gondwana, the Australian Plate separated from Antarctica and began moving northward, and the cratonic margin collided with a complex system of island arcs and plate fragments. This resulted in uplift, volcanism, and terrane accretion along the leading edge of the plate, starting from the mid-Tertiary. Thus, northern New Guinea comprises fold belt mountains and the central highlands, while the southern part of the island consists largely of the stable cratonic platform (Hill & Hall 2003). Island arc systems were forming in nearby Southeast Asia and may have provided dispersal pathways for elements of the Oriental and Papuan biotas. Polhemus (1996) noted that taxa may disperse along arcs that are subsequently fragmented, producing disjunct distributions. The fidelity of taxa to their ancestral but now accreted terranes (see discussion in Heads 2002) is difficult to assess for most Diptera. However, Clarke *et al.* (2004) noted that the greatest diversity of *Bactrocera* Macquart and *Dacus* Fabricius species (Tephritidae) in Papua New Guinea is associated with northern geological elements (the New Guinea orogen, accreted terranes, and off-shore islands) rather than on the southern Australian craton. This possibly supports the concept of terrane fidelity, where species remain associated with their ancestral habitat and do not disperse widely. During the Quaternary, fluctuations in sea level occurred, at which time New Guinea was twice joined with Australia across the Torres Strait and had land connections with adjacent islands. As a result of this recent land connection, rainforests in northern Cape York Peninsula mark the southernmost limit of many Papuan taxa.

Some New Guinea taxa show an Australian connection, but these are fewer than those with distinct Oriental tropical affinities. McAlpine (1982) cited only a few acalyptrate taxa in New Guinea that are distinctively Australian in origin, such as *Cairnsimyia* Malloch (Heteromyzidae = Heleomyzidae), *Metopochetus* Enderlein (Micropezidae), *Trigonometopus* Macquart (Lauxaniidae), and *Leonophila* Guérin-Méneville and *Microepicausta* Hendel (Platystomatidae). He pointed out that acalyptrate families that are characteristic of temperate southern Australia are largely unknown in New Guinea. However, Bickel (1999a) noted the presence of the primarily temperate southern Australian genus *Nothorhaphium* Bickel (Dolichopodidae) in montane New Guinea above 2,000 m. This suggests that during the Tertiary, while the Australian Plate pushed northward into the tropics and the New Guinea Highlands arose, temperate southern taxa (including many Gondwanan groups such as the tree genus *Nothofagus*) would have tracked the cool moist conditions higher into the mountains, and subsequently become disjunct from their temperate Australian congeners.

Many families in the Papuan region show a strong connection with the tropical Oriental region. Maa (1982) demonstrated that the New Guinea Pupipara (Hippoboscidae, incl. Streblidae and Nycteribiidae; all obligate exoparasites on mammals and birds) have strong affinities with the Oriental Region and negligible Australian/Tasmanian and Polynesian faunal elements. Two lower Brachyceran genera, *Rachicerus* Walker (Xylophagidae), with six species confined to New Guinea (Nagatomi 1970), and *Solva* Walker (Xylomyidae), with nine species in the Papuan subregion, including three shared with Southeast Asia and one also in northern Queensland (Daniels 1977), are both Old World taxa whose distribution reaches its limit in New Guinea. The acalyptrate family Gobryidae is another predominately Oriental taxon that reaches it limit in New Guinea.

New Guinea appears to be the center of radiation for the family Platystomatidae, with 30 known genera (some highly diverse). Generic richness drops eastward from New Guinea, although New Caledonia has a high generic endemicity and species richness (McAlpine 2001). One of the more striking radiations is the genus *Achias* Fabricius, with 97 endemic New Guinea species (and three from northern Queensland), and which includes the remarkable stalk-eyed fly *A. rothschildi* Austen with a head width of 55 mm (McAlpine 1994). The tribe Phytalmiini (Tephritidae), including the six species of the remarkable antler flies of the genus *Phytalmia* Gerstaecker, is another group centered in New Guinea, with a single species in northern Queensland (McAlpine & Schneider 1978).

2.2 Solomon Islands

The Solomon Islands comprise a double chain of rugged, mostly elongate islands between New Guinea and Vanuatu. Craig *et al.* (2006) provided a good summary of the complex geology of this archipelago. Briefly, the islands arose as a result of the interaction of the Australian Plate with the

Ontong Java Plateau, a huge subterranean mass of basalt and sediment lying to the northeast on the Pacific Plate. Initially, as the Australian Plate was moving north, the Pacific Plate was being subducted under it, forming the Vitiaz Trench. However, as the massive Ontong Java Plateau approached the Trench, the subduction was reversed, causing the Australian Plate to move under the Pacific Plate. This produced a complex of distinct geological provinces within the Solomon Islands, with varied uplift and volcanism.

The Solomon Islands Diptera are very poorly documented, and apart from a few groups of economic importance, our taxonomic knowledge of most families comprises isolated descriptions of lowland species from limited collections. However, based on the Simuliidae, Craig *et al.* (2006) suggested the fauna had a western Papuan/Oriental origin and probably arrived in the archipelago relatively recently, in the last few million years.

2.3 Fiji and Vanuatu

These two archipelagoes are located in the tectonically active zone between the Australian and Pacific Plates, and each comprises a complex of volcanics, pyroclastic sediments and limestone reef deposits developed since the early Cenozoic. Repeated volcanism and tectonic activity, combined with uplift and erosion, have produced a complicated geology. A subduction zone was formed at the Vitiaz Trench, and Fiji was situated on an island arc system (the Vanuatu-Fiji-Lau-Tonga Ridge) that migrated eastward. Vanuatu is thought to have been close to Fiji until the mid-Miocene (*ca.* 20 Mya). Most authors suggest an early to mid-Miocene (10–14 Mya) emergence for Viti Levu (references in Evenhuis & Bickel 2005).

Revision of recent collections of Dolichopodidae have shown a close relationship between Fiji and Vanuatu, with, for example, two clades in the *Amblypsilopus pulvillatus* group including species from both archipelagos (Bickel 2006b).

Interestingly, the black fly subgenus *Simulium* (*Hebridosimulium* Grenier & Rageau) has greatly diversified on Vanuatu, with some 11 species occurring in the archipelago, whereas in Fiji there is only one species, *Simuliium* (*H.*) *lacinatorum* Edwards, widely distributed on the five main islands (Craig *et al.* 2007). The absence of speciation within this subgenus on Fiji remains unclear, especially since Fiji has a great range of fluvial habitats that might be expected to facilitate speciation.

In contrast to New Caledonia (see below), Belkin (1962) noted that the unusual Culicidae fauna of Fiji consists of 'primitive, clearly marked, relict endemics having no close relatives anywhere in the South Pacific.' He suggested that its relatives were in the Indomalayan and Afrotropical areas, and that these elements entered Fiji via a northern route (island arc) with subsequent isolation. Since many of these mosquitoes are container or tree hole breeders, as opposed to ground pool breeders, a dispersal via land connections is seen as most likely.

Fiji has some remarkable long-distance zoogeographic connections. The best known are the Fijian iguanas, with two extant species and fossils also known from both Fiji and Tonga. They are isolated in the Southwest Pacific and have their closest relatives in the Neotropics (Keogh, et al. 2008). However, recent entomological surveys have revealed additional disjunct patterns. Taveuni, a large and high but relatively young (2–3 My) island, has a species of *Mythicomyia* Coquillet (Mythicomyiidae) belonging to a Neotropical species group (N.L. Evenhuis, unpubl.), and a species of *Simulium* (*Psilopelmia* Enderlein) (Simuliidae), whose closest relatives occur in Central America (D.A. Craig, unpubl.). Neither Mythicomyiidae, which are characteristic of semiarid regions, nor Simuliidae, with larvae in fast flowing streams, are likely to be good long-distance dispersers. And an undescribed genus of Fijian Keroplatidae is apparently also known only from Madagascar (N.L. Evenhuis, unpubl.).

2.4 New Caledonia

New Caledonia has a rich, endemic biota long considered to be Gondwanan and vicariant with the fauna of southern temperate landmasses. The basic geology of New Caledonia is complex, but it appears to be part of a zone of continental crust along the Norfolk Ridge, sometimes called Zelandia, and involves the breakup and coalescing of small terranes over time, with intermittent land connections (see Pelletier 2006 and Ladiges & Cantril 2007). However, Hall (2001) noted that there is little evidence to suggest New Caledonia was above sea level until the late Eocene. This presents a major problem for the idea that New Caledonia, with its rich endemic flora, had continuous terrestrial habitat since the breakup of Gondwana in the Cretaceous (Grandcolas *et al.* 2008).

Sanmartín & Ronquist (2004) noted that although New Caledonia and New Zealand have a sister-group relationship based on area cladograms, there are few biogeographic sister-taxon relationships between the two landmasses to support this. However, such a pattern is evident in the genus *Paraclius* Loew (Dolichopodidae), where the distinctive *xanthurus* Group comprises seven species in New Caledonia and one from the North Island of New Zealand. The New Zealand species appears to be the sister taxon of one of the New Caledonian species, suggesting that the species group radiated while both ancestral landmasses were in closer physical contact, before the postulated opening of the New Caledonian Basin in the mid-Tertiary, some 40–30 Mya. (Bickel 2008).

Many more biogeographical patterns involving New Caledonian Diptera include Australia. Belkin (1962) noted that the affinities of the New Caledonian Culicidae are almost entirely Australian with rather little local endemicity (45%), in marked contrast to culicid endemicity levels in most other Pacific archipelagoes (average 85%). This gives the impression that New Caledonia has a relatively young mosquito fauna comprising many Australian species colonized by dispersal.

A number of genera appear to have an exclusive Australian/New Caledonian distribution. For example, the dolichopodid genus *Antyx* Meuffels & Grootaert is known only from eastern Australia (3 spp.) and New Caledonia (5 spp.) (Bickel 1999b). Sinclair (2008) recorded *Austrothaumalea* Tonnoir (Thaumaleidae) in New Caledonia and, although the distribution of this genus in South America, Australia and New Zealand suggests it is an old Gondawanan taxon, he argues that the close relationship between the single New Caledonian species and Australian species suggests the ancestor of the former arrived via long-distance dispersal from Australia. Molecular analyses and divergence estimates suggest that the presence of *Anabarhynchus* Macquart (Therevidae) in New Caledonia is the result of dispersal from Australia (C. Lambkin, unpubl.), which is in agreement with the fact that many species of *Anabarhynchus* in Australia occur in coastal habitats and therefore are potentially good dispersers. Further, there is an undescribed species of *Anabarhynchus* in Fiji which she suggests may be the result of dispersal from New Caledonia, similar to the Fijian *Parentia* species (Dolichopodidae) discussed above. [Interestingly, the Fijian *Anabarhynchus* and *Parentia* species are known only from Sigatoka, Viti Levu, and both were collected in the unique vegetated dune habitat found there.]

DIPTERA DIVERSITY: STATUS, CHALLENGES AND TOOLS
(EDS T. PAPE, D. BICKEL & R. MEIER). © 2009 KONINKLIJKE BRILL NV.

2.5 New Zealand

New Zealand has a varied and highly endemic Diptera fauna (Macfarlane & Andrew 2001). Two families are endemic, the sciomyzoid acalyptrate family Huttoninidae with eight species (Harrison 1959) and the bizarre monotypic bat fly family Mystacinobiidae (sometimes regarded as a subfamily of the Calliphoridae) comprising a single species, *Mystacinobia zelandica* Holloway, which is associated with the endemic New Zealand bat family Mystacinidae (also known from Tertiary fossils in Australia) (Holloway 1976).

The New Zealand Diptera are quite distinct from the tropical Pacific fauna and appear to be most closely related to the faunas of southern South America and Australia (e.g., Hennig 1966). A recent semi-popular book (Gibbs 2007) provides an excellent overview of the New Zealand biota, while McDowall (2008) reviews the biogeography.

The breakup of Gondwana and the northern movement of ancestral New Zealand terranes occurred in the Cretaceous, along with the formation of the Tasman Sea, some 110–85 Mya (Stevens 1985). New Zealand and New Caledonia apparently remained closer than now until a gap between the landmasses developed about 40–30 Mya. During the Oligocene, some 26–38 Mya, New Zealand is thought to have been greatly reduced in landmass and extensively submerged. Some authors consider it to have been completely submerged, leading to the hypothesis that all of New Zealand's extant terrestrial biota arrived *via* long-distance dispersal since the mid-Tertiary, presumably from Australia (Waters & Craw 2006, Goldberg *et al.* 2008).

However, there is good evidence among the Diptera to suggest that New Zealand existed as continuous land, however reduced in area, throughout the Tertiary. Brundin (1966), for example, in his study of the phylogenetic relationships of the trans-Antarctic subfamilies Podonominae and Aphroteniinae (Chironomidae), found that, *a*) the sister group of a New Zealand group occurred in South America only, or both South America and Australia–Tasmania, and, *b*) there are no direct relationships between New Zealand and Australia. Further, Munroe (1974) noted that Australian and New Zealand species of the genus *Australosymmerus* Freeman (Ditomyiidae) have a sister-group relationship with Chilean species rather than with one another. The trans-Antarctic tracks of both families suggest long-term survival in New Zealand, since dispersal would have shown sister-taxon relationships with Australia, not South America.

Although many Diptera genera occur in both New Zealand and Australia, few are known well enough to either support or exclude the possibility of long-distance colonization of New Zealand from Australia. For example, Taylor *et al.* (2007) described the first New Zealand species of *Fergusonina* Malloch (Fergusoninidae) along with a new species of its associated nematode *Fergusobia* from *Metrosideros* (Myrtaceae) galls in New Zealand. The Fergusoninidae are widely diversified in Australia, and their larvae have an obligate association with nematodes that help them form galls in the shoot buds of various myrtaceous genera on which they feed. Taylor *et al.* (2007) set out three possible biogeographic hypotheses regarding the relationship between the Australian and New Zealand *Fergusonina*. Two hypotheses postulate long-distance dispersal from Australia to New Zealand along with plant host switching, while the third is a vicariant hypothesis involving the speciation of the New Zealand *Fergusonina* during the early divergence of *Metrosideros* from a myrtaceous ancestor, before the separation of New Zealand and Australia, some 80 Mya. The authors favour the dispersal models in light of the postulated recent evolution of the Fergusoninidae.

3. Summary

This short summary of Dipteran distribution in the Southwest Pacific has been designed to show a range of patterns and presumed biogeographic processes. Indeed, some of the patterns can be interpreted in various ways, depending on the quality of the evidence and, more frequently, the predilection of the worker. However, the baseline of data is ever expanding with new geological data and hypotheses, systematic revisions (still woefully inadequate considering the size of the undescribed Diptera fauna), and supplementary evidence from other taxonomic groups. That does not necessarily make the task easier, as Nature is 'messy', and there will always be examples to counter any hypothesis as well as enigmatic distributions that defy reasonable explanation. Adding to this, extinction has exerted its confounding impact, leaving gaps in the distribution patterns. All this, of course, is part of the appeal of biogeography.

Acknowledgements

Neal Evenhuis and Dan Polhemus provided constructive comments on an earlier version of this manuscript. Doug Craig and Christine Lambkin provided useful information. Michael Elliott constructed Fig. 9.1 using ArcMap, version 9.3. This article is Contribution Number 2008-006 to the Fiji Arthropod Survey, funded in part by NSF grant DEB-0425790.

References

Belkin, J.N. (1962) *The Mosquitoes of the South Pacific (Diptera, Culicidae)*. Vols 1–2, University of California Press, Berkeley & Los Angeles, 608 pp.

Bickel, D.J. (1996) Restricted and widespread taxa in the Pacific: Biogeographic processes in the fly family Dolichopodidae (Diptera). Pages 331–346 *in*: Keast, A. & Miller, S.E. (eds), *The origin and evolution of Pacific Islands biotas, New Guinea to eastern Polynesia: patterns and processes*. SPB Academic Publishing, Amsterdam.

Bickel, D.J. (1999a) Australian Sympycninae II: *Syntormon* Loew and *Nothorhaphium* gen. nov., with a treatment of the Western Pacific fauna, and notes on the subfamily Rhaphiinae and *Dactylonotus* Parent (Diptera: Dolichopodidae). *Invertebrate Taxonomy* 13: 179–206.

Bickel, D.J. (1999b) Australian *Antyx* Meuffels & Grootaert and the New Caledonian connection (Diptera: Dolichopodidae). *Australian Journal of Entomology* 38: 168–175.

Bickel, D.J. (2002) The Sciapodinae of New Caledonia (Diptera: Dolichopodidae). *Zoologia Neocaledonica* 5, *Mémoires du Muséum national d'Histoire naturelle Serie A: Zoologie* 187: 11–83.

Bickel, D.J. (2006a) *Parentia* (Diptera: Dolichopodidae) from Fiji: a biogeographic link with New Caledonia and New Zealand. *Bishop Museum Occasional Papers* 89: 45–50. [*In*: Evenhuis, N.L. & Bickel, D.J. (eds), *Fiji Arthropods V.*]

Bickel, D.J. (2006b) The *Amblypsilopus pulvillatus* species group (Diptera: Dolichopodidae: Sciapodinae), a radiation in the western Pacific. *Bishop Museum Occasional Papers* 90: 51–66. [*In*: Evenhuis, N.L. & Bickel, D.J. (eds), *Fiji Arthropods VI.*]

Bickel, D.J. (2008) The Dolichopodinae (Diptera: Dolichopodidae) of New Caledonia, with descriptions and records from Australia, New Zealand and Melanesia. *Zoologia Neocaledonica* 6, *Mémoires du Muséum national d'Histoire Serie A: Zoologie* 197: 13–48.

Brundin, L. (1966) Transantarctic relationships and their significance, as evidenced by chironomid midges, with a monograph of the Podonominae and Aphroteniinae and the austral *Heptagyia*. *Kungliga Svenska Vetenskapsakademiens Handlingar* 11: 1–472.

Clarke, A.R., Balagawi, S., Clifford, B., Drew, R.A.I., Leblanc, L., Mararuai, A.N., McGuire, D., Putulan, D., Romig, T., Sar, S. & Tenakanai, D. (2004) Distribution and biogeography of *Bactrocera* and *Dacus* species (Diptera: Tephritidae) in Papua New Guinea. *Australian Journal of Entomology* 43: 148–156.

Craig, D.A., Currie, D.C., Hunter, F.F. & Spironella, M. (2007) A taxonomic revision of the southwestern Pacific subgenus *Hebridosimulium* (Diptera: Simuliidae: *Simulium*). *Zootaxa* 1380: 1–90.

Craig, D.A., Currie, D.C. & Joy, D.A. (2001) Geographical history of the central-western Pacific black fly subgenus *Inseliellum* (Diptera: Simuliidae: *Simulium*) based on a reconstructed phylogeny of the species, hot spot archipelagoes and hydrological considerations. *Journal of Biogeography* 28: 1101–1127.

Craig, D.A., Englund, R.A. & Takaoka, H. (2006) Simuliidae (Diptera) of the Solomon Islands: new records and species, ecology, and biogeography. *Zootaxa* 1328: 1–26.

Cranston, P. (2005) Biogeographic patterns in the evolution of Diptera. Pages 274–311 in: Yeates, D. & Wiegmann, B. (eds), *The Evolutionary Biology of Flies*. Columbia University Press, New York.

Craw, R.C., Grehan, J.R. & Heads, M.J. (1999) *Panbiogeography — tracking the history of life*. Oxford University Press, Oxford, 240 pp.

Daniels, G. (1977) The Xylomyidae (Diptera) of Australia and Papua New Guinea. *Journal of the Australian Entomological Society* 15: 453–60.

de Queiroz, A. (2004) The resurrection of oceanic dispersal in historical biogeography. *Trends in Ecology and Evolution* 20: 68–73.

Dyce A.L. (2001) Biogeographic origins of species of the genus *Culicoides* (Diptera: Ceratopogonidae) in the Australasian Region. *Arbovirus Research in Australia*: 8: 133–140.

Evenhuis, N.L. & Bickel, D.J. (2005) The NSF-Fiji Terrestrial Arthropod Survey: overview. *Bishop Museum Occasional Papers* 82: 3–26. [*In:* Evenhuis, N.L. & Bickel, D.J. (eds), *Fiji Arthropods I.*]

Evenhuis, N.L. (ed.) (1989) *Catalog of the Diptera of the Australasian and Oceanian regions*. Bishop Museum Special Publication 86; Bishop Museum, Honolulu and E. J. Brill; 1155 pp. [Online, updated version available at http://hbs.bishopmuseum.org/aocat/aocathome.html]

Farrow, R.A. (1984) Detection of transoceanic migration of insects to a remote island in the Coral Sea, Willis Island. *Australian Journal of Zoology* 9: 253–272.

Gibbs, G. (2007) *Ghosts of Gondwana — the history of life in New Zealand*. Craig Potton, Auckland, 232 pp.

Goldberg, J., Trewick, S.A. & Paterson, A.M. (2008) Evolution of New Zealand's terrestrial fauna: a review of molecular evidence. *Philosophical Transactions of the Royal Society, Series B* 1508: 3319–3334.

Grandcolas, P., Murienne, J., Robillard, T., Desutter-Grandcolas, L., Jourdan, H., Guilbert, H. & Deharveng, L. (2008) New Caledonia: a very old Darwinian island? *Philosophical Transactions of the Royal Society, Series B* 1508: 3309–3317.

Gressitt, J.L. (1961) Problems in the zoogeography of Pacific and Antarctic insects. *Pacific Insects Monograph* 2: 1–94.

Gressitt, J.L. (ed.) (1982) *Biogeography and ecology in New Guinea*. Vols 1–2 Monographiae Biologicae 42; W. Junk, The Hague, 983 pp.

Hall, R. (2001) Cenozoic reconstruction of SE Asia and the SW Pacific: changing patterns of land and sea. Pages 35–56 *in*: Metcalfe, I., Smith, J.M.B., Morwood, M. & Davidson, I.D. (eds), *Faunal and floral migrations and evolution in SE Asia-Australasia*. A.A. Balkema (Swets & Zeitlinger Publishers), Lisse.

Harrison, R.A. (1959) Acalypterate Diptera of New Zealand. *Bulletin New Zealand Department of Scientific and Industrial Research* 128: 1–382.

Heads, M. (2002) Birds of paradise, vicariance biogeography and terrane tectonics in New Guinea. *Journal of Biogeography* 29: 261–283.

Heads, M. (2006) Seed plants of Fiji: an ecological analysis. *Biological Journal of the Linnean Society* 89: 407–431.

Hennig, W. (1966) The Diptera fauna of New Zealand as a problem in systematics and zoogeography. *Pacific Insects Monographs* 9: 1–81.

Hill, K.C. & Hall, R. (2003) Mesozoic-Cenozoic evolution of Australia's New Guinea margin in a west Pacific context. Pages 256–290 *in*: Hillis, R.R. & Muller R.D. (eds), *Evolution and Dynamics of the Australian Plate*. Geological Society of Australia Special Publication 22.

Holloway, B.A. (1976) A new bat-fly family from New Zealand (Diptera: Mystacinobiidae). *New Zealand Journal of Zoology* 3: 279–301.

Holzapfel, E.P. & Harrell, J.C. (1968) Transoceanic dispersal studies of insects. *Pacific Insects* 10: 115–153.

Keast, A. (1996) Pacific biogeography: patterns and processes. Pages 477–512 *in*: Keast, A. & Miller, S.E. (eds), *The origin and evolution of Pacific Islands biotas, New Guinea to eastern Polynesia: patterns and processes*. SPB Academic Publishing, Amsterdam.

Keast, A. & Miller, S.E. (eds) (1996) *The origin and evolution of Pacific Islands biotas, New Guinea to eastern Polynesia: patterns and processes*. SPB Academic Publishing, Amsterdam; 531 pp.

Keogh, J.S., Edwards, D.L., Fisher, R.N. & Harlow, P.S. (2008) Molecular and morphological analysis of the critically endangered Fijian iguanas reveals cryptic diversity and a complex biogeographic history. *Philosophical Transactions of the Royal Society, Series B* 1508: 3413–3426.

Keppel, G., Hodgskiss, P.D. & Plunkett, G.M. (2008) Cycads in the insular South-west Pacific: dispersal or vicariance? *Journal of Biogeography* 35: 1–12.

Kroenke, L.W. (1996) Plate tectonic development of the western and southwestern Pacific: Mesozoic to the present. Pages 19–34 *in*: Keast, A. & Miller, S.E. (eds), *The origin and evolution of Pacific Islands biotas, New Guinea to eastern Polynesia: patterns and processes*. SPB Academic Publishing, Amsterdam.

Ladiges, P.Y. & Cantrill, D. (2007) New Caledonia–Australian connections: biogeographic patterns and geology. *Australian Systematic Botany* 20: 383–389.

Lewis, D.J. & Dyce, A.L. (1989) Taxonomy of Australasian Phlebotominae (Diptera: Psychodidae) with revision of genus *Sergentomyia* from the region. *Invertebrate Taxonomy* 2: 755–804.

Maa, T.C. (1982) On the zoogeography of New Guinean Diptera Pupipara. Pages 699–708 *in*: Gressitt, J.L. (ed.), *Biogeography and Ecology of New Guinea*. Vol 2,. Monographiae Biologicae 42; W. Junk, The Hague.

Macfarlane, R.P. & Andrew, I.G. (2001) New Zealand Diptera diversity, biogeography and identification: a summary. *Records of the Canterbury Museum* 15: 33–72.

Mackerras, I.M. (1961) The zoogeography of western Pacific Tabanidae (Diptera). *Pacific Insects Monographs* 2: 101–106.

McAlpine, D.K. (1982) The acalyptrate Diptera with special reference to the Platystomatidae. Pages 659–672 *in:* Gressitt, J.L. (ed.), *Biogeography and Ecology of New Guinea*. Vol 2; Monographiae Biologicae 42; W. Junk, The Hague.

McAlpine, D.K. (1994) Review of the species of *Achias* (Diptera: Platystomatidae). *Invertebrate Taxonomy* 8: 117–281.

McAlpine, D.K. (2001) Review of the Australasian genera of signal flies (Diptera: Platystomatidae). *Records of the Australian Museum* 53: 113–199.

McAlpine D.K. & Schneider, M.A. (1978) A systematic study of *Phytalmia* (Diptera: Tephritidae), with the description of a new genus. *Systematic Entomology* 3: 159–175.

McCarthy, D. (2005) Biogeographical and geological evidence for a smaller, completely-enclosed Pacific basin in the late Cretaceous. *Journal of Biogeography* 32: 2161–2177.

McDowall, R.M. (2008) Process and pattern in the biogeography of New Zealand – a global microcosm? *Journal of Biogeography* 35: 197–212.

Miller, S.E. (1996) Biogeography of Pacific insects and other terrestrial invertebrates: A status report. Pages 463–476 *in*: Keast, A. & Miller, S.E. (eds), *The origin and evolution of Pacific Islands biotas, New Guinea to eastern Polynesia: patterns and processes*. SPB Academic Publishing, Amsterdam.

Mueller-Dombois, D. & Fosberg, F.R. (1998) *Vegetation of the tropical Pacific Islands*. Springer Verlag, Berlin, 731 pp.

Munroe, D.D. (1974) The systematics, phylogeny, and zoogeography of *Symmerus* Walker and *Austrosymmerus* Freeman (Diptera: Mycetophilidae: Ditomyiinae). *Memoirs of the Entomological Society of Canada* 92: 1–183.

Nagatomi, A. (1970) Rachiceridae (Diptera) from the Oriental and Palearctic regions. *Pacific Insects* 12: 417–466.

Neall, V.E. & Trewick, S.A. (2008) The age and origin of the Pacific islands: a geological overview. *Philosophical Transactions of the Royal Society, Series B* 1508: 3293–3308.

Peck, S.B. (1994a) Aerial dispersal of insects between and to islands in the Galápagos Archipelago, Ecuador. *Annals of the Entomological Society of America* 87: 218–224.

Peck, S.B. (1994b) Sea-surface (pleustron) transport of insects between islands in the Galapagos Archipelago, Ecuador. *Annals of the Entomological Society of America* 87: 576–582.

Pelletier, B. (2006) Geology of the New Caledonia region and its implications for the study of the New Caledonian biodiversity. *Documents Scientifiques et Techniques de l'Institut de Recherche pour le Développement Nouméa*, II 7: 17–30.

Polhemus, D.A. (1996) Island arcs, and their influence on Indo-Pacific biogeography. Pages 51–66 *in*: Keast, A. & Miller, S. (eds), *The origin and evolution of Pacific Island biotas, New Guinea to eastern Polynesia: Patterns and processes*. SPB Academic Publishing bv, Amsterdam.

Polhemus, D.A. (2007) Tectonic geology of Papua. Pages 137–164 *in*: Marshall, A.J. & Beehler, B. (eds). *Ecology of Papua, Part 1*. Periplus Editions (HK) Ltd., Singapore.

Quate, L.W. & Quate, S.H. (1967) A monograph of Papuan Psychodidae, including *Phlebotomus* (Diptera). *Pacific Insects Monograph* 15: 1–216.

Sanmartín, I. & Ronquist, F. (2004) Southern hemisphere biogeography inferred by event-based models: plant versus animal patterns. *Systematic Biology* 53: 216–243.

Sinclair, B.J. (2008) A new species of *Austrothaumalea* from New Caledonia (Diptera: Thaumaleidae). *Zoologia Neocaledonica 6, Mémoires du Muséum national d'Histoire Serie A: Zoologie* 197: 277–281.

Stevens, G. (1985) *Prehistoric NewZealand*. Heinemann Reed, Auckland, 128 pp.

Taylor, G., Davies, K., Martin, N. & Crosby, T. (2007) First record of *Fergusonina* (Diptera: Fergusoninidae) and associated *Fergusobia* (Tylenchida: Neotylenchidae) forming galls on *Metrosideros* (Myrtaceae) from New Zealand. *Systematic Entomology* 32: 548–557.

Waters, J. M. & Craw, D. (2006) Goodbye Gondwana? New Zealand Biogeography, Geology, and the Problem of Circularity. *Systematic Biology* 55: 351–356.

Wilson, E.O. (1959) Adaptive shift and dispersal in a tropical ant fauna. *Evolution* 13: 122–144.

SECTION II:
DIPTERA BIODIVERSITY:
CASE STUDIES, ECOLOGICAL
APPROACHES AND ESTIMATION

CHAPTER TEN

WHY *HILARA* IS NOT AMUSING: THE PROBLEM OF OPEN-ENDED TAXA AND THE LIMITS OF TAXONOMIC KNOWLEDGE

DANIEL BICKEL

Australian Museum, Sydney, Australia

INTRODUCTION

Recent concern over biodiversity loss has highlighted our ignorance about life on this planet. In particular, we do not even know how many species exist, or how many there are likely to be. Somewhere between 1.5–1.8 million plant and animal species have been described (this number is uncertain partially due to potential synonymy), but estimates of the actual number of living species range from 5–100 million (Stork 1997). Such estimates are based on varying methods of calculation, different species concepts, and trophic models, but a figure close to 10 million species has been used in many recent papers.

Various projects have been proposed to electronically document all species in a unified format, the most recent being the *Encyclopedia of Life* (EOL), which plans a separate web page for each described species on the planet (Leslie 2007). Other projects, such as the defunct *All Species Foundation*, called for 'the discovery, identification, and description of all remaining species on Earth within one human generation—25 years' (Berger 2005). This unrealistically optimistic project, while acknowledging the small number of practicing taxonomists (6,000–7,000) and the rather low output of new species described each year (15,000), naively proposed that new tools and computer technology could overcome any problems. Previously, E.O. Wilson (2000, 2003a) stated 'to describe and classify all of the surviving species of the world deserves to be one of the great scientific goals of the new century.' I do not question Wilson's enthusiasm and his ground-breaking achievements in many areas of biology (including Wilson 2003b, his magisterial monograph treating the New World ant genus *Pheidole* with 624 species, 337 newly described, this alone being far more than the individual output of most full-time taxonomists), but I doubt we can come even close to describing the Earth's biota, and reaching even 3 million described species in next century would itself be a major accomplishment.

To begin with, the 1.5–1.8 million described species (those to be awarded individual EOL web pages), include all the larger, prominent species (vascular plants and vertebrates), a general baseline of taxa (butterflies, mollusks, ants, algae, etc.), taxa of direct economic or medical interest (agricultural pests, mosquitoes, etc.), and random groups studied out of personal curiosity (for example, the 10,890 species & subspecies in the cranefly family Tipulidae described by C.P. Alexander; see Oosterbroek 2007). The remaining 5-100 million undescribed taxa comprise mostly invertebrates, fungi, and prokaryotes. Many of these are 'orphan taxa', groups that are totally neglected or lack current taxonomists (Janzen & Hallwachs 1994). For example, in many arthropod groups, a taxonomic 'impediment' exists at the level of genus, which means specimens cannot be identified much beyond family level. Quite simply, appropriate keys do not exist, old descriptions are unusable, and there are insufficient specialist taxonomists in the world. This problem is not confined to rarely encountered or obscure taxa. In highly biodiverse regions of the world, some of the most abundant families taken in routine trapping cannot be identified much beyond family level.

Further, many taxa are 'open-ended' by which I mean they are highly speciose, cosmopolitan or spanning several biogeographic regions, and that based on known local faunas, their true diversity cannot even be accurately estimated. Apart from being speciose, they often remain uncollected or unsorted from mass trapping residues, are unrecorded for vast regions where they undoubtedly occur, and often lack clear generic definition. The total species numbers of such open-ended taxa are as uncertain as the estimates for the total world biota cited above. The effort needed to complete the taxonomic study of such groups is not a simple matter of applying computer technology, as will be discussed later.

This paper reviews examples of open-ended taxa in the Diptera drawn from a range of families and genera. This is by no means a complete conspectus, and specialists will be able to suggest additional examples. There are many large (>1,000 spp.), widely distributed and often paraphyletic genera in the Insecta alone, not to mention other classes. Similar open-ended taxa occur in the Coleoptera, Hymenoptera, Lepidoptera, Hemiptera, Acari, Crustacea, Nematoda, etc. Also, the discussion below is confined to species based on morphology, as species based on molecular differences are another dimension, outside the scope of the traditional taxonomy discussed here. [For an example of the species richness that might be discerned using molecular techniques, see Condon *et al.* (2008). Here six sympatric cryptic species of *Blepharoneura* Loew (Tephritidae) were discovered in the floral calyces of *Gurania spinulosa* (Cucurbitaceae) in eastern Ecuador. They were ecologically and molecularly distinct, but diagnostic morphological characters to separate the species were not discovered.]

1. Examples of Open-Ended Taxa in the Diptera

1.1 Family Cecidomyiidae

Adult Cecidomyiidae are fragile flies that are often abundant in trap samples (their numerous body fragments might even be regarded as a major thickening agent in the Malaise trap 'soup'). Although commonly called 'gall midges,' cecidomyiids have a wide range of larval habits. Much of the taxonomic attention given to the family is related to pest management, although the vast majority of species have no economic impact. From a total of 5,451 described species, Gagné (2007) listed the number of Cecidomyiidae known to occur in each zoogeographical realm: Palaearctic 3,057, Nearctic 1,169, Neotropical 533, Oriental 335, Australasian 247, Afrotropical 165, and Oceanian 32. The geographical numbers show a common pattern among many poorly known Diptera families, with the Palaearctic region having most species, reflecting a long history of research in a settled and accessible region. By itself, Germany has 838 cecidomyiids (Jong 2004), more than any other zoogeographic realm except the Nearctic. If the more than 800 species from a recently glaciated Central European country is a glimpse of potential cecidomyiid diversity, then the true world total, considering the rich floras, varied habitats and ancient landscapes of the tropics and southern hemisphere, must be at least 1–2 orders of magnitude greater. This family is truly immense and open-ended, but unlikely to attract casual entomologists, not the least because species are so small and fragile. Further, as Gagné noted, much of the named fauna is poorly described and needs revision in a wider context to have any meaning. For further discussion, see Espíritu-Santo & Fernandes (2007) on the diversity of gall-inducing insects in general based on ecological considerations.

1.2 Family Ceratopogonidae

Culicoides Latreille. *Culicoides* is a large and difficult genus of some 1,200 described species (Borkent & Wirth 1997), and it includes many species of medical and veterinary importance. The bite of some species causes irritation and allergic reactions in humans, and more importantly, *Culicoides* spp. are proven vectors of arboviruses, protozoa and filarial worms, including Bluetongue virus, a serious disease of sheep and cattle.

Despite its economic importance, this genus is poorly known. For example, Dyce *et al.* (2007) published a photographic atlas showing the diagnostic wing patterns for all the Australasian *Culicoides* species, of which 145 were described and 120 undescribed. This means that 40% of the known Australasian species are represented only by code names and will probably remain in this informal taxonomy for some time. Nevertheless they are identifiable and of practical use to medical entomologists.

Although considerable funding has been available to study and describe potential disease vectors, large and complex genera such as *Culicoides* can still present major problems. Dyce *et al.* (2007) suggested that the publication of their atlas would serve as a framework on which to fine-tune species identification using molecular techniques. Possibly so, but this serves to emphasize the importance of basic morphological identification before molecular work can proceed.

1.3 Family Corethrellidae

Corethrella Coquillett. The Corethrellidae, or frog-biting midges, comprise a single genus, *Corethrella,* with both extant and fossil species. The family is predominantly tropical, and females are known to feed on the blood of male frogs, being attracted to them by their mating calls. The entire family was treated by Borkent (2008), and of the 97 extant species, 52 were newly described. Of these, 38 species are known from Costa Rica, a country that has been intensively collected, by both general Malaise sampling and with frog-call traps. Based on the species collected and regions not visited within Costa Rica, Borkent suggests 35–50 additional species will be found in that small Central American country. Clearly a large undescribed/ uncollected fauna exists elsewhere, not only in the Neotropics but in the even more poorly known Old World tropics. Here is an example of a fauna whose true diversity cannot even be estimated without a specific collecting technique that targets an aspect of its feeding behaviour.

1.4 Family Dolichopodidae

Subfamily Sciapodinae. The Sciapodinae comprise graceful, metallic coloured flies frequently seen resting on leaves. The subfamily is particularly rich in the tropics and subtropics, and this has been demonstrated by mass trapping. Based on available collections, Bickel (2002) revised the Sciapodinae of New Caledonia and treated 55 species, 53 newly described, only two having extralimital distributions. Since that revision, access to Malaise trap collections from Province Nord have revealed an additional 28 species, 25 new endemics and three described from extralimital areas. Therefore, the current total of Sciapodinae for New Caledonia is 83 species, 78 endemic. Of these, 38% of the species are known from a single site and 65% from three or fewer sites. The Fiji Islands have a similar richness, although the Fiji Group has a distinctly different geological history and biota to New Caledonia. Based on a systematic Malaise trapping program (Evenhuis & Bickel 2005), 95 species of Fijian Sciapodinae have been discovered, nine of which were previously described, and only two having wider Pacific distributions (see Bickel 2005, 2006). Such diversity has been revealed by long term trapping, and most species are known only from their type locality. If the faunas of the relatively small, low-elevation landmasses of New Caledonia and Fiji are exemplars of possible richness in Melanesia, then based on Sciapodinae from the relatively small collections taken in topographically complex New Guinea and Solomon Islands, the western Melanesia would have a huge undescribed fauna, currently uncollected.

Campsicnemus Haliday. This genus has undergone a large endemic radiation in the Hawaiian Islands, in a manner analogous to that in the better known Hawaiian Drosophilidae. There are 163 described Hawaiian *Campsicnemus* with some 100–150 known additional species that await description. Many species are known only from a single site which suggests high local endemicity, with more species undoubtedly awaiting collection from areas of difficult access. And one pan trap sample from Hawai'i yielded 11 species of *Campsicnemus*, suggesting a high level of potential sympatry (Evenhuis 2003).

1.5 *Family Drosophilidae*

Cladochaeta Williston. Grimaldi & Nguyen (1999) revised the New World genus *Cladochaeta*, whose larvae are parasitic on spittlebug nymphs, and treated 119 species, 105 newly described. They speculated on the actual diversity of the genus based on species distributions and the geographical coverage of their study collections. Systematic collecting protocol using Malaise traps had been used at only two sites in Costa Rica. These two sites were about 50 km apart and yielded 11 species each, with none shared between the sites, suggesting a much larger uncollected diversity in Costa Rica alone. Using a simple species-area relationship based on the two best sampled countries, the 35 known species from Costa Rica and Panama, they then discussed how many species might be present in other Neotropical subregions. For all of Central America they estimated 645 species (only 64 currently known) and noted that even if a conservative estimate of 300 species was used for all of Central America, this would result in an 'astronomical estimate' of 1,700 South American species (only 45 currently known). Although such speculation might be considered unfounded, this must be balanced against the fact that 28% of *Cladochaeta* species are known only from a single specimen, the holotype. As well, many Neotropical subregions are poorly sampled for *Cladochaeta* (and indeed for Diptera generally), including the highlands of central and southern Mexico, Andean forests from Colombia to Bolivia, western Venezuela, the Atlantic coastal forest of Brazil, as well as isolated massifs and unique biotopes.

Hawaiian Drosophilidae. The endemic Hawaiian Drosophilidae provide a spectacular example of adaptive radiation, with 560 described endemic species in two genera, *Drosophila* Fallén (414 described and 175 known but undescribed species), and *Scaptomyza* Hardy (146 described and 200 known but undescribed species) (P.J. O'Grady, unpubl.). The two genera have evolved with the changing environment of the Hawaiian chain, and successive population bottlenecks and fragmentation has lead to the evolution of hundreds of species. Although this radiation has been the subject of intense study by researchers in evolution, behaviour, and genetics, it is estimated that little more than half the fauna has been described, and that the actual number could approach 1,000 species (Magnacca *et al.* 2008). This estimate is based not only on undescribed species in collections (many of which comprise difficult species groups, especially in *Scaptomyza*), but also the rate at which new species are encountered in

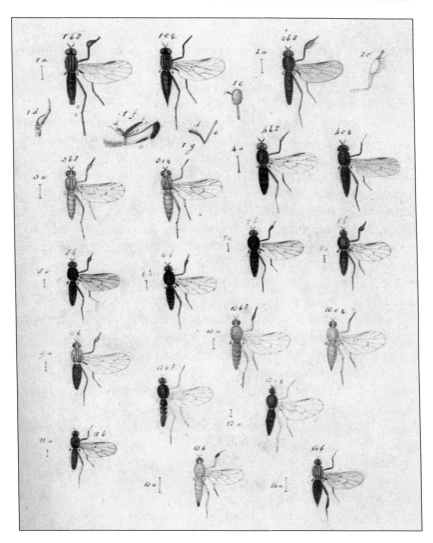

Figure 10.1. European species of *Hilara,* drawn by J.W. Meigen (1818–1838)
(from Morge 1975).

the field. For example, Magnacca (unpubl.) recently found some 15 un-described *Drosophila* species in rearings from a well-collected area in the Hawai'i Volcanoes National Park.

1.6 Family Empididae

Hilara Meigen. *Hilara* is a cosmopolitan genus defined in part by a strong *de novo* apomorphy, the swollen basitarsus on male leg I which contains silk glands (Fig. 10.1). The silk is exuded and used to wrap prey to act as a nuptial gift in attracting females. Male *Hilara* are commonly seen cruis-ing back and forth in loose swarms over pools and streams throughout the world, especially in temperate and upland regions. This dance-like swarming behaviour is the source of the generic name, from the Greek *hilaros*, meaning happy or cheerful. However, this genus does not inspire such feelings in the taxonomist.

The number of species of *Hilara* described from each zoogeographical realm is presented here: Palaearctic 184 species (with 90 in Germany, 85 in the Czech Republic) (Chvála & Wagner 1989), Nearctic 50 species, with 48 additional unpublished eastern species (Roach 1971) and 100 known but undescribed western species (P.H. Arnaud, Jr., unpubl.); Neotropical 22 species, mostly from southern Chile and Argentina (Collin 1933); Ori-ental 44 species, including 25 sympatric endemic species described from Kambaiti, Burma (see discussion under *Prosciara* Frey, below) (Smith 1975); Australasian 21 species, mostly New Zealand and Tasmania (Smith 1989); Afrotropical 15 species, mostly South Africa (Smith 1969). Only the European fauna has been well studied (see Chvála 2005) and shows the *potential* size of other faunas if they were equally well known. Other regions have species described from only a few localities, or in the case of North America, known and with large institutional collections but not formally described.

The generic limits of *Hilara* itself are not clear, as the genus shows end-less variation based on suites of characters. Some local groups that could be considered *Hilara* (*sensu lato*) have been defined as small genera, leav-ing a generalized and highly speciose cosmopolitan residue. The overall similar appearance of the genus is such that if a species like the European type species, *Hilara maura* (Fabricius, 1777), occurred in Australia, it would not surprise me — few such cosmopolitan taxa have biogeographic provenance associated with distinctive synapomorphies (if indeed that is always possible). *Hilara* is undoubtedly paraphyletic, but that is not a problem at this stage.

One can also refer to a '*Hilara* group of genera', comprising most of the tribe Hilarini (Empididae: Empidinae) and all having the male fore basitarsus I swollen and presumably housing silk glands. This group would include some ten genera (Bickel, 1996), but additional genera await description from the Southern Hemisphere. However, genera that appear to be well defined in one region may not hold equally well elsewhere. For example, the predominately southern temperate genus *Hilarempis* Bezzi with some 110 described species is separated from *Hilara* by having vein Sc incomplete, while *Hilara* as defined always has a complete Sc. Although this character appears to have some use in separating groups of species, I have seen this vein being complete and incomplete among closely related Australian species.

The *Hilara*-group is rarely captured in Malaise or water traps, and therefore must be hand collected, primarily sweeping overhanging stream vegetation and netting males over water (a pleasant form of fly fishing!). The Australian fauna is predominately southern temperate, and phenology is important, with drylands species flying in cooler, damp weather. For example, *Hilara* and *Hilarempis* are totally undescribed for Western Australia, but there is a particularly rich and largely uncollected early spring fauna in the southwest. Some 200 *Hilara*-group species are now known from southeastern Australia. I have collected up to 10 *Hilara*-group species together at a stream, and undescribed species are discovered on almost every field trip. I estimate a minimum of 400 species exist in Australia — almost all of them undescribed. Since the group requires hand collecting, dedicated collector-taxonomists are needed to document the fauna. Considering how few areas globally, which have been adequately sampled — if at all — I have little doubt that the world diversity for the *Hilara*-group could easily reach some 3,000–4,000 morphospecies.

Elaphropeza Macquart. The genus *Elaphropeza* comprises small, predaceous, mostly yellow-coloured empidid flies, and it is largely pan-tropical in distribution. Shamshev & Grootaert (2007) revised the Oriental fauna and described 51 new species, bringing the total for that region to 130 species. However, their study focussed primarily on freshly collected material from Singapore gathered by frequent sweep netting and a complete annual cycle with Malaise traps at eight stations. This revealed a remarkable local diversity in the genus, with 544 samples comprising 987 individuals belonging to 52 species, 48 newly described, with the other four previously described species from Java and the Philippines. Even with such intensive collecting effort, 33% of the local Singapore species were only represented by singletons and doubletons, suggesting that a large number of local species still remain undiscovered. Although some additional Singapore species may be more widespread, this remains to be documented. However, so few species have been described from other diverse Oriental countries, and if so many species can co-exist on a highly impacted island like Singapore, undoubtedly many more species await discovery, and with similar suites of undescribed sympatric species.

1.7 Superfamily Mycetophiloidea

The Mycetophiloidea consists of nine families that favour moist forest environments, and five of these families, the Bolitophilidae, Diadocidiidae, Ditomyiidae, Keroplatidae and Mycetophilidae, collectively known as 'fungus gnats,' are particularly diverse in the Holarctic Region where their species richness can be extraordinary. A single Malaise and window trap operated for a year at one site in boreal-oceanic, old-growth deciduous forest in Norway yielded 23,000 specimens of fungus gnats, comprising 315 species, ten of which were new to science (Kjærandsen & Jordal 2007). This included 49 sympatric species in the genus *Mycetophila* Meigen. The Scandinavian fungus gnat fauna would have survived repeated glaciations and probably quickly re-colonised northern Europe during inter-glacial periods. Repeated and periodic isolations in glacial refugia might even partly explain the richness of the fauna. However, this diversity could only have been documented in a country where there are both interested researchers and an extensive knowledge of the regional fauna (e.g. Kjærandsen *et al.* 2007).

1.8 Family Mycetophilidae

Manota Williston. Focused collecting and species-descriptions from a small area can give a hint as to the potential size of a taxon, and also show that regional checklists of described species often fail to reflect true diversity. This is demonstrated in the cosmopolitan genus *Manota*, which prior to 2005 comprised 33 species: Afrotropical 18, Australian 5, Nearctic 1, Neotropical 3, Oriental 3, and Palaearctic 3. Then Jaschhof & Hippa (2005) described 27 new species from Costa Rica alone, suggesting a much larger Neotropical fauna. This was followed by the remarkable study of Hippa

(2006), who described 27 sympatric species of *Manota* from one month of Malaise trapping within a plot less than 1 km^2 in secondary lowland rainforest near Kuala Lumpur, Malaysia. Hippa noted that even more species were present in the samples but were too poorly preserved to describe. Based on this single site alone, it is not possible to know whether these Malaysian *Manota* species are local endemics or occur more widely in the region (the three previously described Oriental species were from Taiwan and Sri Lanka). However, one might expect similar levels of sympatry elsewhere in the Indo-Malaysian region, with potentially hundreds of undescribed species. *Manota* must therefore be regarded as an open-ended taxon, especially in the tropics. This is further supported by preliminary data from the Afrotropical region. Hippa (2008), for example, described five species of *Manota* from Madagascar based on a small sample of only eleven specimens.

1.9 Family Sciaridae

The Sciaridae, or dusky fungus gnats, are one of the most abundant Diptera families in forested habitats, and species are often seen hovering in mating swarms near tree trunks. The larvae feed variously on detritus, decaying plant tissue and fungi, while the short-lived adults take only nectar or other liquids. Sciaridae were the most abundant arthropod family in a tree trunk sticky trap survey in New South Wales, Australia (Bickel & Tasker 2004), and it comprised 48% of all specimens, being the dominant taxon at most sites from all seasons, based on more than 103,000 arthropod specimens in 215 families. Yet despite their abundance, little of the Australian fauna is known. Some 65 species of Australian Sciaridae have been described, mostly between 1888–1890 by Frederick Skuse, who placed them in the then vaguely defined type-genus, *Sciara* Meigen. There has been little taxonomic work since and there are no keys to the Australian fauna. By comparison, some 728 species are described for the Palaearctic region (with more than 400 species in Germany alone), keys are available, and there are several active workers, who estimate the Palaearctic fauna to be closer to 1,100 species (Menzel & Mohrig 1997), not including cryptic species complexes. The Sciaridae are thus an orphan taxon in Australia, and specimens cannot be identified much beyond family level.

Prosciara Frey. The Sciaridae can show extraordinary levels of sympatry, as in the genus *Prosciara*, where 43 species have been described from Kambaiti, northeast Burma, out of 79 species total known from the entire Indomalayan region (Hippa & Vilkamaa 1991; Vilkamaa & Hippa 1996). And it is worth noting that this Kambaiti biodiversity hot spot (all collected by René Malaise, of Malaise trap fame in 1934) also has eight sympatric species of the sciarid genus *Keilbachia* Mohrig (Vilkamaa *et al.* 2006). Further sampling would be needed to know if similar levels of sympatry and local endemicity are found through the complex of forested ranges along the Himalayan front.

1.10 Family Phoridae

Megaselia Rondani. The cosmopolitan genus *Megaselia* is diverse in both species number and larval lifestyles, and it is possibly the largest genus of living organisms, but its actual size can only be hinted at. Of all insect genera, it possesses the greatest diversity of larval habits. This includes feeding on micro-organisms in aquatic habitats, dung, carrion (including human corpses), fungi and plants, and some *Megaselia* species are also known to be kleptoparasites, predators, parasitoids and parasites (Disney 1994).

There are some 1,500 described species of *Megaselia*, but Disney (1994 & unpubl.) estimates this to be only 10% of the true world fauna. Some 400 species are known from Europe, including 246 from Germany and 230 from the United Kingdom (Weber & Prescher 2004, Disney 1989). An example of local sympatry in Europe is the 56 *Megaselia* species recorded from Buckingham Palace Garden (Disney 2001). Further, Bonet *et al.* (2006) reported that Malaise trap sampling in burnt hemiboreal forest near Stockholm yielded 331 species of *Megaselia* of which 112 species were newly recorded for Sweden. In such a genus, unless a fauna has been consistently worked by specialists, published species numbers in catalogs do not begin to reflect true regional diversity. For example, Tasmania has 39 (Disney 2003) and New Guinea has 21 (Evenhuis 1989) known species of *Megaselia*, each less than a garden fauna in England. However, the following species numbers from focused study of circumscribed areas give a hint of possible numbers on continental landmasses: Arabian Peninsula: 84; Greenland 6 (including one above the Arctic Circle); Azore Islands 15; Seychelle Islands: 29 (Disney, pers. comm.). And the Kambaiti, Burma site (discussed above under *Prosciara*) has 35 sympatric species, apparently all local endemics (Delfinado *et al.*1975). Clearly *Megaselia* is an open-ended, mega-genus.

Table 10.1. The Papuan Psychodidae treated by Quate & Quate (1967)

	New Species	Described Species
Lowland Endemic [< 500 m] (1 site only)	58	2
Upland Endemic [> 500 m] (1 site only)	68	1
Lowland (2-4 sites)	12	1
Upland (2-4 sites)	20	–
Lowland Widespread Papuan	7	3
Upland Widespread Papuan	26	–
Lowland & Upland, with wide extralimital distribution	–	30

1.11 Family Psychodidae

Based on some 20,000 specimens, Quate & Quate (1967) treated 228 species of Psychodidae from the Papuan subregion, with 191 species newly described. This is a superb revision, a model of concise description, with detailed line drawings laid out in atlas form for comparison of closely related species. Their remarks on both species and higher taxa clearly lay out problems of taxonomic interpretation and intraspecific variation, provide essential summaries of species bionomics, and reflect deep knowledge and experience with a large and complex family. They estimated that the 228 species was but a small part of the total, possibly not more than 25% of the actual Papuan fauna. How did they reach this conclusion? I have summarized their data in Table 10.1.

In interpreting species locality data for Table 10.1, the lowland (<500 m) and upland (>500 m) categories were arbitrarily established, and although some distributions include both, each species was assigned to only one category. Nevertheless, this summary is representative of the distributions. The following points can be made:

> 129 species or almost 57% of the total Papuan psychodid fauna are known only from a single site. This high level of local endemism suggests the entire Papuan subregion is under collected for Psychodidae.

> Fig. 10.2 shows the collecting locales for the material used by the Quates. Although not shown in this outline map, the island of New Guinea is extraordinarily rugged and difficult to access. The collecting effort indicated seems almost puny in comparison to the immensity of the to-

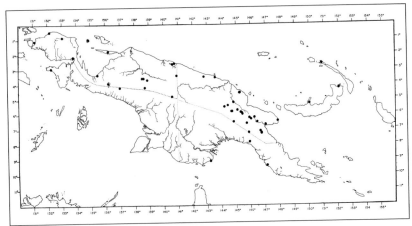

Figure 10.2. Collecting localities for Psychodidae in the Papuan Subregion
(from Quate & Quate 1967; reproduced courtesy of Bishop Museum Press).

pography, and entire mountain ranges and major drainages were unsampled.

All species with wide extralimital distributions were previously described, and often occurred across the Orient and/or Australasia.

Intensive collecting at single sites yielded large and diverse samples, suggesting possible levels of sympatry for the Psychodidae. The two richest
sites were in Indonesian New Guinea:

> 73 species: Kebar Valley, Vogelkopf, 550 m, 4–31.I.1962, L. & S.
> Quate

> 55 species: Sibil Valley, Star Mountains, 1245 m, X–XI.1961, L. & S.
> Quate

It is no coincidence that both of the richest sites were collected by specialist psychodid workers, Laurence and Stella Quate, who used hand
collecting in addition to Malaise and light traps. This supports the axiomatic rule that specialist collecting will greatly increase species diversity, even though attendance of many such family level specialists at
such remote sites is rarely possible.

With such levels of single site endemicity (57%) and local sympatry up to
73 species per site, the Quates' suggestion that they had treated some 25%
of the Papuan psychodid fauna was highly conservative. The true fauna
of such a vast and geographically complex region might be several thousands, essentially open-ended.

DIPTERA DIVERSITY: STATUS, CHALLENGES AND TOOLS
(EDS T. PAPE, D. BICKEL & R. MEIER). © 2009 KONINKLIJKE BRILL NV.

2. Discussion

The examples above treat large, open-ended Diptera taxa from several standpoints:

Taxonomic, with examples of open-ended taxa at the level of family, subfamily and genus.

Spatial, with examples of open-ended taxa at the level of the world, region, and local site.

Comparison with well-studied faunas. Here the large number of species in the Palaearctic fauna (and in particular northern Europe) show the potential size of the fauna when it is well studied. The large species lists obtained from small European countries (e.g. the species numbers for German Cecidomyiidae, *Hilara* and *Megaselia,* and Norwegian and Swedish Mycetophiloidea) are a reflection of the many interested biologists and naturalists working in a settled and accessible landscape, where both the species and their natural history are relatively well-known. However, such detail could never be obtained for wild, remote or even partially settled regions in the humid tropics because the biota is just too complex, apart from logistical and cultural considerations. Paradoxically, perhaps a region has to be tamed and heavily impacted before its biota can become well known. Then only when the remaining natural fragments are saved as nature reserves can they be the subject of thorough investigation. Based on the *Elaphropeza* example and other studies, Singapore has the right combination of interested biologists and small accessible nature reserves, and with its position in the Oriental tropics (rather than recently glaciated northern Europe) it has the potential to become a global invertebrate inventory 'hotspot.'

The large number of congeneric species that can occur sympatrically at one site. Apart from questions of ecological niche separation, such numbers suggest a genus has undergone extensive speciation, either 'explosively' or slowly accumulating species over time, and this is likely to have occurred throughout much of its range. *Manota* in lowland Malaysia, Hawaiian Drosophilidae and European *Megaselia* are examples of high levels of congeneric sympatry.

Large numbers of singletons and doubletons in samples. Traditionally this has been used as an ecological measurement based on the numerical abundance per species (e.g. Novotný & Basset 2000), but it can equally well be used as the number of site occurrences per species in taxonomic works. Thus, the 57% of the total Papuan psychodid fauna known only from a single site is an indication of undercollecting and the existence of a large number of undocumented species.

Local endemicity is hard to discern without widespread regional collecting to determine complementarity, or the index of species shared between sites. In many instances, a large number of locally sympatric species is correlated with a large number of local endemics, but this must be used with caution. With *Manota* it would not be surprising to find similar levels of sympatry but with totally different suites of endemic species in both lowland Malaysia and lowland Kalimantan. On the other hand, most *Megaselia* from the Buckingham Palace Gardens have much wider distributions.

Although we cannot say what we do not know, the Diptera examples presented here strongly suggest that immense faunas exist that are both uncollected and undescribed.

2.1 Is the naming of all taxa possible?

The description of the entire world's species is quite simply impossible. Outside of specific goal-directed projects (mosquitoes, aphid pests, etc.), taxonomic description has traditionally been haphazard, based on the interests and passions of individual taxonomists. Some of this has been supported by grants and salaries, and some by personal funds. I don't know how this can be otherwise since individual passion appears to be the main driver in taxonomy. The Tipulidae *sensu lato* would be a much smaller family without the more than 10,800 species described by C.P. Alexander. Such output was not specified in any job description or even expected, but it was a reflection of individual interest and energy. Can we ask for even 500 described species as a total career output from any professional taxonomist? In such a case the taxonomist would need to be directed to a group requiring work. However, taxonomists have varied personalities, interests, and professional and personal commitments, and for most, 'burnout' would come well before 500 species (for other problems with taxonomists, see Evenhuis 2007).

If Grimaldi & Nguyen (1999) were only able to treat 119 species of *Cladochaeta* in their revision, who is willing or able, or more importantly, even interested in describing the more than 2,000 remaining species they estimate to exist in nature? Further, to accomplish this, who would be able to collect new material in the field, or indeed, sort them out from the many jars of Malaise trap soup sitting on the shelves of various institutions?

And consider the 750 (but undoubtedly many more) Papuan psychodid species that Quate & Quate (1967) estimated to be undocumented. Based on Fig. 10.1, few species were described from the Fly River drainage, and that would be a good region to start a two-year sampling program, with intensive collecting perhaps at 400 m intervals. Even if such a program were funded and the massive samples were sorted to family by trained parataxonomists, who would describe the species? Apart from a major lack of funding for traditional alpha taxonomy (see Flowers 2007), such large and open-ended taxa characteristically have few devoted specialists, and those that exist may already be committed to other projects.

2.2 Who wants to know and why?

Some taxa are of greater interest than others, and there are priorities in taxonomic focus and productivity. For example, ants are a keystone group in ecological processes and are an important group for the investigation of sociality. Many people study ants and identify ants, and there are excellent keys and treatments of the world genera. The Formicidae are not an 'open-ended' taxon and could be completely (95% +) described to species level in the next 25 years, with interactive keys, molecular barcodes, etc. So a taxonomic work describing 20 new species in an ant genus from Laos will generate much more interest than a paper describing 20 new species of Laotian *Megaselia*. This is the way things are — hierarchies of interest exist even among the small fraction of the population who study biology and natural history.

Why should all species on the planet have names, apart from the impossibility of achieving this task and most people not really caring? The conservation argument is often put forward, in that we need to know what we want to conserve, or that to know species and their phylogeny will help us make conservation decisions. Quite frankly, we already know enough based on vertebrates, vascular plants, and biotic communities to make these choices — the impediment is not the lack of knowledge about small flies, but the lack of political will, increasing human population pressure, cultural attitudes and economic issues. Madidi National Park, comprising some 1,890,000 hectares on the Amazonian slopes of the Bolivian Andes, was established in 1995. Yet any biological foundation for its establishment would have been the plant and vertebrate life (even then only as fragmentary knowledge), and other considerations, such as watershed protection, lobbying, political concerns, etc. might have had more impact

DIPTERA DIVERSITY: STATUS, CHALLENGES AND TOOLS
(EDS T. PAPE, D. BICKEL & R. MEIER). © 2009 KONINKLIJKE BRILL NV.

in decision-making. It goes without saying that almost all of the Cecidomyiidae in that reserve would be both uncollected and undescribed, and would comprise thousands of new species. Do they *need* to be described? Should it be a priority, any more than describing many equally unknown families? And who would fund a team to describe the gall midges of Bolivia? Possibly someone might take an interest in the group, and spend the rest of their life describing Andean cecidomyiids, but this would be a matter of personal interest and energy (as with C.P. Alexander and the Tipulidae), and not any governmental or institutional directive.

2.3 What is needed?

To say that the world's Diptera can never be fully described does not mean I am defeatist or a philistine out to wreck the basis for justifying taxonomy. On the contrary, I love these large, open-ended taxa and their infinite variety. Nor will this stop taxonomy from continuing on its rather haphazard way, and new species from all groups will continue to be described, as they always have been.

However, different taxa require different approaches. Description of most Hawaiian drosophilid species, for example, might be achieved and would help answer interesting evolutionary questions. By contrast, the description of isolated new species in a complex, unrevised genus contributes little and may actually be a hindrance to future revisionary work. In the Diptera, summaries of generic level data and keys are the greatest need, and outside of the Holarctic region, many taxa cannot be identified beyond the level of family. The publication of the *Manual of Nearctic Diptera* (McAlpine 1981, 1987) set a standard for further such works, and the forthcoming *Manual of Central American Diptera* (Brown *et al.*, in press), will be truly ground-breaking in providing the first set of generic keys and taxonomic summaries for a tropical country. Such works require extensive coordination and commitment from the few taxonomists who have experience with such complex faunas. Interactive keys such as Hamilton *et al.* (2006) are another useful means for disseminating such taxonomic data. As focus is shifting away from traditional morphological taxonomy, there is even greater need to deal with the world's Diptera fauna at the generic level before the expertise that exists is lost.

3. Summary

Proposals to taxonomically describe all life on Earth fail to understand the diversity of highly speciose taxa. The true size of these groups often cannot even be accurately estimated, let alone having all species described. However, this should not be discouraging. Large, open-ended taxa with their immense species richness are in themselves wonderful. The fact that their species can never be individually named is not the issue. After all, they have existed and evolved for millions of years without names, and their continued survival in large intact communities is the important thing, nothing else. Our inability to encompass them is secondary.

Acknowledgements

R.H.L. Disney, N.L. Evenhuis, J. Kjærandsen, K.N. Magnacca and P.M. O'Grady provided useful comments and information. I thank the Bishop Museum Press for permission to reproduce Fig. 10.2.

REFERENCES

Berger, J.K. (2005) Mission possible: ALL Species Foundation and the call for discovery. *Proceedings of the California Academy of Sciences* 56: (Supplement 1): 114–118.

Bickel, D.J. (1996) *Thinempis*, a new genus from Australia and New Zealand (Diptera: Empididae), with notes on the tribal classification of the Empidinae. *Systematic Entomology* 21: 115–128.

Bickel, D.J. (2002) The Sciapodinae of New Caledonia (Diptera: Dolichopodidae). Zoologia Neocaledonica, *Mémoires de Museum National d'Histoire Naturelle* 187: 11-83.

Bickel, D.J. (2005) *Plagiozopelma* (Diptera: Dolichopodidae: Sciapodinae) from Fiji, Vanuatu and the Solomon Islands. Fiji Arthropods I. *Bishop Museum Occasional Papers* 82: 47–61.

Bickel, D.J. (2006) The *Amblypsilopus pulvillatus* species group (Diptera: Dolichopodidae: Sciapodinae), a radiation in the western Pacific. Fiji Arthropods VI. *Bishop Museum Occasional Papers* 90: 51–66.

Bickel, D.J. & Tasker, E.M. (2004) Tree trunk invertebrates in Australian forests: conserving unknown species and complex processes. Pages 888–899 *in*: Lunney, D. (ed.), Conservation of Australia's Forest Fauna (2nd ed.) Royal Zoological Society of New South Wales, Mosman, Australia.

Bonet, J., Ulefors, S.-O., Viklund, B. & Pape, T. (2006) Mass sampling, species richness and distribution of scuttle flies (Diptera: Phoridae, *Megaselia*) in a wildfire affected hemiboreal forest. Page 27 *in*: Suwa, M. (ed.), Abstract volume, 6th International Congress of Dipterology, 23–28 September 2006, Fukuoka, Japan.

Borkent, A. (2008) The Frog-Biting Midges of the World (Corethrellidae: Diptera). *Zootaxa* 1804: 1–456.

Borkent, A. & Wirth, W.W. (1997) World species of biting midges (Diptera: Ceratopogonidae). *Bulletin of the American Museum of Natural History* 233: 1–257.

Brown, B., Borkent, A., Cumming, J.M., Wood, D.M., Woodley, N.E. & Zumbado, M. (eds) (in press) *Manual of Central American Diptera*. Vol. 1. NRC Press, Ottawa, 752 pp.

Chvála, M. (2005) The Empidoidea (Diptera) of Fennoscandia and Denmark IV Genus *Hilara*. *Fauna Entomologica Scandinavica* 40: 1–234.

Chvála, M. & Wagner, R. (1989) Family Empididae. Pages 228–336 *in*: Soós, Á. & Papp, L. (eds.) *Catalogue of Palaearctic Diptera. Volume 6. Therevidae – Empididae*. Elsevier Science Publishers, Amsterdam.

Collin, J.E. (1933) Empididae, Part 4, *Diptera of Patagonia and South Chile*. British Museum (Natural History), London, 334 pp.

Condon, M., Adams, D.C., Bann, D, Flaherty, K, Gammons, G, Johnson, J, Lewis, M.L., Marsteller, S, Scheffer, S.J., Serena, F. & Swensen, S. (2008) Uncovering tropical diversity: six sympatric cryptic species of *Blepharoneura* (Diptera: Tephritidae) in flowers of *Gurania spinulosa* (Cucurbitaceae) in eastern Ecuador. *Biological Journal of the Linnaean Society* 93: 779-797.

Delfinado, M.D., Hardy, D.E. & Teramoto, L. (1975) Family Phoridae. Pages 261–292 *in*: Delfinado, M.D. & Hardy, D.E. (eds.) *A Catalog of the Diptera of the Oriental Region. Volume II. Suborder Brachycera through division Aschiza, suborder Cyclorrhapha.* University Press of Hawaii, Honolulu.

Disney, R.H.L. (1989) Scuttle Flies — Diptera Phoridae Genus *Megaselia*. *Handbooks for the Identification of British Insects* 10(8): 1–155.

Disney, R.H.L. (1994) *Scuttle Flies: The Phoridae.* Chapman & Hall, London, 467 pp.

Disney, R.H.L. (2001) The scuttle flies (Diptera: Phoridae) of Buckingham Palace Garden. *Supplement to the London Naturalist* 80: 245–258.

Disney, R.H.L. (2003) Tasmanian Phoridae (Diptera) and some additional Australasian species. *Journal of Natural History* 37: 505–639.

Dyce, A.L., Bellis, G.A., & Muller, M.J. (2007) *Pictorial Atlas of the Australasian Culicoides Wings (Diptera: Ceratopogonidae).* Australian Biological Resources Study, Canberra, 88 pp.

Espíritu-Santo, M. M. & Fernandes, G.W. (2007) How many species of gall-inducing insects are there on Earth, and where are they? *Annals of the Entomological Society of America* 100: 95–99.

Evenhuis, N.L. (1989) Family Phoridae. Page 422–431, *in* Evenhuis, N.L. (ed.) *Catalog of the Diptera of the Australasian and Oceanian regions.* Bishop Museum Special Publication 86; Bishop Museum, Honolulu and E. J. Brill, Leiden.

Evenhuis, N.L. (2003) Review of the Hawaiian *Campsicnemus* species from Kaua'i (Diptera: Dolichopodidae), with key and descriptions of new species. *Bishop Museum Occasional Papers* 75: 1–34.

Evenhuis, N.L. (2007) Helping Solve the "Other" Taxonomic Impediment: Completing the Eight Steps to Total Enlightenment and Taxonomic Nirvana. *Zootaxa* 1407: 3–12.

Flowers, R.W. (2007) Comments on "Helping Solve the 'Other' Taxonomic Impediment: Completing the Eight Steps to Total Enlightenment and Taxonomic Nirvana" by Evenhuis (2007). *Zootaxa* 1494: 67–68.

Gagné, R.J. (2007) Species numbers of Cecidomyiidae (Diptera) by zoogeographical realm. *Proceedings of the Entomological Society of Washington* 109: 499.

Grimaldi, D.A. & Nguyen, T. (1999) Monograph on the spittlebug flies, genus *Cladochaeta* (Diptera, Drosophilidae, Cladochaetini). *Bulletin of the American Museum of Natural History* 241, 326 pp.

Hamilton, J.R., Yeates, D.K., Hastings, A., Colless, D.H., McAlpine, D.K., Bickel, D., Cranston, P.S., Schneider, M.A., Daniels, G. & Marshall, S. (2006) *On The Fly: The Interactive Atlas and Key to Australian Fly Families.* (CD ROM). Australian Biological Resources Study/Centre for Biological Information Technology (CBIT).

Hippa, H. (2006) Diversity of *Manota* Williston (Diptera: Mycetophilidae) in a Malaysian rainforest, with description of twenty-seven new sympatric species. *Zootaxa* 1161: 1–49.

Hippa, H. (2008) Notes on Afrotropical *Manota* Williston (Diptera: Mycetophilidae), with the description of seven new species. *Zootaxa* 1741: 1–23.

Hippa, H. & Vilkamaa, P. 1991. The genus *Prosciara* Frey (Diptera, Sciaridae). *Entomologica Fennica* 2: 113–155.

Janzen, D.H. & Hallwachs, W. (1994) *All taxa biodiversity inventory (ATBI) of terrestrial systems: a generic protocol for preparing wildland biodiversity for non-damaging use.* Draft report of an NSF Workshop, 16–18 April 1993, Philadelphia, Pennsylvania.

Jaschhof, M & Hippa, H. (2006) The genus *Manota* in Costa Rica (Diptera: Mycetophilidae). *Zootaxa* 1011: 1–54.

Jong, H. de 2004. Cecidomyiidae. *In*: Jong, H. de (ed.), *Fauna Europaea, Diptera: Nematocera*, version 1.1, available online at http://www.faunaeur.org (last accessed 31 May 2007).

Kjærandsen, J. & Jordal, J.B. (2007) Fungus gnats (Diptera: Bolitophilidae, Diadocidiidae, Ditomyiidae, Keroplatidae and Mycetophilidae) from Møre og Romsdal. *Norwegian Journal of Entomology* 54: 147–171.

Kjærandsen, J., Hedmark, K., Kurina, O., Polevoi, A., Økland, B. & Götmark, F. (2007) Annotated checklist of fungus gnats from Sweden (Diptera: Bolitophilidae, Diadocidiidae, Ditomyiidae, Keroplatidae and Mycetophilidae). *Insect Systematics and Evolution Supplements* 65: 1–128.

Leslie, M. (2007) The Ultimate Life List. *Science* 316 (5826): 818.

Magnacca, K.N., Foote, D. & O'Grady, P.M. (2008) A review of the endemic Hawaiian Drosophilidae and their host plants. *Zootaxa* 1728: 1–58.

McAlpine, J.F. (ed.). (1981) *Manual of Nearctic Diptera.* Volume 1. Biosystematics Research Centre, Research Branch. Agriculture Canada. Monograph No. 28. Ottawa, pp 1–674.

McAlpine, J.F. (ed.) (1987) *Manual of Nearctic Diptera.* Volume 2. Biosystematics Research Centre, Research Branch. Agriculture Canada. Monograph No. 28. Ottawa, pp. 675–1332.

Menzel, F. & Mohrig, W. (1997) Family Sciaridae. Pages 51–69 *in*: Papp, L. & Darvas, B. (eds), *Contributions to a Manual of Palaearctic Diptera.* Vol. 2. Science Herald, Budapest.

Morge, G. (1975) Dipteren-Farbtafeln nach den bisher nicht veröffentlichen Original-Handzeichnungen Meigens: "Johann Wilhelm Meigen: Abbildung der europaeische zweiflügeligen Insekten nach der Natur" Tafel 1–80. *Beiträge zur Entomologie* 25: 383–500.

Novotný, V. & Basset, Y. (2000) Rare species in communities of tropical insect herbivores: pondering the mystery of singletons. *Oikos* 89: 564–572.

Oosterbroek, P. (2007) *Catalogue of the Craneflies of the World (Diptera, Tipuloidea: Pediciidae, Limoniidae, Cylindrotomidae, Tipulidae).* Online at http://nlbif.eti.uva.nl/ccw/manual.php.

Quate, L.W. & Quate, S.H. (1967) A monograph of Papuan Psychodidae, including *Phlebotomus* (Diptera). *Pacific Insects Monograph* 15: 1–216.

Roach, W. K. (1971) *A revision of the genus* Hilara *in Eastern North America*. University Microfilms, Ann Arbor (Michigan), 271 pp.

Shamshev, I.V. & Grootaert, P. (2007) Revision of the genus *Elaphropeza* Macquart (Diptera: Hybotidae) from the Oriental Region, with a special attention to the fauna of Singapore. *Zootaxa* 1488: 1–64.

Smith, K.G.V. (1969) The Empididae of southern Africa (Diptera). *Annals of the Natal Museum* 19: 1–342.

Smith, K.G.V. (1975) Family Empididae. Pages 185–211 *in:* Delfinado, M.D. & Hardy, D.E. (eds), *A Catalog of the Diptera of the Oriental Region. Volume II. Suborder Brachycera through division Aschiza, suborder Cyclorrhapha*. University Press of Hawaii, Honolulu.

Smith, K.G.V. (1989) Family Empididae. Pages 382–392 *in:* Evenhuis, N.L. (ed.), *Catalog of Australasian and Oceanian Diptera*. Bishop Museum Press, Honolulu.

Stork, N.E. (1997) Measuring global biodiversity and its decline. Pages 41–68 *in:* Reaka-Kudla, R. L, Wilson, D.E. & Wilson, E.O. (eds), *Biodiversity II: Understanding and protecting our Biological Resources*. Joseph Henry Press, Washington, DC.

Vilkamaa, P. & Hippa, H. (1996) Review of the genus *Prosciara* Frey (Diptera, Sciaridae) in the Indo-malayan region. *Acta Zoologica Fennica* 203: 1–57.

Vilkamaa, P., Komarov, L.A., & Hippa, H. (2006) The genus *Keilbachia* Mohrig (Diptera: Sciaridae) in a biodiversity hot spot: new sympatric species from Kambaiti, Burma. *Zootaxa* 1123: 39–55

Weber, G. & Prescher, S. (2004) Phoridae. *In:* Pape, T. (ed.), *Fauna Europaea*, version 1.1, available online at http://www.faunaeur.org (last accessed 31 May 2007).

Wilson, E.O. (2000) A global biodiversity map. *Science* 289: 2279.

Wilson, E.O. (2003a) The encyclopedia of life. *Trends in Ecology & Evolution*. 18: 77–80.

Wilson, E.O. (2003b) Pheidole *in the New World, a dominant, hyperdiverse ant genus*. Harvard University Press, Cambridge (Massachussetts), 794 pp.

CHAPTER ELEVEN

DIPTERA AS ECOLOGICAL INDICATORS OF HABITAT AND HABITAT CHANGE

MARC POLLET

Royal Belgian Institute of Natural Sciences, Brussels, Belgium
& University of Ghent, Belgium.

INTRODUCTION

Since the introduction of the term biodiversity, both the genetic, species and ecosystem diversity has drawn a great deal of attention. One of the aspects of biodiversity taxonomists are most familiar with is species richness, which is particularly high in insects. In fact, invertebrates and insects in particular make up about 62% and 54% (or 950,000 species) respectively of the total named biodiversity (Hammond 1992). About 1.75 million species have been described thus far but the estimates of the real diversity range from 5 (Hodkinson & Casson 1991) to 30 million (Erwin 1982). It is thus very unlikely that the major part of, let alone the entire, species diversity will be characterized very soon, especially as long as a solution for the taxonomy impediment seems not easily found.

Species description and identification is an important part of, e.g., phylogenetic and biogeographic research, but it is only a first step in bio-indication and site assessment studies. Indeed, in order to use biotic elements as ecological or bio-indicators it is essential that the ecology (or at least the habitat affinity) of the selected species is documented, i.e., their relationship with their environment and their response to changes in it. And at present, it must unfortunately be concluded that for most taxa and in most biogeographical realms, these data are largely, if not entirely, lacking. Western Europe appears to be the only region that makes an exception to this rule, which is largely due to centuries of intense and large-scale sampling by professional and amateur biologists.

Not all taxa are equally fit to serve as ecological indicators and a number of conditions must be fulfilled before a taxon can be considered suitable (Speight 1986). Nowadays, invertebrates are widely regarded as more informative ecological indicators than, e.g., mammals or vascular plants due to their generally higher species diversity and abundances, and their more subtle responses to environmental changes. Ground beetles (Coleoptera: Carabidae) and spiders (Araneae) are well known ecological indicators (Maelfait *et al.* 1989), but very few dipteran taxa have been used for this purpose due to the lack of ecological data (Hövemeyer 2000). Only larval Chironomidae are known to make part of a traditional set of taxa to monitor the quality of waterways (e.g. Resh & Rosenberg 1984).

Dolichopodidae or long-legged flies are rather easily recognized as a taxon by their generally metallic green body colour, their long legs (hence the name), slightly protruding proboscis and a typical, simplified venation. They are found in every terrestrial and semi-aquatic habitat but are especially prominent in moist places like humid forests, salt marshes, wetlands and shallow banks of rivers and ponds. Some species (*Medetera* spp., *Neurigona* spp., *Sciapus* spp.) mainly occur on vertical surfaces. Except for *Thrypticus*, both adults and larvae are considered predatory on soft-bodied invertebrates (Laurence 1951, Smith & Empson 1955, Smith 1959, White 1976, Schlee 1977). Especially during the last 24 years, both intensive field surveys and large scale projects in western Europe contributed tremendously to the knowledge of their ecology and distribution. On the basis of these data, long-legged flies have proved to be as suitable as ecological indicators as the more traditionally used invertebrates. Moreover, Speight (1986) already acknowledged this taxon as an 'auxiliary group' for wetlands, together with Sciomyzidae, Stratiomyidae, Tabanidae and Zygoptera (Odonata). To provide evidence for the usefulness of Dolichopodidae as ecological indicators, in the present paper the western European fauna is checked against the bio-indicator criteria listed by the latter author, and it is illustrated with the results of some field experiments.

CRITERIA

Criteria for bio-indication are of taxonomic, biogeographical, biological or logistic nature.

1. Taxonomic Criteria

1.1 Identification of the species should be done without undue effort

Most species can be readily identified with the current, though outdated and incomplete, keys in combination with more recently published revisions of, e.g., *Achalcus* Loew (Pollet 1996), *Gymnopternus* Loew (Pollet 1990), *Sciapus* Zeller (Meuffels & Grootaert 1990) and some very useful but, unfortunately, unpublished keys to *Argyra* Macquart and *Dolichopus* Latreille (Meuffels, unpubl. data). Assis Fonseca (1978) is restricted in use to the British Isles and western Europe, and although the dolichopodid chapters by Stackelberg and Negrobov in the Lindner series cover the entire Palaearctic, they only treat the Dolichopodinae (Stackelberg 1930, 1933, 1934, 1941, 1971), Medeterinae (Negrobov & Stackelberg 1971, 1972, 1974a,b; Negrobov 1977), Hydrophorinae (Negrobov 1977, 1978a, 1979a,b), and Rhaphiinae (Negrobov 1979b). This leaves Parent (1938) the most useful key for the western European fauna, but there is an urgent need for an update. Nevertheless, western Europe is the only region with keys to its entire dolichopodid fauna.

1.2 Nomenclature must be reasonably stable

Apart from some long pending problems, listed by Collin (1940) but recently treated by Pollet (unpubl. data) in the frame of the *Fauna Europaea* project (see further below), and some recent generic changes (Chandler 1998) this criterion is largely met with in the region under consideration.

1.3 Taxonomic literature should be available in the form of holistic [= comprehensive] reviews

This is certainly not the case for the entire fauna, but the subfamilies listed above and genera like *Syntormon* Loew (Negrobov 1975), *Xanthochlorus* Loew (Negrobov 1978b, Chandler & Negrobov 2008), some *Chrysotus* Meigen species groups (Negrobov 1980, Negrobov *et al.* 2000), *Gymnopternus*, *Sciapus*, *Achalcus* and *Micromorphus* Mik (Negrobov 2000) have been thoroughly revised during the last 25 years.

1.4 Taxonomic literature should be published in widely used, international languages [English, French, Spanish, German]

Most papers, both old and recent, are, indeed, published in the before mentioned languages. Russian dolichopodid workers like Dr O. Negrobov and Dr I. Grichanov have published many papers in Russian, especially the former author, but fortunately part of these publications have systematically been translated into English. The latter author currently publishes nearly exclusively in English, but deals mainly with either the African or the North European fauna.

2. Biogeographic Criteria

2.1 Reliable national species lists must be available

In this respect, the *Fauna Europaea* project (1.iii.2000–27.ii.2004) has made a considerable contribution (http://www.faunaeur.org). Funded by the European Commission this project aimed at producing a Web-based information infrastructure of all European terrestrial and freshwater animals including both taxonomic and distributional data. As one of the 17 Diptera Brachycera taxonomists involved, I dealt with the family Dolichopodidae. The first research results presented by Pollet & Pape (2002) revealed that even in Europe certain regions were strongly undersampled. Less than 5 species appear to be recorded from Cyprus, Malta, Sardinia, Iceland and Andorra, and the national lists of Portugal and Luxemburg merely consist of 23 and 24 species respectively. On the contrary, dolichopodid faunas of western European countries like Belgium (297 species), France (382 spp.), Germany (365 spp.), Great Britain (285 spp.), The Netherlands (243 spp.) and of nearly all Fennoscandian countries (Finland: 234 spp., Norway: 216 spp., Sweden: 322 spp.) much better reflect the real biodiversity.

2.2 Regional distribution data must be available

To my knowledge, detailed distribution maps on a country scale are currently available only for Belgium (Pollet 2000) and for Nearctic *Medetera* (Bickel 1985). Distribution maps of Dolichopodidae of The Netherlands have been generated recently (see Pollet *et al.* 2003) but not published yet. Pollet (2000) clearly illustrates that even in a small country like Belgium, many species show a very restricted geographical distribution. Examples are (thalasso-) halophilous species like *Aphrosylus* spp. (*celtiber* Haliday, *ferox* Haliday), *Hydrophorus oceanus* (Macquart), *Machaerium maritimae* Haliday and *Muscidideicus praetextatus* (Haliday), which are entirely confined to the coastal region, whereas the psammophilous *Medetera micacea* Loew and *M. plumbella* Meigen exclusively occur in the sandy, more inland dune region and heathlands of Flanders (northern Belgium). Moreover, derelict humid heathlands in the westernmost part of the country and the large eastern heathlands house specific and, in part, different dolichopodid faunas with, e.g., *Telmaturgus tumidulus* (Raddatz) and *Rhaphium fascipes* (Meigen) only recorded from the west, and *Rhaphium longicorne* (Fallén) restricted to the east.

2.3 World range data must be available

Most probably due to their pronounced stenotopic behaviour, only a few species are recorded from several biogeographical realms. And species with a seemingly worldwide distribution like *Micromorphus albipes* (Zetterstedt) most probably involve records based on misidentifications. In the updated *Catalog of American Dolichopodidae north of Mexico*, Pollet *et al.* (2004) only list 55 Holarctic species with a predominance of *Dolichopus* (n = 20) and *Rhaphium* Meigen (n = 15) species. Most of the species show a northern (boreal) distribution range or occur in coastal areas, which immediately explains their circumpolar distribution. Only a few species (e.g., *Medetera truncorum* Meigen, *Thambemyia borealis* (Takagi)) seem to be introduced in North America as a result of human activities. This databased catalog by Pollet *et al.* (2004) and subsequent online North American checklist (Brooks *et al.* 2005) together with the *Fauna Europaea* database (see Pollet 2004) present an excellent and continuously updated source for Holarctic range assessment.

3. Biological Criteria

3.1 The species' habitat must be defined

Although miscellaneous ecological papers occasionally contain site descriptions and apart from the excellent though empirical paper by Emeis (1964), in-depth ecological research on dolichopodid flies is currently conducted only in northern Germany (Meyer and co-workers) and Belgium (Pollet and co-workers). Thanks to the combined efforts of both teams, a large number of western European species has been ecologically characterized. Whereas Meyer mainly focuses on saltmarshes and adjacent habitats in Schleswig-Holstein (see, e.g., Meyer *et al.* 1995, Meyer & Heydemann 1990), Pollet and co-workers have investigated both forests (Pollet *et al.* 1986; Pollet & Grootaert 1987, 1991), heathlands (Pollet *et al.* 1989), marshlands (Pollet & Decleer 1989, Pollet 1992a,b), coastal habitats

(Pollet & Grootaert 1995, 1996), and grasslands (Pollet 2001). Since 1997, eight ecological projects funded by the Belgian or Flemish government have been conducted, yielding valuable additional data on species habitat affinity and distribution (see further below).

3.2 The taxa under consideration must be better agents for detection of site attributes than vertebrates or vascular plants
See discussion.

3.3 A selection of sets of characteristic species must be possible for all available habitats and the taxa [= sets of selected species] must represent each major trophic group in each habitat type
During the sampling campaigns in Belgium, ecological information was collected for every single sample site. Systematically, this information was converted into a macro- and microhabitat code which allowed for the analysis of species distributions over the different macrohabitats — and microhabitats, if relevant (see Pollet & Grootaert 1998, Pollet 2000). In this way, most of the species could be assigned to ecological groups (e.g., species typical for forests, marshlands, riparian habitats, heathlands and moors, saltmarshes, etc.) which roughly correspond with the sets Speight (1986) refers to.

As all dolichopodid species are considered to be predaceous on soft-bodied invertebrates, this taxon does not fit the second part of this criterion. As a matter of fact, neither do the above mentioned groups (Carabidae, Areaneae, chironomid larvae). This requirement therefore only seems to be fulfilled by combining taxonomic groups that belong to different trophic groups.

4. Logistic Criteria

4.1 All taxonomic groups are collected using a single sampling technique
As for most frequently flying insects, Malaise traps are by far the most yielding sampling technique for Dolichopodidae, both in terms of species and specimens (Pollet & Grootaert 1987). Nevertheless, white and yellow water (or pan) traps also seem to be very efficient and have particular advantages to the former trap type in ecological research (Pollet 1989, Pollet & Grootaert 1994). Species complementarity between Malaise and raised (at 60 cm height), white water traps reaches 30% or more (Pollet & Groot-

aert 1987), which underlines the relevance of their simultaneous application in surveys. On the other hand, arboreal *Medetera*, *Neurigona*, and *Sciapus* species are definitely most attracted to blue and bluish green (and related) colours (Pollet & Grootaert 1987, 1994).

4.2 Total number of species should not outnumber 1000

As revealed by the *Fauna Europaea* project, the European dolichopodid fauna comprises 790 dolichopodid species, but more species can be expected, especially in the Mediterranean basin. At this moment, nearly 20 species from France, Greece, Switzerland and Sweden await description (M. Pollet, unpubl.). In western Europe, most species have been recorded from France, Germany, the Czech Republic (333 spp.) and Sweden, but even with intensive and large-scale sampling, it is very unlikely that these national lists will double.

4.3 Number of taxonomic groups should be limited

See discussion.

5. Sensitivity to Environmental Alterations

The biological criteria as discussed before include knowledge of the habitat requirements of the species, which also forms the basis for the construction of species sets per habitat type. However, species sensitivity to environmental changes was not included explicitly, although this feature is considered essential for the purpose of bio-indication. This sensitivity has been demonstrated in several field experiments in Flanders since the early 1980'ies and two examples are presented here.

5.1 Effects of flooding in reed marshes of the De Blankaart Nature Reserve (Woumen, Belgium)

The Blankaart NR (80.5 ha) is located at 2.6 m above sea level in the valley of the river Ijzer in the westernmost province of Belgium (West-Vlaanderen). It mainly consists of a lake (50 ha) fed by several brooks and bordered by reed marshes (20 ha) with willow carrs at their outermost rims. A ruderalised reed marsh site (Decleer 1990) with a thick litter layer was sampled by pitfall traps during three years. Six pitfall traps were in operation during the 1984, 1985 and 1988 seasons and emptied at fortnightly intervals. However, only yields collected during 27.iv–15.ix.1984,

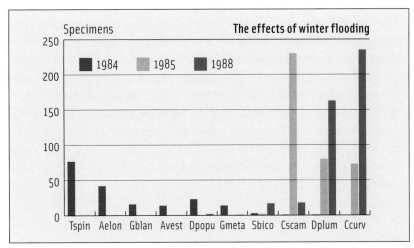

Figure 11.1. The effects of winter flooding (1984–1985) on dolichopodid communities of reedmarsh habitats in De Blankaart Nature Reserve (Belgium). Species codes: Tspin: *Teuchophorus spinigerellus*; Aelon: *Argyra elongata*; Gblan: *Gymnopternus blankaartensis*; Avest: *Argyra vestita*; Dpopu: *Dolichopus popularis*; Gmeta: *Gymnopternus metallicus*; Sbico: *Syntormon bicolorellum*; Cscam: *Campsicnemus scambus*; Dplum: *Dolichopus plumipes*; Ccurv: *Campsicnemus curvipes*.

19.iv–6.ix.1985 and 23.iv–11.ix.1988 were included in the analysis. During the winter of 1984–1985, flooding in the reed marshes of The Blankaart was exceptionally strong and effects on the dolichopodid fauna were measured by comparing the annual pitfall trap catches.

Although overall yields increased from 221 (1984) to 396 (1985) and 460 specimens (1988), species richness and diversity was by far highest in 1984 (21 species, Margalef's diversity index: 3.70) and lowest immediately following the flooding (1985: 11 species, diversity index: 1.67). By 1988, species richness (15 spp.) and diversity index (2.28) seemed to be almost restored, but proved, however, to be exclusively due to the introduction or higher abundances of woodland (*Dolichopus claviger* Stannius, *Gymnopternus celer* (Meigen), *Syntormon bicolorellum* (Zetterstedt)), riparian (*Campsicnemus curvipes* (Fallén), *C. picticornis* (Zetterstedt), *Dolichopus nubilus* Say) and eurytopic species (*D. plumipes* (Scopoli), *D. ungulatus* (Linnaeus)). None of the true reed marsh-inhabiting species (*Teuchophorus spinigerellus* (Zetterstedt), *Argyra elongata* (Zetterstedt), *A. vestita* (Wiedemann), *Gymnopternus blankaartensis* (Pollet)) that accounted for

2/3 of the dolichopodid fauna in 1984 recovered after the flooding (Fig. 11.1). This can be explained by the fact that the soil and litter layer might be crucial for the larval development in these species. Woodland species, on the other hand, appeared to recolonise the reed marsh slowly, which was most probably due to the fact that the slightly more elevated willow carrs — considered the source habitat for these species — were not so heavily affected by the flooding. Undoubtedly the most successful ecological group proved to be the riparian, mud-preferent species, which were obviously favoured by the thick layer of mud that was retained after the flooding.

5.2 Effects of afforestation on heath-like highway verges (Waasmunster, Belgium)

During the late 1960'ies the construction of the highway E17 through a sandy hill at Waasmunster created extensive slopes facing north and south (Desender *et al.* 1987). The largely bare soil surfaces were planted with heather (*Calluna vulgaris*), which was managed and maintained until about 1985 by periodical removal of tree seedlings. When management came to an end in the late 1980'ies, natural afforestation took place, especially in the higher parts of the slopes. In 1985, three sites on both the north-facing (A, B, C) and south-facing slope (D, E, F) were sampled by pitfall traps. Three traps were installed on 21.vi.1985 at the top (A, D), in the middle (B, E) and at the bottom (C, F) of each slope and emptied at fortnightly intervals until 3.x.1985. During 16.vi–6.x.2000 an identical sampling campaign was conducted to assess the effect of the afforestation on the invertebrate community.

Next to the eurytopic ecological group, which accounted on average for about 45% of the dolichopodid fauna, in 1985 one grassland species (*Chrysotus cilipes* Meigen) represented an important faunal element on the north-facing slope, at least in the top and central site (Fig. 11.2). The heath-like grassland apparently offered cool and relatively humid but sunny conditions favoured by this species. On the drier south-facing slope and the north-facing slope bottom, psammophilous species (*Medetera micacea, Sciapus contristans* (Wiedemann)) comprised on average more than 50% of the fauna. Forest species, on the contrary, were only found in the top and central sites in very low abundances. By 2000, forest species (mainly *Sciapus platypterus* (Fabricius), *Xanthochlorus tenellus* (Wiedemann)) clearly outcompeted the eurytopic species in the top and central

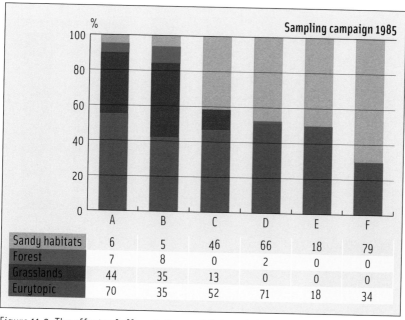

	A	B	C	D	E	F
Sandy habitats	6	5	46	66	18	79
Forest	7	8	0	2	0	0
Grasslands	44	35	13	0	0	0
Eurytopic	70	35	52	71	18	34

Figure 11.2. The effects of afforestation on dolichopodid communities of highway verges in Waasmunster (Belgium) evaluated from the 1985 sampling campaign. Sampling sites: top (A), central (B) and bottom (C) of north-facing slope, and top (D), central (E) and bottom (F) of south-facing slope. Numbers indicate actual number of specimens for all species of the respective ecological group.

sites, where they represented more than 95% and at least 60% of the fauna on the north-facing and south-facing slope respectively (Fig. 11.3). Grassland species disappeared almost entirely, whereas psammophilous species appeared to sustain only in the bottom sites.

6. How to Use Ecological Indicators in Nature Conservation?

Despite the current lack of ecological knowledge on some species-rich genera (e.g., *Medetera*), at present a large number of western European Dolichopodidae are sufficiently known in this respect. Together with the fact that most dolichopodid species appear to react very pronouncedly to environmental changes as shown before, non-stochastic changes in population densities over long periods of time are very likely to be related to changes in their environment. Since both rarity and changes in relative rarity were considered valuable attributes in environmental studies,

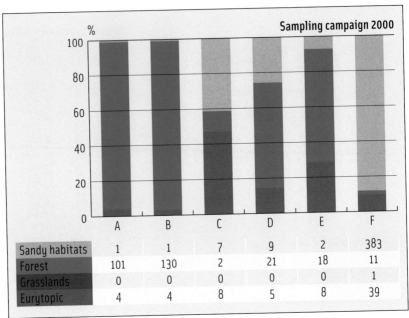

	A	B	C	D	E	F
Sandy habitats	1	1	7	9	2	383
Forest	101	130	2	21	18	11
Grasslands	0	0	0	0	0	1
Eurytopic	4	4	8	5	8	39

Figure 11.3. The effects of afforestation on dolichopodid communities of highway verges in Waasmunster (Belgium) evaluated from the 2000 sampling campaign. See Fig. 11.2 for explanation of sampling sites A–F and numbers.

it was decided to compile all ecological and distributional information on Dolichopodidae in Belgium in the frame of a Red Data Book of Flanders (Pollet 2000). In this way, long-legged flies became a standard agent in nature conservation and site quality assessment studies together with ground beetles, spiders, butterflies and empidid flies. In Pollet (2000), (current) rarity is expressed as the percentage of the sampled UTM 5km squares where the species is found (since 1981). Changes of this rarity in time are calculated between two time periods (1850–1980; 1981–1997) with a subequal number of UTM 5km squares investigated (171; 167). On the basis of both estimations, species with a sufficient ecological characterization were assigned to Red Data Book categories, as shown in Table 11.1. Only 39 species (mostly *Medetera*) were placed in the category 'Insufficiently known'. Furthermore, the Red Data Book includes data sheets for 95 or 36.5% of the 260 Flemish species with information on identification, biogeographical distribution, ecology, status, threats, protection measures and relevant literature references.

Table 11.1. Number of species of Dolichopodidae per Red Data Book category in Flanders (by strict application of rarity and trend criteria).

Current rarity *	Extinct in Flanders	Very rare	Rare	Fairly rare	Not rare	No. species
% sampled UTM 5km squares	0%	1–<2%	2–<5%	5–<10%	≥10%	
Trend (degree of decrease)						
76–100%	Extinct in Flanders (22)	Critically endangered (2)	Endangered (–)	Vulnerable (–)	Nearly threatened (–)	24
51–75%	–	Endangered (4)	Endangered (1)	Vulnerable (1)	Nearly threatened (–)	6
26–50%	–	Vulnerable (8)	Vulnerable (6)	Vulnerable (2)	Nearly threatened (–)	16
≤52%	–	Very rare (33)	Rare (36)	Fairly rare (33)	Safe/ Low risk (73)	175
Insufficiently known	–	–	–	–	–	39
No. species	**22**	**47**	**43**	**36**	**73**	**260**

* estimated during 1981–1997

The assignment to a particular Red Data Book category enabled the calculation of a nature conservation value for every species (Pollet & Grootaert 1999). In order to incorporate entire dolichopodid faunas in site quality assessment studies, an index was subsequently designed that differed from the Species Quality Index (SQI) by Foster (1987) by including not only species richness and rarity but also information on abundances and recent decline. This Site Conservation Quality Index (SCQI) was first applied in a comparative study of grassland and reed marsh communities (Pollet 2001) where it proved to be highly significantly positively correlated with other site quality criteria such as species richness, diversity and rarity. Since its construction, SCQI has confirmed its usefulness as an 'all-in' conservation tool on many other occasions (e.g., nature conservation projects in Flanders).

7. Discussion

As mentioned before, Speight (1986) suggested that Dolichopodidae could be used as bio-indicators but only as an auxiliary group in wetlands. On the basis of 24 years of intensive and large-scale sampling in Belgium, we can conclude that this statement is much too narrow in scope. Dolichopo-did communities are, indeed, poorer in species and can reach consider-ably lower overall abundances in dry habitats as compared to humid en-vironments (e.g., Pollet & Grootaert 1996) but this does not render them inadequate as ecological indicators in this type of habitat. Local changes in light intensity, soil humidity, extent of muddy soil surfaces and con-stitution of litter layer have a strong effect on the structure of dolichopo-did communities (see above, Pollet *et al.* 1986, Pollet 1992a), but it can be assumed that dolichopodids are also suitable to detect the impact of (more) global change(s). In fact, dry habitats show a distinctly different fauna than humid ones even at the level of genera (M. Pollet, unpubl.) and for this purpose alone, dry habitats should also be included in monitoring programs.

Despite their usefulness as ecological indicators, dolichopodid flies do not seem to represent all relevant site attributes (see criterion 3.2). Conse-quently, it is highly recommended that inventories combine several sam-pling techniques and include other invertebrate taxa to reconstruct the most reliable picture of the invertebrate communities under investigation. A first candidate taxon within the order Diptera would seem to be Em-pididae (*sensu lato*) as, contrary to Dolichopodidae, species appear to be highly dependent on higher level landscape elements and habitat mosaic for their survival (Delettre *et al.* 1992, 1998; Pollet & Grootaert 1996). And even on a micro-scale, species of both taxa react entirely differently to a sampling setup as shown by Pollet & Grootaert (1994). Other candidates are Stratiomyidae (larval habitats) and Sciomyzidae (snail hosts). Toge-ther with deer flies (Tabanidae), the latter two fly families should only be considered as 'auxiliary', especially in western Europe, as they lack the high species richness of Dolichopodidae and Empididae *s.l.* Beyond Dip-tera, the most obvious choice of invertebrate ecological indicators are un-doubtedly ground-beetles, as their ecology is elaborately studied and they represent a major invertebrate taxon in most terrestrial habitats (Thiele 1977, Turin 2000).

Despite the fact that the habitat affinity of a large number of western European dolichopodid species has been determined, some knowledge gaps still remain: (i) Dolichopodidae have been reported from only 40% of the UTM 5km squares in Flanders and 21 or more species have been collected in only 85 squares. This most probably implies that the remaining 189 squares are considerably undersampled; (ii) 22 species have not been collected since 1981 and were, according to IUCN-standards, consequently considered extinct in Flanders. However, in a number of cases the pre-1981 capture sites have not recently been investigated, which might possibly reveal their continued presence; (iii) species with larval development in rotholes and sapruns on trees (*Systenus* spp., *Australachalcus melanotrichus* (Mik), *Hercostomus nigrilamellatus* (Macquart)) have not been systematically sampled for, and as a result no reliable statements about their rarity and decline can be made; (iv) very little is known about the biotic and abiotic factors that determine arboreal *Medetera* faunas, which renders them as yet unsuitable for bio-indication. Nevertheless, due to their affinity for tree trunks they might provide important information on the condition of these substrates/plants and the environmental factors that affect them; (v) dolichopodid populations have rarely been investigated for more than one season (most notable exceptions are presented above), but monitoring is essential to gather information on annual population fluctuations and the factors that govern them.

The type of sampling technique is highly depending upon the general aim of the survey as well as the focal taxa. Malaise traps are mainly recommended for faunistic inventories (Pollet & Grootaert 1987) and only useful in ecological research if the sampling period does not cover the entire season as in the ALAS IV surveys (http://www.evergreen.edu/user/serv_res/research/arthropod). Otherwise, yields of several traps per habitat might prove nearly impossible to process in proper time. If gathering information on the habitat preference of species is the main goal of the survey, sets of white (or yellow) pan traps at soil surface level are best used (Pollet & Grootaert 1987, 1994). Finally, raised blue or bluish green pan traps appear to be most effective in collecting arboreal dolichopodid faunas.

It cannot be denied that the concept described above is hardly applicable for any other (dolichopodid) fauna than that of western Europe. Not only is the ecology of most non-European species not or hardly known, but Red Data Books have only a very local relevance (Pollet 2000). However, this must not be considered a strong enough argument to exclude Dolichopodidae from bio-indication in other biogeographical regions. As a matter of fact, by the end of the 1970'ies, only data from about 5,000 Belgian dolichopodid specimens from museum collections were available, mostly void of any information on their collecting sites. Massive, large-scale inventories combined with detailed ecological research conducted in Belgium since the early 1980'ies have provided us with data that allow for a reliable estimation of rarity and decline. It should be clear that collecting site information and storing it in a systematic and accessible way is a crucial step in this process (Pollet & Grootaert 1998, Pollet 2000). As they share about the same dolichopodid faunas as Belgium and hence species with a known ecology, building a similar data set for other, comparable-sized western European countries or regions like The Netherlands (see Pollet *et al.* 2003), southern England, northern France or Germany, Luxemburg or Switzerland only requires large scale sampling (assessment of current distribution and rarity) and the examination of old museum collections (assessment of decline). In other reasonably well-sampled temperate regions like, e.g., Ontario and Quebec with 212 and 215 species recorded respectively thus far (Pollet *et al.* 2004), the before mentioned actions should be accompanied by ecological research in at least the major biomes. It is expected that the number of undescribed species discovered during sampling or in existing collections will definitely be higher than in western Europe, but it is not likely that this will slow down the process considerably. A totally different situation arises in the tropics, even in countries with about the same size of Belgium like Costa Rica and other Central American countries (Morris 1986). The Neotropical catalog by Robinson (1970) lists only 95 Costa Rican species, but the real dolichopodid diversity in this country may be as high as 700 (D. Bickel, pers. comm.). And this should still be considered a very conservative estimate. This huge species richness together with the lack of historical records makes it impossible to build a Red Data Book in an acceptable period of time. Indeed, in order to use the species as potential ecological indicators, they must be described first, which means that taxonomic research deserves the highest priority here. At the same time, however, sufficient attention should be drawn to the collection and storage of relevant information on the sampling sites (habitat and soil type, altitude, humidity, etc.) which will provide important clues to explain or predict their biogeographical distribution and, as such, the rarity of each species. And every additional sampling campaign adds new information which allows continued re-assessment of relevant nature conservation attributes.

The ecology of most western European dolichopodid species is generally considered to be sufficiently well known, but still remains not entirely understood. As described before, eurytopic, mud-preferent species are obviously favoured by flooding in reed marshes and forest-inhabiting species by afforestation, but it is not clear what exactly offers these species the significant advantage against other species. Only in-depth, autecological, biological, physiological and/or ethological research might provide more insight in these matters. However, as this type of 'archaic' natural history research competes poorly with current 'hot' topics in environmental biology (e.g., biodiversity) and systematics (e.g., molecular phylogeny), dolichopodid natural history (and that of most other invertebrates) most probably will never be explained properly ... but perhaps that is what makes it so irresistible!

Acknowledgements

I am greatly indebted to all those people who have provided me with countless samples of dolichopodid flies since the 1980'ies. Most of them had to be very patient to receive any feedback on their deliveries but ultimately their efforts appear to be rewarded here. Special thanks are due to Dr Konjev Desender (Dept of Entomology, Royal Belgian Institute of Natural Sciences — RBINS, Brussels, Belgium), who passed away much too soon in September 2008. Konjev not only supervised my Ph.D. research on carabid beetles, but also showed me how excellent ecological research is done. The second person I am truly grateful to is Dr Patrick Grootaert (Dept of Entomology, RBINS). Patrick provided me with the ultimate stimulus to make the switch from carabid beetles to long-legged flies and was the force behind the Belgian large-scale sampling program for Diptera in the 1980'ies. The numerous discussions I enjoyed with both Konjev and Patrick over the past decades were invaluable for my own research. Thanks are also due to Mr Kris Decleer (Institute for Forest and Nature Research — INBO, Brussels, Belgium), who provided me with the De Blankaart samples. This chap-

ter is a contribution of the Department of Entomology of the RBINS
and of the Research Group Terrestrial Ecology (TEREC) of the Ghent
University, Ghent, Belgium.

References

Assis Fonseca, E.C.M. (1978) Diptera Orthorrhapha Brachycera Dolichopodidae.
    Handbooks for the Identification of British Insects 9(5): 1–90.

Bickel, D.J. (1985) A Revision of the Nearctic *Medetera* (Diptera: Dolichopodidae).
    U.S. Department of Agriculture, Technical Bulletin 1692, 109 pp.

Brooks, S.E., Cumming, J.M. & Pollet, M.A.A. (2005) Checklist of the Dolichopodi-
    dae *s.str.* (Diptera) of America north of Mexico (1st Edition). PDF document, 18 pp.
    Available at http://www.nadsdiptera.org/Doid/Checklist/Dolichopodidae Check-
    list.pdf.

Chandler, P.J. (ed.) 1998) Checklists of Insects of the British Isles (New Series). Part 1:
    Diptera. *Handbooks for the Identification of British Insects* 12(1): 1–234.

Chandler, P.J. & Negrobov, O.P. (2008) The British species of *Xanthochlorus* Loew,
    1857 (Diptera, Dolichopodidae), with description of two new species. *Dipterists
    Digest* 15: 29–40.

Collin, J.E. (1940) Critical notes on some recent synonymy affecting British species of
    Dolichopodidae (Diptera). *Entomologist's monthly Magazine* 76: 261–271.

Decleer, K. (1990) Experimental Cutting of Reedmarsh Vegetation and its Influence
    on the Spider (Araneae) Fauna of the Blankaart Nature Reserve, Belgium. *Biologi-
    cal Conservation* 52: 161–185.

Delettre, Y., Morvan, N., Tréhen, P. & Grootaert, P. (1998) Local biodiversity and
    multi-habitat use in empidoid flies (Insecta: Diptera, Empidoidea). *Biodiversity
    and Conservation* 7: 9–25.

Delettre, Y., Tréhen, P. & Grootaert, P. (1992) Space heterogeneity, space use and
    short-range dispersal in Diptera: A case study. *Landscape Ecology* 6: 175–181.

Desender, K., Van Kerckvoorde, M. & Mertens, J. (1987) Habitat characteristics and
    the composition of the carabid beetle fauna on motorway verges across a hill on
    sandy soil. *Acta Phytopathologica Entomologica Hungarica* 22: 341–347.

Emeis, W. (1964) Untersuchungen über die ökologische Verbreitung der Dolichopo-
    diden (Ins. Dipt.) in Schleswig-Holstein. *Schriften des Naturwissenschaftlichen
    Vereins für Schleswig-Holstein* 35: 61–75.

Erwin, T.L. (1982) Tropical forests: their richness in Coleoptera and other arthropod
    species. *Coleopterists Bulletin* 36: 74–75.

Foster, G.N. (1987) The use of Coleoptera records in assessing the conservation value
    of wetlands. Pages 8–17 *in*: Luff, M.L. (ed.), *The use of invertebrates in site assess-
    ment for conservation*. University of Newcastle upon Tyne, Newcastle.

Hammond, P.M. (1992) Species inventory. Pages 17–39 in: Groombridge, B. (ed.), *Global biodiversity, status of the Earth's living resources*. Chapman and Hall, London.

Hodkinson, I.D. & Casson, D. (1991) A lesser predilection for bugs: Hemiptera (Insecta) diversity in tropical rain forests. *Biological Journal of the Linnean Society* 43:101–109.

Hövemeyer, K. (2000) Ecology of Diptera. Pages 437–489 in: Papp, L. & Darvas, B. (eds), *Contributions to a Manual of Palaearctic Diptera (with special reference to flies of economic importance)*. Vol. 1, *General and Applied Dipterology*. Science Herald, Budapest.

Laurence, B. (1951) The prey of some tree trunk frequenting Empididae and Dolichopodidae (Dipt.). *Entomologist's monthly Magazine* 87: 166–169.

Maelfait, J.-P., Desender, K. & Baert, L. (1989) Some examples of the practical use of spiders and carabid beetles as ecological indicators. Pages 437–442 in: Wouters, K. & Baert, L. (eds), *Invertébrés de Belgique. Comptes rendus du symposium "Invertébrés de Belgique" 25–26 Nov. 1988*. Institut Royal de Sciences Naturelles de Belgique, Bruxelles, Belgium.

Meuffels, H.J.G. & Grootaert, P. (1990) The identity of *Sciapus contristans* (Wiedemann, 1817) (Diptera: Dolichopodidae), and the revision of the species group of its relatives. *Bulletin de l'Institut Royal des Sciences Naturelles de Belgique, Entomologie* 60: 161–178.

Meyer, H., Fock, H., Haase, A., Reinke, H.D. & Tulowitzki, I. (1995) Structure of the invertebrate fauna in salt marshes of the Wadden Sea coast of Schleswig-Holstein influenced by sheep-grazing. *Helgoländer Meeresuntersuchungen* 49: 563–589.

Meyer, H. & Heydemann, B. (1990) Faunistisch-ökologische Untersuchungen an Dolichopodiden und Empididen (Diptera — Dolichopodidae u. Empididae, Hybotidae) in Küsten- und Binnenlandbiotopen Schleswig-Holsteins. *Faunistisch-Ökologische Mitteilungen* 6(3–4): 147–172.

Morris, M.G. (1986) The scientific basis of insect conservation. Pages 357–367 in: Velthuis, H.H.W. (ed.), Proceedings of the 3rd European Congress of Entomology, Amsterdam, The Netherlands, 24–29 August 1986, Part 3.

Negrobov, O.P. (1975) A review of the genus *Syntormon* Meigen (Diptera, Dolichopodidae) from Palearctic. *Ent. Obozr.* 54(3): 652–664.

Negrobov, O.P. (1977) Dolichopodidae. *In*: Lindner, E. (ed.), *Die Fliegen der Palaearktischen Region*, Lieferung 316: 347–386, plates CVIII–CL. Nägele & Obermiller, Stuttgart.

Negrobov, O.P. (1978a) Dolichopodidae. *In*: Lindner, E. (ed.), *Die Fliegen der Palaearktischen Region*, Lieferung 319: 387–418, plates CLI–CLXXIII. Nägele & Obermiller, Stuttgart.

Negrobov, O.P. (1978b) Revision of species from *Xanthochlorus* Lw. genus (Diptera, Dolichopodidae). *Vestnik Zoologii* [1978](2): 17–26.

Negrobov, O.P. (1979a) Dolichopodidae. *In*: Lindner, E. (ed.), *Die Fliegen der Palaearktischen Region*, Lieferung 321: 419–474, plates CLXXIV–CLXXXVIII. Nägele & Obermiller, Stuttgart.

Negrobov, O.P. (1979b) Dolichopodidae. *In*: Lindner, E. (ed.), *Die Fliegen der Palae-arktischen Region*, Lieferung 321: 475–530, plates CLXXXIX–CCVII. Nägele & Obermiller, Stuttgart.

Negrobov, O.P. (1980) A revision of Palearctic *Chrysotus* Mg. (Diptera, Dolichopodi-dae). I. *Ch. cilipes* Mg. and *Ch. laesus* Wied. groups. *Ent. Obozr.* 59(2): 415–420.

Negrobov, O.P. (2000) Revision of the Palaearctic species of the genus *Micromorphus* Mik (Diptera: Dolichopodidae). *International Journal of Dipterological Research* 11: 19–26.

Negrobov, O.P. & Stackelberg, A.A. (1971) Dolichopodidae. *In*: Lindner, E. (ed.), *Die Fliegen der Palaearktischen Region*, Lieferung 284: 238–256, plates XIII–XXVIII. Nägele & Obermiller, Stuttgart.

Negrobov, O.P. & Stackelberg, A.A. (1972) Dolichopodidae. *In*: Lindner, E. (ed.), *Die Fliegen der Palaearktischen Region*, Lieferung 289: 257–302, plates XXIX–XLIV. Nägele & Obermiller, Stuttgart.

Negrobov, O.P. & Stackelberg, A.A. (1974a) Dolichopodidae. *In*: Lindner, E. (ed.), *Die Fliegen der Palaearktischen Region*, Lieferung 302: 303–324, plates XLV–LXXVI. Nägele & Obermiller, Stuttgart.

Negrobov, O.P. & Stackelberg, A.A. (1974b) Dolichopodidae. *In*: Lindner, E. (ed.), *Die Fliegen der Palaearktischen Region*, Lieferung 303: 325–346, plates LXXVII–XVII. Nägele & Obermiller, Stuttgart.

Negrobov, O.P., Tsurikov, M.N. & Maslova, O.O. (2000) Revision of the Palaearctic species of the genus *Chrysotus* Mg. (Diptera, Dolichopodidae). III. *Entomolog-icheskoe Obozrenie* 79(1): 227–238.

Parent, O. (1938) Diptères Dolichopodidae. *Faune de France* 35: 1–720.

Pollet, M. (1989) A technique for collecting quantitative data on Dolichopodidae and Empididae. *Empid and Dolichopod Study Group Newssheet* 6: 5–7.

Pollet, M. (1990) Phenetic and ecological relationships between species of the subge-nus *Hercostomus* (*Gymnopternus*) in western Europe with the description of two new species (Diptera: Dolichopodidae). *Systematic Entomology* 15: 359–383.

Pollet, M. (1992a) Impact of environmental variables on the occurrence of dolichopo-did flies in marshland habitats in Belgium (Diptera: Dolichopodidae). *Journal of Natural History* 26: 621–636.

Pollet, M. (1992b) Reedmarshes: a poorly appreciated habitat for Dolichopodidae. *Dipterists Digest* 12: 23–26.

Pollet, M. (1996) Systematic revision and phylogeny of the Palaearctic species of the genus *Achalcus* Loew (Diptera: Dolichopodidae) with the description of four new species. *Systematic Entomology* 21: 353–386.

Pollet, M. (2000) A documented Red List of the dolichopodid flies (Diptera: Dolichopodidae) of Flanders (in Dutch with English summary). *Communications of the Institute of Nature Conservation* 8, 190 pp. Brussels.

Pollet, M. (2001) Dolichopodid biodiversity and site quality assessment of reed marshes and grasslands in Belgium (Diptera: Dolichopodidae). *Journal of Insect Conservation* 5: 99–116.

Pollet, M. (2004) Dolichopodidae. *In*: Pape, T. (ed.), *Fauna Europaea: Diptera, Brachycera*. Fauna Europaea version 1.1. Available at http://www.fauneur.org.

Pollet, M., Brooks, S.E. & Cumming, J.M. (2004) Catalog of the Dolichopodidae (Diptera) of America north of Mexico. *Bulletin of the American Museum of Natural History* 283: 1–114.

Pollet, M. & Decleer, K. (1989) Contributions to the knowledge of dolichopodid flies in Belgium. III. The dolichopodid fauna of the nature reserve 'Het Molsbroek' at Lokeren (Prov. Eastern Flanders) (Diptera: Dolichopodidae). *Phegea* 17: 83–90.

Pollet, M. & Grootaert, P. (1987) Ecological data on Dolichopodidae (Diptera) from a woodland ecosystem. I. Colour preference, detailed distribution and comparison between different sampling techniques. *Bulletin de l'Institut Royal des Sciences Naturelles de Belgique, Entomologie* 57: 173–186.

Pollet, M. & Grootaert, P. (1991) Horizontal and vertical distribution of Dolichopodidae (Diptera) in a woodland ecosystem. *Journal of Natural History* 25: 1297–1312.

Pollet, M. & Grootaert, P. (1994) Optimizing the water trap technique to collect Empidoidea (Diptera). *Studia dipterologica* 1: 33–48.

Pollet, M. & Grootaert, P. (1995) The Dolichopodid fauna of coastal habitats in Belgium. *Bulletin et Annales de la Société royale belge d'Entomologie* 130: 331–344.

Pollet, M. & Grootaert, P. (1996) An estimation of the natural value of dune habitats using Empidoidea (Diptera). *Biodiversity and Conservation* 5: 859–880.

Pollet, M. & Grootaert, P. (1998) From systematic and ecological databases to Red Data Books and systematic research. Abstracts of the 4th International Congress of Dipterology, 6–13 September 1998, Oxford, UK: 171–172.

Pollet, M. & Grootaert, P. (1999) Dolichopodidae (Diptera): poorly known but excellent agents for site quality assessment and nature conservation. *Proceedings of the section Experimental and Applied Entomology of the Netherlands Entomological Society (N.E.V.)* 10: 63–68.

Pollet, M., Grootaert, P. & Meuffels, H. (1989) Relationships between habitat preference and distribution of dolichopodid flies in Flanders (Dipt., Dolichopodidae). *Verhandelingen van het Symposium "Invertebraten van België"* [1989]: 363–371.

Pollet, M., Meuffels, H. & Grootaert, P. (2003) Status of Dolichopodidae of the Flemish Red Data Book in the Netherlands (Insecta: Diptera). Pages 61–68 *in*: Reemer, M., van Helsdingen, P.J. & Kleukers, R.M.J.C. (eds.), *Changes in ranges: invertebrates on the move*. Proceedings of the 13th International Colloquium of the European Invertebrate Survey, Leiden, 2–5 September 2001. European Invertebrate Survey – The Netherlands, Leiden.

Pollet, M. & Pape, T. (2002) Databasing European Dolichopodidae (Diptera) in the frame of Fauna Europaea. Page 195 *in*: Abstract Volume, 5th International Congress of Dipterology, Brisbane, Australia, 29 September–4 October 2002.

Pollet, M., Verbeke, C. & Grootaert, P. (1986) Verspreiding en fenologie van Dolichopodidae in een bosbiotoop te Wijnendale (West-Vlaanderen). *Bulletin et Annales de la Société royale belge d'Entomologie* 122: 285–292.

Resh, V.H. & Rosenberg, D.M. (eds) (1984) *The Ecology of Aquatic Insects.* Praeger Publishers, CBS Inc., New York, 625 pp.

Robinson, H. (1970) 40. Family Dolichopodidae. Pages 1-92 *in*: Papavero, N. (ed.), *A catalogue of the Diptera of the Americas south of the United States.* Universidade de São Paulo, Museu de Zoologia, São Paulo.

Schlee, D. (1977) Chironomidae als Beute von Dolichopodidae, Muscidae, Ephydridae, Anthomyiidae, Scatophagidae und anderen Insecta. *Stuttgarter Beiträge zur Naturkunde, Series A (Biologie)* 302: 1–22.

Smith, K.G.V. (1959) A note on the courtship and predaceous behaviour of *Neurigona* species (Dipt., Dolichopodidae). *Entomologist's monthly Magazine* 95: 32–33.

Smith, K.G.V. & Empson, D.W. (1955) Note on the courtship and predaceous behaviour of *Poecilobothrus nobilitatus* L. (Dipt. Dolichopodidae). *British Journal of Animal Behaviour* 3: 32–34.

Speight, M.C.D. (1986) Criteria for the selection of insects to be used as bio-indicators in nature conservation research. Pages 485–488 *in*: Velthuis, H.H.W. (ed.), Proceedings of the 3rd European Congress of Entomology, Amsterdam, The Netherlands, 24–29 August 1986, Part 3.

Stackelberg, A.A. (1930) Dolichopodidae. *In*: Lindner, E. (ed.), *Die Fliegen der Palaearktischen Region*, Lieferung 51: 1–64, plates I–II. Nägele & Obermiller, Stuttgart.

Stackelberg, A.A. (1933) Dolichopodidae. *In*: Lindner, E. (ed.), *Die Fliegen der Palaearktischen Region*, Lieferung 71: 65–128, plates III–IV. Nägele & Obermiller, Stuttgart.

Stackelberg, A.A. (1934) Dolichopodidae. *In*: Lindner, E. (ed.), *Die Fliegen der Palaearktischen Region*, Lieferung 82: 129–176. Nägele & Obermiller, Stuttgart.

Stackelberg, A.A. (1941) Dolichopodidae. *In*: Lindner, E. (ed.), *Die Fliegen der Palaearktischen Region*, Lieferung 138: 177–224, plates V–XII. Nägele & Obermiller, Stuttgart.

Stackelberg, A.A. (1971) Dolichopodidae. *In*: Lindner, E. (ed.), *Die Fliegen der Palaearktischen Region*, Lieferung 284: 225–238. Nägele & Obermiller, Stuttgart.

Thiele, H.U. (1977) *Carabid beetles in their environments.* Springer Verlag, Berlin, Heidelberg, New York, 369 pp.

Turin, H. (2000) De Nederlandse Loopkevers : verspreiding en oecologie (Coleoptera: Carabidae). *Nederlandse fauna*, 3. Stichting Uitgeverij van de Koninklijke Nederlandse Natuurhistorische Vereniging, Leiden, 666 pp.

White, O.M. (1976) On the Feeding Habits of Four Species of Adult Dolichopodidae (Diptera). *The Entomologist's Record and Journal of Variation* 88: 94–96.

CHAPTER TWELVE

BIODIVERSITY RESEARCH BASED ON TAXONOMIC REVISIONS – A TALE OF UNREALIZED OPPORTUNITIES

Torsten Dikow[1], Rudolf Meier[2], Gaurav G. Vaidya[2] & Jason G. H. Londt[3]

[1]*Cornell University, Ithaca, USA and
American Museum of Natural History, New York, USA;*
[2]*National University of Singapore, Singapore;*
[3]*Natal Museum, Pietermaritzburg, South Africa and
University of KwaZulu-Natal, Pietermaritzburg, South Africa*

INTRODUCTION

If we were to ask the average biologist or university administrator about his or her opinion on the relative importance of revisionary taxonomy and biodiversity research, the vast majority would consider taxonomy uninteresting or even unnecessary, while many would find biodiversity research interesting and important. After all, we live at a time when we do not know even within an order of magnitude how many species exist on our planet, where they are found, and whether they are threatened by extinction. Yet, here we will argue that these positions are incongruous, because without the help from revisionary taxonomy, biodiversity and conservation research will remain restricted to less than 10% of the known species diversity; i.e., mostly vascular plants, butterflies, mammals, and birds. Some will argue that the remaining 90% can be safely ignored because they are less glamorous and deserve less attention. Glamour may be important for conservation organizations when collecting donations from the public and arguing for the conservation of an area, but when it comes to scientific research in biodiversity and conservation biology it is important to also consider invertebrate species. It also appears from the literature, that biodiversity researchers are not avoiding invertebrate data because they are regarded as unimportant. Instead, such data are generally regarded as unavailable. Here we will demonstrate how specimen lists from taxonomic revisions will provide this much needed access. We will furthermore argue that using these data will not only help biodiversity research. Analyzing specimen lists from revisions also creates new opportunities for taxonomists, who are now living in a scientific environment where they need to make their research more relevant to a larger audience and need to produce more publications with high immediate impact (Wheeler 2004).

The focus of this chapter is not yet another discussion of what biodiversity means, whether the inclusion of invertebrates is desirable, or how biodiversity can be preserved. Instead, we will use several Diptera examples for discussing how invertebrate data can be incorporated into quantitative biodiversity research. We will demonstrate how specimen data from taxonomic revisions can be used (1) to compare the species richness and levels of endemism of two or more areas, (2) to provide quantitative data for Red Lists of endangered species, and (3) to estimate the full species diversity in a clade. We will start by pointing out how much data are already available in the taxonomic literature and end with discussing the problems with using specimen data. Throughout the chapter we will use examples from the Asilidae (Diptera: Brachycera: Asiloidea). Robber flies are a diverse group of predatory insects (some 7,000 described species) that mostly catch other insects on the wing. The largest species diversity is found in arid and semiarid regions all over the world. In contrast to the many invertebrate groups that are only collected by a dedicated group of specialists, asilids are popular with amateur and professional entomologists alike. Many robber flies are large or conspicuous due to their habit of resting on exposed vegetation or on the ground. Furthermore, catching asilids poses a nice challenge gladly taken up by many collectors. Due to the combination of these factors, Asilidae collections are unusually large and collected by a diverse group of entomologists thus creating a more random specimen sample than is available for most other groups of invertebrates.

In discussing the use of specimen data from taxonomic revisions, we will rely on two data sets. One covers a large proportion of the sub-Saharan Asilidae. Londt (1977–2002, 37 publications) and Dikow (2000–2003, 3 publications) have revised and described 724 species of the approximately 1,500 described Afrotropical robber flies and we compiled 21,505 specimen records from these taxonomic revisions. The second data set was compiled for a revision of the Danish Asilidae fauna. This data set is unusual in that it contains 4,300 specimens for a relatively small fauna of only 30 species.

1. Taxonomic Revisions: How Much Information is Available?

The scientific literature contains a vast number of taxonomic revisions and most include specimen data. For example, in a recent search Meier & Dikow (2004) found that the *Zoological Record* listed more than 2,300 taxonomic revisions that were published between 1990–2002 and Gaston (1991) documented that more than 10,000 new species in the four hyper-diverse insect 'orders' were described between 1986–1989. In order to be able to produce distribution or range maps and to analyze the phenol-ogy or seasonality of species, taxonomists routinely collect data from the specimen labels. These labels often include, for example, locality, date/year of collection, collector, altitude, and ecological information. The data are generally available within the revision, sometimes in smaller font, as an appendix, or as an electronic supplement. We will here promote the use of these data and believe that they are preferable over data obtained by the currently more popular approach of digitizing label information from museum specimens:

1. The data from revisions are readily available and do not require specimen label digitization by non-specialists; they are thus more cost-effective and of higher quality because an expert can avoid many data transfer errors.

2. The specimens were identified by the best expert in the field; i.e., the taxonomist who is carrying out the taxonomic revision. He is often the only expert in the world who can correctly identify closely related species. Misidentifications which can be common in museum collections (Meier & Dikow 2004) are thus, as much as possible, avoided.

3. Specimens from many collections contribute label data to the speci-men list in taxonomic revisions thus maximizing specimen coverage for a particular group.

4. Specimens in the 'unsorted' drawers are more likely to be included because specialists tend to sort through or borrow unsorted material from a large range of institutions.

5. Given the large number of published revisions, millions of label data are already available and there is no need to wait for museum digitization projects to be completed.

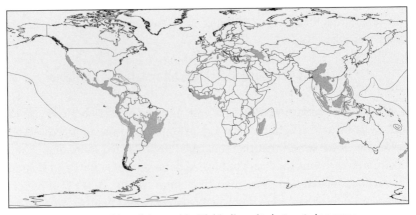

Figure 12.1. Map of the world with biodiversity hotspots in orange.

2. Use of Specimen Data for Comparing the Biodiversity of Conservation Areas

One of the main goals in conservation biology is to optimize the selection of conservation areas given that only very limited financial resources are available for protecting biodiversity. Special attention is usually given to areas with outstanding biodiversity whereby the latter is assessed using a variety of different criteria. In the older literature, raw species counts were often used, which had the undesirable effect that many species with small ranges were not covered by the selected regions. Today, more attention is paid to a variety of other criteria. For example, Myers *et al.* (2000) focused on maximizing the number of endemic species in habitats that have already lost most of their natural vegetation. In 1991, Humphries *et al.* argued for a combined approach including species richness, phylogenetic diversity, complementarity, and taxonomic distinctness, and Vane-Wright *et al.* (1991) proposed a taxic diversity measure based on complementarity analysis of faunas and floras. One year later, Platnick (1992) discussed the comparison of species richness and species composition (overlap) between three allopatric areas and favored an approach that would maximize the preservation of species assemblages. All these techniques have one element in common. They are critically dependent on having species distribution data. This also explains why invertebrates have been largely ignored in these analyses because it is commonly assumed that distribution data are not available.

Diptera Diversity: Status, Challenges and Tools
(eds T. Pape, D. Bickel & R. Meier). © 2009 Koninklijke Brill NV.

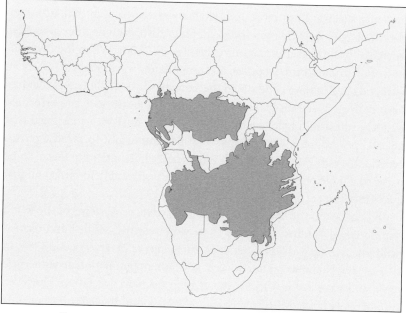

Figure 12.2. Map of sub-Saharan Africa with Wilderness Areas
(Congo Basin and Miombo-Mopane) in orange.

2.1 *Assessing the validity of Myers* et al.*'s biodiversity hotspots for Diptera*

Many biologists have pondered the question where on our planet we find the highest and most threatened animal and plant diversity. Botanists were the first to provide answers because they have long accumulated detailed information on the distribution of vascular plant species published in numerous regional floras. Analyses of these data revealed global diversity centers (Barthlott *et al.* 1996) and biodiversity hotspots (Myers *et al.* 2000; Fig. 12.1). Barthlott *et al.* (1996) evaluated the number of species within a specific area of 10,000 km² and distinguished diversity regions by isotaxas; i.e. lines of equal species richness. Myers *et al.* (2000) went beyond just mapping species richness and also incorporated a conservation angle by considering species endemism and habitat loss. Myers *et al.* designated areas as biodiversity hotspots if they harbored at least 1500 endemic species of vascular plants (0.5% of the described species diversity on our planet) and if 70% of the habitat in the area had already been lost through human interference. Twenty-five biodiversity hotspots were originally defined and today these cover a combined area of only 1.4% of the Earth's land masses. Yet, they contain 44% of all flowering plants as endemic species. Even more remarkable was the finding that, although these hotspots were defined based on plant distributions, they were also hotspots for terrestrial tetrapods (Myers *et al.* 2000). When the distribution of mammals, birds, reptiles, and amphibians were mapped, 35% of all tetrapod species were endemic to the same biodiversity hotspots that had been defined based on plant distributions. The hotspots also performed well in protecting phylogenetic diversity. Some authors had argued that not all species are equal and that it is also important to protect phylogenetic diversity. Fortunately, the biodiversity hotspots were found to be home to a large proportion of the phylogenetic diversity in birds and primates (Sechrest *et al.* 2002).

In contrast to the biodiversity hotspots, some ecosystems of the world remain largely undisturbed and these were recently defined and termed 'wilderness areas' (Mittermeier *et al.* 2003; Fig. 12.2). They occupy 44% of all terrestrial habitats on Earth, have low human population densities (<5/ km²), and have lost less than 30% of their original vegetation. However, despite being large, Mittermeier *et al.* (2003) found that they harbor only 18% of all plant and 10% of all terrestrial vertebrate species as endemics; i.e., they are not very effective in providing a safe haven for a large proportion of vascular plant and terrestrial tetrapod species.

These studies have been very influential in defining conservation priorities and much funding is now channeled into the protection of biodiversity hotspots regardless of the fact that all this research only considers much less than 10% of the global species diversity. The obvious question is whether invertebrate diversity is also concentrated in these hotspots defined based on plants. In order to answer this question, we decided to test the biodiversity hotspots and wilderness areas in sub-Saharan Africa for robber flies based on all available specimen data from taxonomic revisions. We plotted the distribution of the 21,505 specimens representing 724 species of Afrotropical Asilidae on a map of sub-Saharan Africa (Fig. 12.3). Overall, 1,727 unambiguous localities were included (= geographic co-ordinates are known). As is evident from the map, the most comprehensive revisionary research has been conducted on the fauna of the Republic of South Africa, which was the focus of Jason Londt's studies. South Africa is also the home of two biodiversity hotspots *sensu* the original circumscription of the biodiversity hotspots by Myers *et al.* (2000; Fig. 12.3). These are the Succulent Karoo (SK) on the Atlantic west coast stretching north to southwestern Namibia and the Cape Floral Region (CFR)

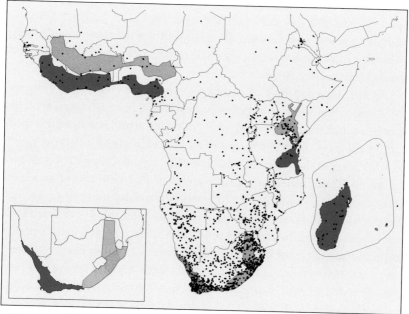

Figure 12.3. Map of sub-Saharan Africa with collecting localities (circles), Biodiversity hotspots in blue, and 'imaginary hotspots' in orange. Inset with detailed map of southern Africa with combined Cape Floral Region and Succulent Karoo biodiversity hotspot and two 'imaginary hotspots' ESA and SEA.

coinciding with the Mediterranean-type climate in southwestern South Africa. Other hotspots tested in our study are the Eastern Arc Mountains in Tanzania and Kenya and the Guinea Forests comprising the tropical rainforest belt along the Atlantic Ocean coast of western Africa (Fig. 12.3). Although Madagascar constitutes yet another sub-Saharan hotspot, it is here excluded because of its island status.

We find that of the 724 robber-fly species in this study, 295 (41%) are resident in at least one of the biodiversity hotspots, which combined occupy 8.5% of the surface area of sub-Saharan Africa. This number is unexpectedly high compared to the corresponding numbers for vascular plants (42.2%) and terrestrial vertebrates (26.4%). Note that the latter two numbers are actually overestimates because in the case of species overlap between several biodiversity hotspots, species are double-counted, which makes the performance of the hotspots for Asilidae even more impressive. However, when we only consider the endemic species among the hotspot residents, we find that only 149 (20.6%) fall into this category. This is a considerably lower proportion of endemics than the hotspot endemicity values for vascular plants (Africa: 42%, global: 44%) and terrestrial verte-brates (Africa: 29%, global: 35%).

At this stage one may be inclined to reject the biodiversity hotspots for Asilidae, but this first impression deceives. In evaluating an area, the absolute numbers may give the wrong impression and it is equally impor-tant to compare the performance of an area earmarked for conservation to other areas of the same size and shape that are not proposed for conser-vation. We thus created 'imaginary hotspots' that were chosen based on three criteria: (1) identical size to a real biodiversity hotspot, (2) similar geographic location, and (3) similar sampling intensity. We rejected ran-dom area selection because it would likely yield areas of incomparable sampling intensity. For the same reason, only 1–2 comparison areas were here identified for each hotspot. For the Eastern Arc and the Guinea For-ests, we tested areas of identical size and shape situated north of the origi-nal hotspot (Fig. 12.3). For the southern African hotspots, we combined the adjacent hotspots Cape Floral Region (CFR) and Succulent Karoo (SK) and created two 'imaginary hotspots' in eastern South Africa (ESA — Eastern South Africa and SEA — Southeastern South Africa; Fig. 12.3). These were of identical size, positioned at similar latitude as the combined original hotspots, and included an area that had been particularly well sampled by the staff of the Natal Museum (Pietermaritzburg, South Af-rica). For each imaginary hotspot, we established the number of collecting events and counted the number of endemic and resident species and gen-era (subtracting species overlap; Table 12.1). For the ESA and SEA imagi-nary hotspots, we used the mean number of resident endemic species for comparison with the combined real hotspot. For the combined CFR and SK hotspots, we used the sum of endemic species for both hotspots, which underestimates the correct number because some species that are not en-demic for either hotspot may be endemic for the combined area.

When using the imaginary hotspots as point of reference, we find that the real hotspots perform extremely well for Asilidae. The real hotspots house 57% more species than are present in the imaginary hotspots (284 vs. 181 species). More remarkably yet, the levels of endemism in the real hotspots are elevated by 424% (140 vs. 33 species) over the level in the comparison areas. Furthermore, the biodiversity hotspots contain three endemic genera while endemism at the generic level is absent in the imaginary hotspots (Table 12.1). In order to rule out that the performance differences are due to unequal sampling, we obtained unbiased values through data-set resampling. The imaginary hotspots had lower numbers of collecting events than the corresponding real hotspots (Eastern Arc, real: 76 vs. imaginary: 62; Guinea Forests, real: 42 vs. imaginary: 25; combined CFR and SK, real: 1,822 vs. imaginary ESA: 984 and SEA: 1,261). To correct for this bias we rarefied the larger data sets from the real hotspots ten times to match the size of the respective smaller data sets and again determined the average number of resident and endemic species based on the resampled data sets (Table 12.1). The number of resident species and endemic species still remained higher in the biodiversity hotspots in com-

Table 12.1. Comparison of resident and endemic species in biodiversity and 'imaginary' hotspots.

	Biodiversity Hotspots	Biodiversity Hotspots	'Imaginary' Hotspots	Resampled Imaginary Hotspots[3]
	Resident/ Endemic species	Resident/ Endemic genera	Resident/ Endemic species/ Resident genera	Resampled resident/ Endemic species
Madagascar	12/9	9/2	n.a.	n.a.
Eastern Arc	26/3	11/0	31/5/17	23±1.7/2.9±0.3
Guinea Forests	16/3	7/0	11/1/5	10±1.4/2.0±0.8
Succulent Karoo (SK)	163/48	29/0	n.a.	n.a.
Cape Floral Region CFR)	168/51	29/1	n.a.	n.a.
Combined CFR & SK	248/134	33/3	ESA[2]: 147/21/34 SEA[2]: 150/32/33	ESA: 191±5/98±5 SEA: 212±7/113±5
Totals	295/114, 149[1]	53/5	n.a.	n.a.
Totals w/o Madagascar	284/105, 140[1]	53/3	181/33	230±12/ 110±8.8

[1] 114 = Sum of CFR & SK, 149 = Combined CFR & SK

[2] Two imaginary hotspots tested for the combined CFR & SK hotspot

[3] Hotspot dataset resampled to dataset size of imaginary hotspots

parison to the imaginary hotspots (resident: 230 in real hotspots vs. 181 in imaginary => +27%; endemic: 111 in real hotspot vs. 33 in imaginary hotspot => +336%; Table 12.1).

The performance difference between the real hotspots and the comparison areas is mainly due to the contribution of the two southern African hotspots (Cape Floral Region and Succulent Karoo) that contain several large radiations of plant lineages. One of the most speciose asilid genera world-wide, *Neolophonotus* Engel, which is confined to the Afrotropical Region, is especially species-rich in these two hotspots and of the 270 described species 70 are endemic in the hotspots.

The wilderness areas proposed by Mittermeier *et al.* (2003) are a different kind of conservation area that can be tested for invertebrates. These areas are generally very large and the two high-diversity wilderness areas in Africa (Congo Basin and Miombo-Mopane) comprise 11.9% of the surface area of sub-Saharan Africa (Fig. 12.2). On the other hand, they have only 7,900 (15.8%) endemic vascular plant species and 170 (3.9%) endemic terrestrial vertebrate species (Brooks *et al.* 2001). Our Asilidae data indicate a relatively high number of species that reside in the Congo Basin and the Miombo-Mopane woodlands (196 species, 27%), but only a few, 60 species (8%), are endemic to them. Although large in size, the wilderness areas are thus also relatively ineffective in protecting the sub-Saharan robber-fly species diversity. The small number of endemic species is especially surprising given that large areas have a higher chance of harboring a large number of endemics (Brooks *et al.* 2002).

Overall, we find that the biodiversity hotspots that were defined based on plant distribution data perform well not only for terrestrial tetrapods, but also for Asilidae. This correlation is all the more surprising because Asilidae are not phytophagous. For phytophagous insects, a correlation would intuitively have been expected, but robber-fly larvae as well as the imagines are predators with little evidence for prey specialization beyond size. It thus appears more likely that, for example, historical reasons (e.g., geology, fragmentation through climate change) account for the simultaneously high levels of endemism in the biodiversity hotspots in vascular plants, robber flies, and terrestrial vertebrates. Our example here demonstrates how data from taxonomic revisions can be used to test whether existing conservation priority areas have any relevance to invertebrates. Note that these tests were carried out based on published data that are freely available in the literature. Note also that the same set of techniques

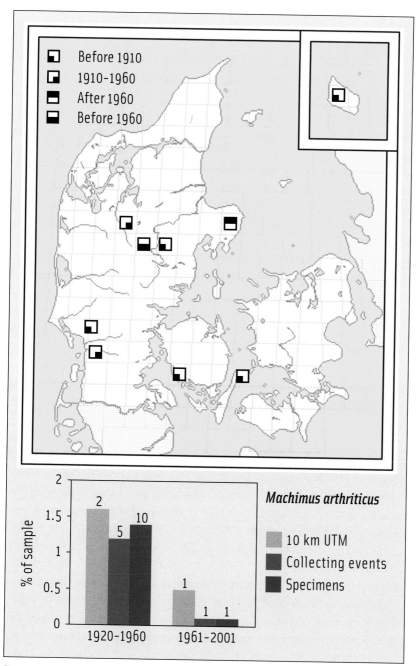

Figure 12.4. Collecting records for *Machimus arthriticus* (Zeller) in Denmark over two time periods (1920–1960; 1961–2001; inset: Bornholm).

can be used for assessing the validity of other conservation areas for invertebrates.

3. Use of Specimen Data for Proposing Red Lists

Red Lists for invertebrates are usually based on the specialists' guesses instead of quantitative data as demanded by the World Conservation Union (IUCN 2001), but see Red Lists for Finland and Sweden (Gärdenfors 2001, Gärdenfors *et al.* 2001). The main perceived problem is the lack of quantitative data for invertebrates, that supposedly prevented biologists from applying the IUCN criteria that require quantitative statements about species abundances, distributions, and/or probabilities of extinction. However, we will argue here that some taxonomic revisions contain enough information. Recently, Larsen & Meier (2004) revised the Danish Asilidae and proposed a Red List based on 4500 specimens for 30 species. The data set was divided into two time periods (1920–1960, 1961–2001) and for each the number of specimens, collecting events, and number of 10 km² UTM grids in which a particular species was found was determined. Changes in abundance were then evaluated after correcting for collecting effort by relating the records for a particular species to the overall collecting activity during the two time periods and the geographic regions that were sampled (for other methods, see Fagan & Kareiva 1997, Ponder *et al.* 2001).

After the correction, we still had to apply the IUCN criteria for regional Red Lists (Gärdenfors 2001, Gärdenfors *et al.* 2001). Of the criteria suggested by IUCN for ranking species, we believe only the 'geographic-range criterion' (criterion B, and rarely category D) can realistically be used for insects. It requires that it is documented that the 'extent of occurrence' or 'area of occupancy' (see IUCN 2001) is smaller than a defined size, whereby different threat categories have different size-thresholds. It is furthermore necessary to document that two of the following three phenomena apply to the species in question: (1) fragmented distribution or existence at few locations, (2) population decline (observed, inferred, or projected), and (3) extreme fluctuations in range or number of populations. The latter criterion probably can not be used for most insect groups, because there are not enough data to document such fluctuations. However, the first two criteria can be applied if enough data are available. We were able to rank all 30 Danish asilid species and found that seven species are now 'critically endangered' (CR), one species is 'endangered' (EN), four species are 'vulnerable' (VU), two species are 'near threatened' (NT), and 16 species are 'least concern' (LC). Below is a typical example for a species assessment:

Machimus arthriticus (Zeller): IUCN Criteria B1+2a,b: extent of occurrence and area of occupancy is probably less than 100 km^2 and 10 km^2 respectively (see Fig. 12.4); a: only one post-1950 locality is known; b: number of populations is in decline; evidence: the species has been collected five times in two 10 km^2 grids between 1920 and 1960. Afterwards, it has only been taken once and there is a decline in the number of UTMs, collecting events, and number of specimens. However, it must be pointed out that collecting in the grids with known occurrence has dramatically decreased after 1960 (1920–1960: 80 events, post-1960: 12). One might thus be inclined to consider the decline a sampling artifact, but between 1900 and 1920 alone the species was known from four additional grids. These grids have 20 additional collecting events without any evidence for *M. arthriticus*. Collecting at the old localities is nevertheless urgently needed to confirm the status of the species.

Proposing Red Lists is increasingly important for invertebrates given that the habitat for many species is quickly vanishing. In order to give any credence to such efforts, it is necessary to base ranking decisions on a quantitative assessment of data. As our example demonstrates, in some cases the data are already available, existing deficiencies can be determined and addressed through analytical techniques or additional collecting. In other cases, a compilation of all data will reveal that there is not sufficient information for an assessment. However, compilation of the existing data will at least reveal where the information gaps are. Some taxonomists may argue that guesses by an expert are sufficient, but ultimately guesses are also based on data, and for those insects where identification requires microscopic study these data will come with label information; i.e., one may as well reveal the quantitative data that support the guesses.

4. Use of Specimen Data for Estimating Clade Species Richness

Many attempts have been made to estimate the number of species on our planet. Vascular plants and vertebrates are comparatively well-known taxonomically and only relatively few additional species are described ev-

DIPTERA DIVERSITY: STATUS, CHALLENGES AND TOOLS
(EDS T. PAPE, D. BICKEL & R. MEIER). © 2009 KONINKLIJKE BRILL NV.

ery year. The same is not true for fungi and algae, and for invertebrates such as insects, crustaceans and nematodes. Here, new field work produces millions of specimens every year with many of these belonging to undescribed species. The estimates for the total number of species on our planet range between 3–80 million species, with the recent proposals favoring from 5–10 million species (Gaston 1991, Ødegaard 2000, Stork 1988). These estimates are based on a variety of techniques. For example, Erwin (1982) and Ødegaard (2000) used the number of known species of plants and estimates of host-specificity for phytophagous insects. Hodkinson & Casson (1991) compared the number of described to the number of undescribed species in a sample of true bugs (Hemiptera) from Sulawesi. As can be seen from these examples, most studies either only concentrate on a single taxon and/or a single sample from field work in tropical forests. The problems for estimating the global species diversity from such a basis are obvious. Estimates based on a few taxa or samples will be inherently unstable because different taxa have very different ecological requirements that result in different species distributions. For example, as mentioned earlier, Asilidae are most speciose in arid and semiarid environments around the world and have therefore their highest species diversity outside of the equatorial belt.

Specimen data from the thousands of taxonomic revisions can help to overcome the taxon bias in previous estimates of the global species richness. Just imagine having estimates of species richness for the thousands of taxa covered in the thousands of taxonomic revisions that have been published in the last decades. These estimates would cover a wide variety of taxa and geographic regions. In order to derive such estimates, we can again make use of the quantitative information inherent in specimen lists. This information can be used in conjunction with statistical species-richness estimation techniques that were first proposed for estimating the number of unknown species in ecological samples (Colwell & Coddington 1994). These methods are designed to estimate the full species diversity in a sample even if only a subset of the species has been collected. But these techniques can also be applied to specimen data from museums and taxonomic revisions (Meier & Dikow 2004, Petersen *et al.* 2003, Soberón *et al.* 2000).

Here, we present two examples for such species-richness estimations based on specimen lists from taxonomic revisions. One is for the predominantly African robber-fly genus *Euscelidia* Westwood (Dikow 2003)

Table 12.2. Summary of species richness estimation of Afrotropical Asilidae.

	Afrotropical Region	Republic of South Africa
Known Species Number	≈ 1500	n.a.
Species from Revisions	710	470
Specimens from Revisions	21058	15467
Singleton Species	110	76
Extrapolated Species Number	997-1094	668-738

and the other for the sub-Saharan robber-fly example used previously. In both cases, we use two non-parametric estimators (ICE – incidence-based estimator and Jack2 – second-order Jackknife, 300 random sample-order runs) as implemented in EstimateS (Colwell 2000). Meier & Dikow (2004) submitted the specimen data from the *Euscelidia* revision to species-richness estimation. The revision recognized 68 species distributed primarily in the Afrotropical Region (55 species) with additional species being found in the Oriental and Palaearctic regions. Overall, some 1,500 specimens had been studied and 14 species are currently only known from the holotype. The non-parametric species-richness estimation techniques indicated that there might still be 36–48 additional, uncollected species (Fig. 12.5a). On the other hand, the fauna of the Republic of South Africa is relatively well-sampled and only 3–5 new species can be expected here (Meier & Dikow 2004). The fauna of the Oriental Region has 11 species, of which 4 are known only from the holotype, and it is therefore the least known fauna and the estimators were unable to suggest an estimate; i.e., too few data points were available for an estimate.

When we apply species-richness estimation techniques to our data set including 710 sub-Saharan robber-fly species with sufficient data for estimation, the results imply that 287–384 species are still waiting to be discovered in sub-Saharan Africa (Fig. 12.5b, Table 12.2). Focusing the attention on the best sampled area, the Republic of South Africa, with 470 species, the number of 198–268 new species is still relatively high (Fig. 12.5c, Table 12.2). This indicates that even after extensive taxonomic work including many targeted field trips conducted by Jason Londt over the past 28 years this highly diverse region is still undersampled. This might not come as a surprise as a single person can hardly be held responsible for revising such a diverse fauna, but it also highlights how far a revisionary effort has come and that more taxonomic work needs to be carried out in order to completely cover the fauna.

But how can we obtain a better idea about the global species richness based on such estimates? Thousands of taxonomic revisions with specimen data can be translated into thousands of point estimates of clade richness for different taxa. Instead of relying on a single taxon for the estimate or relying on only tropical samples, this technique would provide repeatable estimates for many taxa. Given that taxonomic experts tend to revise all species in a taxon and not only the tropical species, we should also include predominantly subtropical and temperate clades in the estimates. Currently, most taxonomic revisions start with discussing the taxonomic history and biology of the revised taxa and then proceed to presenting the main results of the revision. It is here that we believe taxonomists should also estimate the true species richness of the revised clade. Some taxonomists have done so, but for the most part taxonomic revisions lack any quantitative evaluation of the data that have been generated.

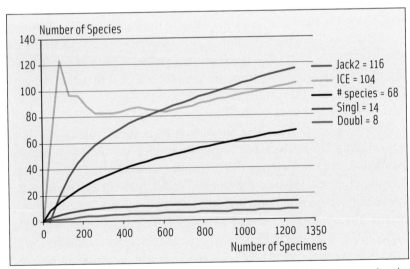

Figure 12.5a. Genus *Euscelidia* Westwood with all species included. Jack2 = second-order Jackknife estimator; ICE = incidence-based estimator; # species = number of observed species; singl = singleton species; doubl = doubleton species.

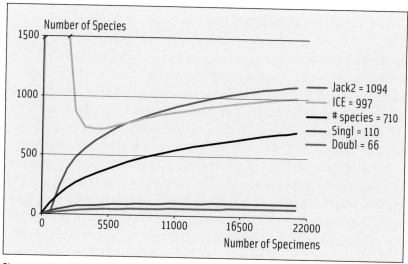

Figure 12.5b. All species of Afrotropical Asilidae data set. Jack2 = second-order Jack-knife estimator; ICE = incidence-based estimator; # species = number of observed species; singl = singleton species; doubl = doubleton species.

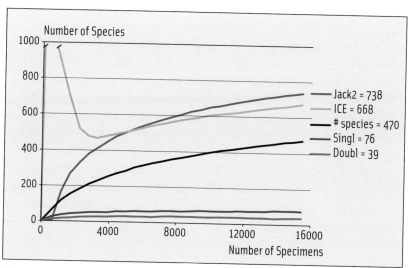

Figure 12.5c. South African species only data set. Jack2 = second-order Jackknife estimator; ICE = incidence-based estimator; # species = number of observed species; singl = singleton species; doubl = doubleton species.

5. Use of Specimen Data: The Numerous Problems

So far we have painted an optimistic picture of how specimen lists from taxonomic revisions can potentially provide the answers to long-standing questions. However, the reality is less favorable, or as Ponder *et al.* (2001) have remarked: 'shortcomings of the data include the ad-hoc nature of the collections, presence-only data, biased sampling, and large collecting gaps in time and space ...'; i.e., specimen lists from taxonomic revisions have many problems and in many cases the data can mostly serve as a baseline for future work only. The most obvious shortcoming is that the specimen samples used in taxonomy are highly non-random in many ways. Taxonomists tend to concentrate on collecting particular species. For example, rare and conspicuous species are over-represented in collections, while very common species are underrepresented. This bias extends to collecting localities, with promising and easily accessible areas being over-collected and 'difficult' and 'uninteresting' areas being essentially unexplored. For example, combining the known collecting localities for many taxa in the Amazon Basin yields a river map for the Amazon and its main tributaries that can be traversed by boat (Heyer *et al.* 1999), and a map with all localities for Papilionidae and Pieridae butterflies in Mexico resembles this country's highway map (Soberón *et al.* 2000). Collectors furthermore do not record if a particular collecting attempt was unsuccessful; i.e., there are no empty pins with label data in museum collections. All these biases are reflected in the specimen lists in taxonomic revisions. The good news is that these lists are more explicit than many other data used in biodiversity research, i.e., some of these biases can be detected and at least partially be removed by techniques such as rarefaction.

Another mixed blessing of revision data is the large proportion of old specimens. Obviously, old specimens are of great importance in documenting that a species has been present in a particular locality and for documenting that a population has continuously occupied a certain area. However, for present-day conservation decisions they are only of limited value.

The last serious problem that we want to mention here is that not all taxonomic revisions are suitable for the kind of quantitative evaluation that we are advocating. There are those that do not include specimen lists because the editors have insisted on the removal of the data from the manuscript. We hope that such cases will become rare and that specimen lists will at least be preserved as supplementary information on journal homepages. Then there are those revisions, for which the specimen lists are no longer available in electronic format or for which the electronic format is unsuitable for databasing. Both are more serious problems than the reader may think. We encountered these problems when preparing our dataset for the sub-Saharan Asilidae. Printed specimen lists only provide the information available on the specimen label so that for data-poor specimens, the number of fields is very different than for data-rich ones. When label data are printed in a running format like it is typical in taxonomic revisions, it becomes near impossible to automatically capture the data. Lastly, many revisions uncover such a large number of new species based on such a small number of specimens that species-richness estimation and species range mapping becomes impossible, e.g., Grimaldi & Nguyen (1999). However, even in these cases we would advocate that taxonomists summarize the numerical data instead of just predicting that there are a large number of undiscovered species in the group.

6. Summary and Conclusions

The specimen information in taxonomic revisions is an outstanding source of information for incorporating hyper-diverse taxa into quantitative biodiversity research. This data source is rich given that more than 2,300 taxonomic revisions have been published between 1990–2002 (Meier & Dikow 2004) and thousands of species descriptions with specimen data have been made available in taxonomic journals.

Here we have described how these data can be used for testing the plant-based biodiversity hotspots with Diptera data. This approach holds considerable promise because it allows for the inclusion of hyper-diverse invertebrate taxa into conservation biology studies. These taxa are otherwise rarely considered. We show that the predatory Asilidae are very speciose in the southern African biodiversity hotspots and thus corroborate the hotspots originally described based on vascular plants. Regardless of the explanation for the observed concordance between Asilidae and plants, our data support initiatives channeling conservation funding into the African biodiversity hotspots (Myers 2003, Reid 1998) and contradicts predictions that restricted-range taxa like invertebrates will be poorly represented by areas selected based on standard indicator taxa (Moore *et al.* 2003, van Jaarsveld *et al.* 1998).

The use of explicit specimen data for compiling Red Lists of endangered species promises to become another important application of specimen data from taxonomic revisions. The case of the Danish Asilidae highlights the suitability of this approach for well-sampled regions of the world and corroborates the notion that human development and high population density may increase extinction risks, even for historically common species. The data for the Danish Red List were compiled from a revision of the Danish fauna and the specimens came from two natural history collections. However, instead of just digitizing the label information, all specimens were re-identified in order to reduce misidentification problems.

Clade species-richness estimation based on specimen data is also a method that can easily be applied to many invertebrate taxa and therefore provides a much better estimate of the total species diversity. Adding species richness estimates of a variety of taxa from different biogeographical regions can then be used to extrapolate the number of species roaming on Earth. The result would be a much more accurate estimate for global species diversity because it would be based on multiple, independent data sources and analyzed in a standardized way. The estimation techniques can also direct future collecting efforts to undersampled areas for which most species are only known from a few specimens (Meier & Dikow 2004).

Specimen data from taxonomic revisions also have numerous shortcomings, however. Overcoming these will be important. Fortunately, specimen data are explicit enough that they can be evaluated for collecting biases that can then be addressed by a variety of techniques including targeted new collecting, the use of comparison areas, species-richness estimation, data-set resampling, and the modeling of species distributions (Colwell & Coddington 1994, Meier & Dikow 2004, Williams *et al.* 2002). We would nevertheless like to point out that not all published revisions are suitable for the approaches outlined above. In particular, more standardized formats for all specimen data would improve usefulness of these data.

The three approaches outlined here may help the beleaguered field of revisionary taxonomy (Godfray 2002, Wheeler 2004) by providing a high-impact use for specimen data generated during relatively low-impact taxonomic revisions. Given the potential of the data, conservation biologists and systematists should start collecting specimen data from revisions into a 'specimen bank' of similar design and accessibility as GenBank. Such a database would quickly evolve into the most comprehensive and reliable global source of data on species distributions for a wide variety of taxa (Godfray 2002).

Acknowledgements

We would like to acknowledge financial support from the European Commission's 'Improving Human Potential' program (Transnational Access to major Research Infrastructures) that supported this research through a grant from the Copenhagen Biosystematics Centre (COBICE) to TD. TD would also like to acknowledge the support from Cornell University and the American Museum of Natural History in form of a graduate student fellowship. Further funding supporting the research came from the Danish Natural Science Research Council (grant nos. 9502155 and 9801904), and FRC grants financed by the Ministry of Education of Singapore (R-377-000-025-112, R-377-000-024-101). We also acknowledge the help from numerous colleagues, who have over the years provided help with carrying out quantitative diversity research on Diptera.

References

Barthlott, W., Lauer, W. & Placke, A. (1996) Global distribution of species diversity in vascular plants: towards a world map of phytodiversity. *Erdkunde* 50: 317–327.

Brooks, T., Balmford, A., Burgess, N., Fjeldså, J., Hansen, L.A., Moore, J., Rahbek, C. & Williams, P. (2001) Toward a blueprint for conservation in Africa. *BioScience* 51: 613–624.

Brooks, T.M., Mittermeier, R.A., Mittermeier, C.G., da Fonseca, G.A.B., Rylands, A.B., Konstant, W.R., Flick, P., Pilgrim, J., Oldfield, S. & Magin, A. (2002) Habitat loss and extinction in the hotspots of biodiversity. *Conservation Biology* 16: 909–923.

Colwell, R.K. (2000) EstimateS: *Statistical estimation of species richness and shared species from samples*, Version 5. User's Guide and application. Available at http://viceroy.eeb.uconn.edu/estimates.

Colwell, R.K. & Coddington, J.A. (1994) Estimating terrestrial biodiversity through extrapolation. *Philosophical Transactions of the Royal Society of London B Biological Sciences* 345: 101–118.

Dikow, T. (2003) Revision of the genus *Euscelidia* Westwood, 1850 (Diptera: Asilidae: Leptogastrinae). *African Invertebrates* 44: 1–131.

Erwin, T.L. (1982) Tropical forests: their richness in Coleoptera and other arthropod species. *Coleopterists Bulletin* 36: 74–75.

Fagan, W.F. Kareiva, P.M (1997) Using compiled species lists to make biodiversity comparisons among regions: a test case using Oregon butterflies. *Biological Conservation* 80: 249–259.

Gärdenfors, U. (2001) Classifying threatened species at national versus global levels. *Trends in Ecology and Evolution* 16: 511–516.

Gärdenfors, U., Hilton-Taylor, C., Mace, G.M. & Rodríguez, J.P. (2001) The application of IUCN Red List criteria at regional levels. *Conservation Biology* 15: 1206–1212.

Gaston, K.J (1991) The Magnitude of Global Insect Species Richness. *Conservation Biology* 5: 283–296.

Godfray, H.C. J. (2002) Challenges for taxonomy. *Nature* 417: 17–19.

Grimaldi, D. & Nguyen, T (1999) Monograph on the spittlebug flies, genus *Cladochaeta* (Diptera: Drosophilidae: Cladochaetini). *Bulletin of the American Museum of Natural History* 241: 1–326.

Heyer, W.R., Coddington, J., Kress, W.J., Acevedo, P., Cole, D., Erwin, T.L., Terry, J.B., Meggers, J., Pogue, M.G., Thorington, R.W., Vari, R.P., Weitzman, M.J. & Weitzman, S.H. (1999) Amazonian biotic data and conservation decisions. *Ciência e Cultura São Paulo* 51: 372–385.

Hodkinson, I.D. & Casson, D. (1991) A lesser predilection for bugs: Hemiptera (Insecta) diversity in tropical rain forests. *Biological Journal of the Linnean Society* 43: 101–109.

Humphries, C.J., Vane-Wright, R.I. & Williams, P.H. (1991) Biodiversity Reserves: Setting new priorities for the conservation of wildlife. *Parks* 2: 34–38.

IUCN – The World Conservation Union (2001) IUCN Red List Categories and Criteria, Version 3.1. IUCN Publications Services, Cambridge, 30 pp.

Larsen, N.M. & Meier, R. (2004) The species diversity, distribution, and conservation status of the Asilidae (Diptera: Insecta) in Denmark. *Steenstrupia* 28: 177–241.

Meier, R. & Dikow, T. (2004) Significance of specimen databases from taxonomic revisions for estimating and mapping the global species diversity of invertebrates and repatriating reliable and complete specimen data. *Conservation Biology* 18: 478–488.

Mittermeier, R.A., Mittermeier, C.G., Brooks, T.M., Pilgrim, J.D., Konstant, W.R., da Fonseca, G.A.B. & Kormos, C. (2003) Wilderness and biodiversity conservation. *Proceedings of the National Academy of Sciences of the United States of America* 100: 10309–10313.

Moore, J.L., Balmford, A., Brooks, T., Burgess, N.D., Hansen, L.A., Rahbek, C. & Williams, P.H. (2003) Performance of sub-Saharan vertebrates as indicator groups for identifying priority areas for conservation. *Conservation Biology* 17: 207–218.

Myers, N. (2003) Biodiversity hotspots revisited. *BioScience* 53: 916–917.

Myers, N., Mittermeier, R.A., Mittermeier, C.G., da Fonseca, G.A.B. & Kent, J. (2000) Biodiversity hotspots for conservation priorities. *Nature* 403: 853–858.

Ødegaard, F. (2000) How many species of arthropods? Erwin's estimate revised. *Biological Journal of the Linnean Society* 71: 583–597.

Petersen, F.T., Meier, R. & Larsen, M.N. (2003) Testing species richness estimation methods using museum label data on the Danish Asilidae. *Biodiversity and Conservation* 12: 687–701.

Platnick, N.I. (1992) Patterns of Biodiversity. Pages 15–24 *in*: Eldredge, N. (ed.), *Systematics, Ecology, and the Biodiversity Crisis*. Columbia University Press, New York.

Ponder, W.F., Carter, G.A., Flemons, P. & Chapman, R.R. (2001) Evaluation of museum collection data for use in biodiversity assessment. *Conservation Biology* 15: 648–657.

Reid, W.V. (1998) Biodiversity hotspots. *Trends in Ecology and Evolution* 13: 275–280.

Sechrest, W., Brooks, T.M., da Fonseca, G.A.B., Konstant, W.R., Mittermeier, R.A., Purvis, A., Rylands, A.B. & Gittleman, J.L. (2002) Hotspots and the conservation of evolutionary history. *Proceedings of the National Academy of Sciences of the United States of America* 99: 2067–2071.

Soberón, J.M., Llorente, J.B. & Oñate, L. (2000) The use of specimen-label databases for conservation purposes: An example using Mexican Papilionid and Pierid butterflies. *Biodiversity and Conservation* 9: 1441–1466.

Stork, N.E. (1988) Insect diversity: facts, fiction and speculation. *Biological Journal of the Linnean Society* 35: 321–337.

van Jaarsveld, A.S., Freitag, S., Chown, S.L., Muller, C., Koch, S., Hull, H., Bellamy, C., Kruger, M., Endrödy-Younga, S., M.W. Mansell & Scholz, C.H. (1998) Biodiversity assessment and conservation strategies. *Science* 279: 2106–2108.

Vane-Wright, R.I., Humphries, C.J. & Williams, P.H. (1991) What to Protect Systematics and the Agony of Choice. *Biological Conservation* 55: 235–254.

Wheeler, Q.D. (2004) Taxonomic triage and the poverty of phylogeny. *Philosophical Transactions of the Royal Society of London B Biological Sciences* 359: 571–583.

Williams, P.H., Margules, C.R. & D.W. Hilbert. (2002) Data requirements and data sources for biodiversity priority area selection. *Journal of Biosciences* 27: 327–338.

SECTION III:

BIOINFORMATICS AND DIPTERAN DIVERSITY

DNA BARCODING AND DNA TAXONOMY IN DIPTERA: AN ASSESSMENT BASED ON 4,261 COI SEQUENCES FOR 1,001 SPECIES

RUDOLF MEIER[1] & GUANYANG ZHANG[2]

[1]*National University of Singapore, Singapore;*
[2]*University of California Riverside, Riverside, California, USA*

INTRODUCTION

Few problems in the biological sciences are older than the crisis in taxonomy. The dipterist Willi Hennig already bemoaned the status of affairs in the 1930s in a high school essay that he wrote at the age of 18 (Schlee 1978). He diagnosed that the main problem was the lack of perceived and/ or actual scientific rigour in the field. Fortunately, he was able to help reversing the trend with regard to higher-level systematics (Meier 2005) by introducing scientific rigor through the redefinition of 'phylogenetic relationship' and 'monophyly' and by insisting that only apomorphic character states can demonstrate monophyly. His work started the revival of phylogenetics, which is now a very productive and highly regarded field (Craft *et al.* 2002). The same has not happened to taxonomy (Wheeler 2004). It is generally acknowledged that taxonomy is important and that our declining skill base for identifying and describing new species is a major problem for science and society (Godfray 2002). However, few solutions are in sight given the scale of the problem and the manpower costs for solving it using the traditional tools.

It thus comes as no surprise that some scientists have suggested DNA-sequence based solutions (Hebert *et al.* 2003a, Tautz *et al.* 2003). The appeal is obvious. The sequencing of standard genes for fresh tissues can be automated and does not require the taxon-specific knowledge of highly trained experts with a PhD degree. However, the proper use of DNA sequences in taxonomy is currently vigorously debated (DeSalle *et al.* 2005, Ebach & Holdrege 2005, Meier *et al.* 2006, Meyer & Paulay 2005, Moritz & Cicero 2004, Prendini 2005, Rubinoff 2006, Sperling 2003, Will *et al.* 2005, Will & Rubinoff 2004). The proposals range from 'DNA barcoding' — here, sequences are only used to identify specimens — to 'DNA taxonomy', where the sequences are also used for species discovery and delimitation (Meier *et al.* 2006, Vogler & Monaghan 2007). Although these are in theory very different proposals (Meier *et al.* 2006, Vogler & Monaghan 2007), the differences are now becoming less distinct with DNA barcoding increasingly being involved also in species discovery (e.g., Hajibabaei *et al.* 2006a, Hebert *et al.* 2004a, Smith *et al.* 2006). In this chapter, we will investigate to what extent DNA sequences may be able to deliver more automated and accurate species identification, discovery, and description. We will not extensively discuss whether DNA sequences are a useful tool when combined with morphological evidence. Several decades of research on 'integrative taxonomy' have left little doubt that taxonomy can benefit tremendously from using a multitude of data sources including DNA sequences (Will *et al.* 2005).

1. History: DNA Sequences in Diptera

The use of DNA sequences for taxonomic purposes in Diptera predates the formal proposal of DNA barcoding and DNA taxonomy by more than one decade. Indeed, for several reasons dipterists were among the first systematists to extensively use DNA sequences for species identification and delimitation. Firstly, Diptera contain a large number of economically important taxa (e.g., hematophagous, phytophagous, and forensically important species) and some are notoriously difficult to identify using traditional methods. For these taxa, DNA sequences were used as an alternative to morphological tools as soon as DNA sequencing became widely available and affordable. Secondly, extensive genetic research on *Drosophila* Fallén and *Anopheles* Meigen made it technically comparatively easy to design primers and obtain sequences for Diptera. Given these factors, it does not surprise that there is an extensive literature on the use of DNA sequences in Diptera taxonomy and for species identification. Particularly extensive was their use in Culicidae in general (Beebe *et al.* 2002, Cockburn 1990, Crabtree *et al.* 1995) and within two genera in particular: *Aedes* Meigen (Cook *et al.* 2005, Patsoula *et al.* 2006), and *Anopheles* (some pre-barcoding publications: Carew *et al.* 2003, Cornel *et al.* 1996, Gunasekera *et al.* 1995, Hill *et al.* 1991, Krzywinski & Besansky 2003, Linton *et al.* 2001, Linton *et al.* 2003, Lounibos *et al.* 1998, Malafronte *et al.* 1997, Malafronte *et al.* 1998, Malafronte *et al.* 1999, Marrelli *et al.* 1999, Marrelli *et al.* 1998, Porter & Collins 1991, Sawabe *et al.* 2003, Sedaghat *et al.* 2003, Torres *et al.* 2000, Xu & Qu 1997). Particularly remarkable is that these techniques were applied very early and to taxonomic problems from different continents. Not surprisingly, similar techniques were more or less simultaneously also developed for other hematophagous nematocerans such as the Ceratopogonidae (Cetre-Sossah *et al.* 2004, Li *et al.* 2003, Pages & Monteys 2005, Ritchie *et al.* 2004), Simuliidae (Ballard 1994, Higazi *et al.* 2000, Tang *et al.* 1996, Tang *et al.* 1995), and Psychodidae (Depaquit *et al.* 2002, Depaquit *et al.* 2004, Ready *et al.* 1997, Testa *et al.* 2002, Xiong & Kocher 1991). Other economically important taxa followed. These included the economically important phytophagous Tephritidae (Basso *et al.* 2003, Douglas & Haymer 2001, Muraji & Nakahara 2002, Nakahara *et al.* 2002, Yu *et al.* 2005, Yu *et al.* 2004) and Agromyzidae (Scheffer *et al.* 2006, Scheffer *et al.* 2001), calyptrates with forensic or veterinary importance (Cai *et al.* 2005, Chen *et al.* 2004, Gleeson & Sarre 1997, Harvey *et al.* 2003, Litjens *et al.* 2001, Noel *et al.* 2004, Otranto & Traversa 2004, Otranto *et al.* 2004, Otranto *et al.* 2005, Saigusa *et al.* 2005, Schroeder *et al.* 2003, Sperling *et al.* 1994, Tenoria *et al.* 2003, Thyssen *et al.* 2005, Wallman & Donnellan 2001, Wells & Williams 2007, Zehner *et al.* 2004), and taxa that are important as freshwater quality bioindicators (Chironomidae: Carew *et al.* 2003, Carew *et al.* 2005, Sharley *et al.* 2004). However, the use of DNA sequences for solving taxonomic problems in dipterans with little economic significance remained rare (but see Milankov *et al.* 2005, Perez-Banon *et al.* 2003, Petersen *et al.* 2007). For these taxa most sequences in GenBank were obtained for phylogenetic and phylogeographic studies.

A wide variety of molecular markers has been used in Diptera, but the best taxon coverage is available for the mitochondrial gene 'cytochrome c oxidase subunit I (COI)', although it is the internal transcribed spacers ('ITS1 & ITS2') that dominate the culicid literature. In writing this chapter, we had the choice to either review the extensive literature on the use of DNA sequences in Diptera taxonomy or to take a fresh look at the evidence by reanalyzing in a consistent manner all available data for one gene. We opted for the latter because much of the literature is so scattered and consists of so many relatively short papers that it is difficult to summarize. A literature review would have been little more than a count of how many times DNA sequences were successfully or unsuccessfully used for taxonomic purposes. We believe that the more interesting question is whether information from DNA sequences alone is sufficient for identifying described species ('DNA barcoding') or delimiting and describing species ('DNA taxonomy').

We thus opted for a comprehensive reanalysis of an extensive data set of COI sequences from Diptera. This data set comprises 4,261 COI sequences from GenBank and covers 1,001 species of Diptera of which 334 are represented by more than one sequence (for similar study on a smaller scale, see Meier *et al.* 2006). We chose COI, because it is the standard gene for DNA barcoding and because it is the gene with the widest taxon coverage in Diptera. One additional advantage of using COI is that the gene codes for a protein and lacks introns. This allowed us to align the sequences based on amino-acid translations using Alignmenthelper (McClellan & Woolley 2004) in conjunction with ClustalW (Thompson *et al.* 1994). The amino-acid alignment was very conserved and lacked indels in all but a few species of Cecidomyiidae and one species of Culicidae, and we thus believe that the DNA sequence alignment is also reliable. In the rest of the chapter, we will only discuss results from sequence comparisons that were based on a minimum overlap of 400 bp.

Before discussing the results of our analysis, we need to briefly address whether it is appropriate to use GenBank data given that some sequences in the database are likely misidentified (Harris 2003, Vilgalys 2003) and these sequences with mistaken identity could potentially lead to an underestimation of the potential for using DNA sequences in taxonomy. However, it is important to remember that the future DNA barcode database will be similar to GenBank in that it is a community effort. Many different researchers will submit sequences and quality control will be difficult. In recognizing this potential problem, the Consortium for the Barcode of Life has recently set rigorous data standards for which sequences can be labeled a 'barcode' in GenBank. These standards are an important step towards proper vouchering and sequence documentation. They will allow for tracing a particular sequence to a voucher specimen including information on collector, identifier, collecting locality, etc. However, these standards can only assure the tractability of a sequence, but do not necessarily improve the quality of the identifications. The latter also remains a significant problem for DNA barcoding initiatives given that museum specimens are targeted as an important source of identified tissues (Hajibabaei *et al.* 2005, Hajibabaei *et al.* 2006c), and a large proportion of museum specimens can be misidentified (Meier & Dikow 2004). For these reasons we believe that GenBank sequences are better suited for testing DNA barcoding than barcode databases. GenBank sequences were submitted by researchers who had a scientific interest in the taxa for which the sequences were submitted. One would think that given this interest, the identification standards are overall high. The other advantage of the existing data in GenBank is taxon coverage. GenBank contains sequences for many species from different continents. Thus, widespread species are more likely to be sampled across their geographic range than in standard DNA barcoding studies that currently still have a narrow taxonomic and geographic scope.

2. DNA Barcoding

The term 'DNA barcoding' was coined by Hebert *et al.* (2003a), who pointed out that the decline in traditional taxonomy is rendering it increasingly difficult to obtain species-level identifications for most taxa and that this problem would worsen in the future because many taxonomists are nearing the end of their career and few young taxonomists are hired. It was concluded that the best solution would be finding new and more cost-effective ways to identify specimens and possibly discover new species (e.g., Hajibabaei *et al.* 2006a, Hebert *et al.* 2004a, Smith *et al.* 2006). Previous research, much of it on Diptera, had shown that many species can be identified based on DNA sequences, and Hebert *et al.* (2003) carried out a pilot study to test whether DNA sequences may be a more general solution to the identification problem. This pilot study was limited in scope, but it was used to support the somewhat extravagant claim that DNA sequences can identify almost all specimens to species, genus, family, and order (Hebert *et al.* 2003a). The extravagancy of the claim proved to be a success with those funding agencies that have a taste for ambitious projects. It was less popular with systematists, whose reactions were more mixed because some of Hebert *et al.*'s (2003a) claims were controversial and subsequent studies revealed that species identification based on DNA sequences is not as straightforward as had been initially suggested (Meier *et al.* 2006, Moritz & Cicero 2004, Rubinoff 2006, Seberg 2004, Sperling

2003, Will *et al.* 2005, Will & Rubinoff 2004). Some of these problems are discussed below.

2.1 The barcoding gap

DNA barcoding aspires to provide a new identification tool for unidentified specimens. It is proposed that for each specimen in need of identification, a ca. 650 bp fragment of COI be sequenced. This sequence is then compared to the sequences in a sequence database consisting of COI sequences of known species identity (barcode database). The process is analogous to a taxonomist trying to identify a specimen by comparing it to a reference collection consisting of identified specimens. Initially he may compare the unidentified specimen (e.g., a *Drosophila melanogaster* Meigen) to all identified specimens in the reference collection, but ultimately the identification problem pairs down to deciding whether the specimen belongs to one of a few, morphologically very similar species (e.g., in the *melanogaster* group of the subgenus *Sophophora* Sturtevant). Determining an unidentified specimen to species is straightforward if the intraspecific variability is small — i.e., the unidentified specimen is a good match to a referenced species — and if the differences between the best-matching species and the next best match is large — i.e., the specimen is a good match to only one referenced species. Translating this procedure into DNA sequence data, this means that it is desirable to have a large 'barcoding gap' (Meyer & Paulay 2005) between the intraspecific and interspecific sequence variability. In the barcoding literature, the interspecific variability is often quantified as the mean interspecific distance between congeneric species (Armstrong & Ball 2005, Ball *et al.* 2005, Barrett & Hebert 2005, Clare *et al.* 2007, Cywinska *et al.* 2006, Hajibabaei *et al.* 2006a, Hajibabaei *et al.* 2006b, Hebert *et al.* 2003a, Hebert *et al.* 2004a, Hebert *et al.* 2003b, Hebert *et al.* 2004b, Hogg & Hebert 2004, Lefebure *et al.* 2006, Lorenz *et al.* 2005, Powers 2004, Saunders 2005, Seifert *et al.* 2007, Smith *et al.* 2005, Smith *et al.* 2006, Ward *et al.* 2005, Zehner *et al.* 2004). However, this is inappropriate as our analogy reveals. For identifying *Drosophila melanogaster* it is irrelevant whether the species differs from the 'average' species in the genus *Drosophila*. Instead, it matters whether *D. melanogaster* is sufficiently distinct from its most similar relative (Cognato 2006, Meier *et al.* 2006, Meyer & Paulay 2005, Moritz & Cicero 2004, Roe & Sperling 2007, Sperling 2003, Vences *et al.* 2005a, Vences *et al.* 2005b). Thus, in studying the barcoding gap, it is necessary

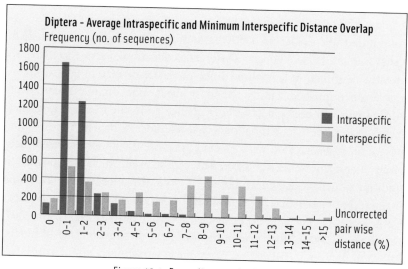

Figure 13.1. Barcoding gap for Diptera.

to quantify the smallest instead of the mean interspecific distances (Meier et al. 2008).

Do we find a barcoding gap between intraspecific and smallest interspecific variability in our Diptera data set? The answer is 'no'. Instead, intraspecific and interspecific variability overlap widely even when the extreme 5% of all intra- and interspecific values are ignored (Fig. 13.1; Meier *et al.* 2006). The overlap ranges from 0.1–7.6% and 25% of all observations for pairwise distances between congeneric sequences fall into this range; i.e., an observed distance of 1% between two sequences has a fair chance of being either intraspecific or interspecific. This ambiguity is problematic for those who want to identify an unidentified specimen based on its COI sequence. An observed distance of 1% between the unidentified sequence and an identified sequence in a barcode database could either indicate that the specimen is con- or allospecific with its best match. Such ambiguity is not rare. For example, 125 or 14.5% of all species in our data set have allospecific, congeneric matches that are 1% or below. Yet, BOLD, 'The Barcode of Life Data System' will identify a sequence to species if the sequence has a match within 1% of an identified DNA barcode. BOLD will then also report this identification with a confidence of 100% (Ratnasingham & Hebert 2007).

2.2 Identification success in Diptera

As pointed out earlier, in DNA barcoding a specimen is identified by matching its COI sequence to a reference collection of sequences that are already identified. At least three different classes of 'matching techniques' can be distinguished: Firstly, the most intuitive technique utilizes overall sequence similarity. Here, the sequence from the query specimen is matched to its best-matching identified reference sequence. For this purpose either uncorrected or corrected pairwise distances are used. Secondly, diagnostic character differences can be used which is analogous to traditional identification techniques. In DNA sequences they could come in the form of particular bases in specific positions or in the form of short DNA sequence motifs that are species-specific (Little & Stevenson 2007; DeSalle et al. 2005). Thirdly, identifications can be based on tree-like identification diagrams. This latter technique is particularly popular in DNA barcoding, where pairwise distances — usually K2P distances — are used to calculate neighbour-joining trees (Hebert et al. 2003a). The position of a query sequence from an unidentified specimen on the NJ tree is then used to infer the species identity of the sequence/specimen. Below, we will evaluate the identification success for Diptera using these three techniques.

2.2.a Species identification using overall similarity

Species identification based on pairwise distances is implemented in a variety of software packages that either work with prealigned or unaligned data (Meier *et al.* 2006, Steinke *et al.* 2005). We are here using software for prealigned data (TaxonDNA: Meier *et al.* 2006). It matches a query sequence by finding those sequences in a barcode database that have the best match based on the smallest pairwise distance. But finding these is only the first step in the identification process. In the next step it is necessary to decide whether the query sequence and its best-matching barcode are conspecific. If so, the query should be assigned the same name as the best-matching barcode; otherwise, the query should remain unidentified. TaxonDNA uses 'best close match' for making this decision. All intraspecific distances in the barcode database are calculated and TaxonDNA determines the distance below which 95% of all intraspecific observations fall. The rationale is that if a query has a match within this interval, then it is likely conspecific.

We can use this technique to establish the identification success for our COI database with 4,261 sequences for 1,001 species. For each sequence, we can pretend that its species identity is unknown. We can then re-identify it using 'best close match'. For each of the 4,261 sequences the possible outcomes are identification success (query is assigned the correct name), no identification (query has no match within the threshold distance), ambiguous (best matching sequences are from multiple species), or misidentified (best matching sequence is allospecific). We find that for our dataset the identification success is relatively low (74.2%). Many queries remain unidentified (7.0%), while the number of ambiguous identifications is moderate (6.2%). More worrying is the relatively large number of misidentified sequences (12.7%). The relatively low identification success is largely due to what we will call 'singleton species'. These are species for which the data set only includes one sequence. Such singleton species cannot be correctly identified once their sequence becomes a query, because it then lacks a conspecific match in the barcode database.

When we only consider the results for sequences from species with at least two sequences in the dataset (3,594 sequences), the identification success is much higher (89.3%), the proportion of unidentified species shrinks considerably (0.2%). The proportion of ambiguous sequences remains largely unaffected (6.7%), while misidentifications become rare (3.9%); i.e., identification success rates of approximately 90% can be achieved with DNA barcodes as long as one does not try to identify sequences from species that lack barcodes in the reference database. These results raise three main questions: First, given that complete barcode databases are important for achieving a high identification success, can these be obtained? Second, is an identification success of 90% sufficient to justify the expense of building a barcoding system? Third, why is the identification success as high as 90% if there is little evidence for a barcoding gap in the data?

With regard to obtaining complete barcode databases, we note that the first COI sequence for Diptera was submitted to GenBank in 1996. After more than 10 years of collecting, COI sequences are now known for approximately 1,000 species. We know from a recently completed catalogue of all described Diptera species, that approximately 153,000 species of Diptera have been named (Evenhuis *et al.* 2008); i.e., 11 years of sequencing yielded barcodes for less than 1% of all described species. Given that there are at least 850,000 additional undescribed species of Diptera, one has to question whether a Diptera barcode database with decent species coverage can be obtained within a reasonable amount of time. The main problem is rare species. Museums are unlikely to contain suitably preserved DNA for these species; i.e., they will have to be collected again and it is difficult to see how this could be accomplished given that collecting is expensive, needs expert manpower, and the subsequent specimen sorting requires so much funding that virtually every major natural history museum has millions of tropical insect specimens that are not even sorted to order. Furthermore, all undescribed species will have to be described before they can be barcoded. Yet, the lack of taxonomists makes it very unlikely that these descriptions will be forthcoming. Overall, it is thus difficult to see how one can obtain the kind of complete barcode coverage that is needed for high identification success. Reason for even more concern about the feasibility of the project arises when the cost of specimen acquisition of correctly identified specimens is considered. Currently, mostly the molecular cost is discussed when projecting a budget for DNA barcoding. However, the main cost will be specimen acquisition and curation (see below).

But is a 90% success rate sufficient to support a DNA barcoding initiative? We would argue that this depends on the uses of the data. For purposes such as forensics or revisionary taxonomy, 90% is not good enough. For example, we doubt that a taxonomic expert who misidentifies 10% of all species/specimens in his taxon would be regarded as being successful, and 90% is also not good enough for using DNA sequences as a 'triage' tool as has been suggested in the barcoding literature (Schindel & Miller 2005). However, 90% is good enough for many other purposes where it is not necessary to identify a specimen to species. For example, for a quarantine officer it may not matter what species of fruit fly he has discovered in a shipment, and a customs officer may only have to identify a biological tissue to genus before he can determine whether it is protected under CITES. Numerous additional examples exist where approximate identifications are sufficient and here DNA barcodes can be very useful. Usually, these cases involve the application of barcodes to an identification problem where the species diversity is limited and the barcode coverage can be nearly complete. Incidentally, these are the uses of molecular markers for species identification that were first championed in the literature (see references above).

Our dataset suggests that even without evidence for a barcoding gap, species identification success can be relatively high. Some readers may be surprised about this combination of no barcoding gap and decent identification success rates. However, in order to obtain high identification success, conspecific sequences only have to be more similar to each other than they are to allospecific sequences. The absolute distances do not matter and there is also no need for a consistent amount of intraspecific and interspecific variability across genera.

2.2.b Species identification using diagnostic characters

As mentioned earlier, an alternative approach to species identification based on DNA sequences would involve the search for species-specific differences in the sequences (Little & Stevenson 2007). This approach differs from those based on pairwise distances in that they take into account where the sequences differ between species. After all, a 1%-difference between two pairs of sequences may involve very different segments of the sequences. One way to investigate whether different species of Diptera have diagnostic differences is through the study of their consensus sequences. Hitherto we had been discussing DNA barcodes for individuals instead of DNA barcodes for species. Genuine species barcodes can be obtained by generating strict consensus sequences for all sequences from individuals of the same species. Clear diagnostic differences between two species are lacking when their consensus sequences are identical (pairwise distance = 0). We used TaxonDNA to investigate how many of the 1,001 species in our data set share strict consensus sequences. We found that this was the case for 70 species (7%). However, the incidence among the species with multiple sequences was much higher (59 of 334: 17.7%); i.e., as sampling improves more species loose their 'unique' barcodes. We thus believe that COI is of relatively low diagnostic value when it comes to distinguishing closely related species.

Maybe we should not be surprised that COI's prospects as a diagnostic tool are not very good. Taxonomists have long known that it can be very difficult to find diagnostic characters that distinguish all closely related species. Therefore, taxonomists use characters coming from a wide variety of character systems. In particular in insects and Diptera, it is generally sexually dimorphic characters that distinguish closely related species and there is general agreement that these are under sexual selection (Arnqvist 1998, Eberhard 1985). However, COI is not subjected to sexual selection and most changes in COI are apparently due to neutral evolution since they occur predominantly in third positions and do not affect amino acid assignments (R. Meier, pers. obs.). We would thus expect differences between species to evolve at a relatively steady rate. This implies that we would expect relatively recently evolved species to share COI barcodes, while old species may have so much variability in COI that a consensus barcode will include few informative sites; i.e., it is unrealistic to expect COI to be of high diagnostic value for distinguishing closely related species.

2.2.c Species identification using NJ trees

For reasons that have yet to be made clear, this is the identification method of choice in most DNA barcoding studies. We can only surmise that it is the appeal of a visual aid such as a tree-like diagram that led to the initial choice of trees as an identification tool. However, from a scientific point of view the use of NJ trees for species identification is hard to justify and the list of problems with identification trees is long (Meier *et al.* 2006, Will & Rubinoff 2004). Hebert *et al.* (2003a) calculates these trees based on K2P distances and regards a sequence to be correctly identified if it clusters on the tree with conspecific sequences; i.e., all sequences for one species form a monophyletic group. This is indefensible. For example, a query sequence that is the sister to all remaining sequences from the same species cannot be identified based on its position on the tree. It could either belong to the same species or to its sister species. The only way to potentially distinguish between the two scenarios is to directly compare sequence divergence and to make a judgement call whether the sequences are sufficiently similar to come from the same species. This is now implicitly acknowledged by 'The Barcode of Life Data System' (BOLD: Ratnasingham & Hebert 2007), which largely relies on direct sequence comparison. But if it is the direct comparison of the sequences that is ultimately needed for matching queries to barcodes, then the additional step of calculating a tree is unnecessary and is a potential source of error. One reason is that as many as a quarter of the currently recognized species are 'paraphyletic' (Funk & Omland 2003), which should not come as a surprise given that the majority of all species concepts do not require species monophyly (Wheeler & Meier 2000). To use a tree-based identification method and requiring monophyly for species identification is thus likely to yield misidentification, and we were thus not surprised to find that for our data set the identification success rate based on pairwise distances is much higher (74.3%) than the identification success based on NJ trees (60.9%; see also Meier *et al.* 2006). Additional problems with NJ trees, for example tie-trees and undesirable taxon-order effects (Meier *et al.* 2006), are known and the use of NJ trees for species identification should thus be abandoned.

3. DNA Taxonomy

DNA barcoding and DNA taxonomy share the belief that DNA sequences will be essential for the future of taxonomy, but they differ with regard to what role the sequences will play. DNA barcodes were at least initially conceived as a tool for species identification although recently species discovery is becoming more important (Hajibabaei *et al.* 2006a, Hebert *et al.* 2004a, Smith *et al.* 2006) and the boundaries between DNA barcoding and DNA taxonomy are more fluid. In DNA taxonomy, DNA sequences were always supposed to provide the main scaffold of a future taxonomy; i.e., they were supposed to be used for revising described and discovering and describing new species. For a taxonomic revision based on DNA sequences, biologists would collect specimens and then sequence a few standard genes for either all material or a few representatives of phenotypically similar 'morphospecies' (Monaghan *et al.* 2005, Pons & Vogler 2005). After collecting the sequences, DNA taxonomists need techniques for grouping DNA sequences into clusters that correspond to species. This step raises a host of conceptual and technical questions. The most intractable conceptual problem is that biologists cannot agree on an universal species concept (Wheeler & Meier 2000); i.e., in discussing different clustering techniques, one would have to test whether they recognize units as species that comply to at least one species concept (Laamanen *et al.* 2003). However, for the purpose of this chapter, we will be more pragmatic and simply ask whether different clustering techniques yield results that are compatible with the species identifications in GenBank.

We would again argue that the techniques used for clustering sequences into 'DNA species' fall into three classes. First, there are techniques based on genetic distance thresholds. Threshold-based arguments are commonly encountered in empirical studies, but they have been largely dismissed in the theoretical literature (Ferguson 2002, Tautz *et al.* 2003). Secondly, authors have argued for assembling sequences into 'species' based on diagnostic differences (e.g., population aggregation analysis: Davis & Nixon 1992). Thirdly, tree-based approaches have been proposed (Brower 1999, Pons & Vogler 2005). We can evaluate all three proposals based on our dataset of COI sequences for Diptera, although most proponents of DNA taxonomy have argued for the use of multiple genes and our test based on a single gene can only be considered preliminary.

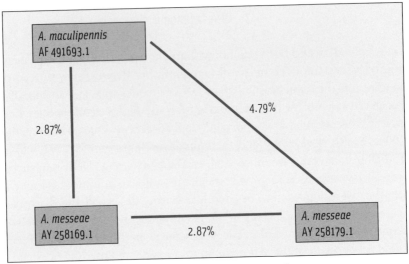

Figure 13.2. Pairwise distances for three *Anopheles* sequences.
All three sequences have one <3% distance, but one pair wise
distance exceeds the threshold of 3%.

3.1 Species delimitation based on overall similarity

The use of genetic distance thresholds is quite popular in the empirical literature. It is not uncommon to find that authors synonymise two species based on genetic identity or near identity (e.g., Hwang *et al.* 2004, Wilkerson *et al.* 2003). In other papers, genetic thresholds are used to estimate the species diversity of a taxon (e.g., Lambert *et al.* 2005) or specimens are labeled as 'cryptic species' because their sequences are unusually divergent from putatively conspecific sequences (e.g., Armstrong & Ball 2005, Ball *et al.* 2005, Barrett & Hebert 2005, Clare *et al.* 2007, Cywinska *et al.* 2006, Hajibabaei *et al.* 2006a, Hebert *et al.* 2004a, Hogg & Hebert 2004, Janzen *et al.* 2005, Lambert *et al.* 2005, Seifert *et al.* 2007, Smith *et al.* 2005, Smith *et al.* 2006, Ward *et al.* 2005). Yet, close inspection reveals that for methodological and biological reasons, threshold-based approaches are seriously flawed (Ferguson 2002, Meier *et al.* 2006). Here, we will only briefly mention the main problems and illustrate with empirical data that pairwise-distance based methods will lead to taxonomic chaos.

The first problem is that the choice of threshold is quite arbitrary (DeSalle *et al.* 2005, Mallet *et al.* 2005) given the wide overlap between intra- and interspecific variability observed for our data set (Fig. 13.1). But potentially more problematic is that it is impossible to consistently apply distance thresholds (Meier *et al.* 2006). The problem is illustrated in Fig. 13.2. For three sequences, it is common to find that two pairwise distances are below a threshold (Fig. 13.2: 3%) while the third exceeds the threshold. How should we classify the three sequences? Two reasonable solutions are conceivable. We can stubbornly maintain the threshold and assign some sequences to two clusters; i.e., the same sequence/specimen would simultaneously belong to two species (Fig. 13.2: *Anopheles messeae* Falleroni, AY258169.1). This solution is hardly acceptable. Alternatively, we can tolerate a threshold violation and create clusters consisting of sequences that have at least one below-threshold match with another sequence in the cluster. The downside of this solution is that the maximum distance within such clusters will exceed the threshold and that clusters have to be recalculated whenever new sequences are added. We would argue that this solution is nevertheless more acceptable because it avoids assigning the same specimen to different DNA species.

We tested how common such threshold violations are in real data sets. Table 13.1 reveals numerous violations with the largest distance within a cluster based on a 3% threshold being as large as 15.8%. But another problem with threshold-based approaches is even more striking. Given that the species-identity of our sequences is known, we can assess whether the clusters assembled based on thresholds conform to traditional species limits. A compatible cluster would only contain all sequences for one species while an incompatible cluster may contain only some of the sequences for one species, or sequences from multiple species. Table 13.1

Table 13.1. Threshold-based taxonomy for 1,001 species of Diptera.

Threshold Distance	No. of clusters	No. of clusters with threshold violations	Maximum pair wise distance	No. of clusters compatible with trad. species (success rate)/profiles with only one species	No. of clusters with split species	No. of clusters with multiple species	Proportion of successful species
1%	1,204	79 (6.6%)	13.9%	739 (61.4%)/1141	402	63	73.8%
2%	1,007	55 (5.5%)	15.8%	703 (69.8%)/916	213	91	70.0%
3%	927	43 (4.6%)	15.8%	662 (71.4%)/819	157	108	66.1%
4%	844	43 (5.1%)	18.5%	604 (71.6%)/724	120	120	60.3%
5%	768	45 (5.9%)	19.3%	544 (70.8%)/638	94	130	54.3%

reveals that regardless of what distance threshold is chosen, a significant number of clusters are incompatible with the traditionally recognized species limits. For example, using a 3% threshold, the sequences for 1,001 species are grouped into 927 clusters. Of these only 71.4% correspond to traditional species; i.e., only the sequences for 66.1% of our 1,001 species are in species-specific clusters. The remaining clusters either lump species and/or place sequences from the same species into different clusters. Our results suggest that if a threshold-based taxonomy was adopted (3% threshold), 33.9% of all 'traditional' species would have to be redescribed. One main reason for proposing DNA-based approaches to taxonomy is to overcome the taxonomic impediment and to speed up species discovery and description. Yet, our result implies that adopting a DNA sequence clustering approach would at least initially slow down progress in taxonomy because many species would have to be redescribed.

3.2 Species delimitation based on diagnosis

In 1992, Davis and Nixon described a method called 'populations aggregation analysis (PAA)' that is used for locating diagnostic differences between populations that deserve species status under the phylogenetic species concept (Wheeler & Platnick 2000). In Brower's (1999) more formal description the PAA consists of five steps: (1) identification of individual organisms as representatives of local populations; (2) identification of attributes useful for comparing populations; (3) tabulation of attributes for a local population to develop a population's attribute profile; (4) comparison of profiles among populations to distinguish between characters (=fixed differences within populations) and traits (=not fixed); (5) aggregation of those populations into a single species which are not distinguished by characters. PAAs can be conducted for DNA sequence data and would potentially yield phylogenetic species that are entirely based on DNA sequences (but see Brower 1999). As pointed out earlier, here we do not want to argue whether the phylogenetic species concept is a good choice (see Willmann & Meier 2000), neither can we conduct PAAs on our data because we lack the necessary locality information for the specimens whose sequences had been deposited in GenBank. But we can test whether COI sequences are diagnostic for species as they are recognized today. As documented earlier, there is good evidence that they are not. A significant number of species share barcodes (7%) and the probability of sharing increases as the taxon sampling improves (7%=>17.7%). However, the power of our test for diagnostic character differences is limited because we cannot determine whether the lack of diagnostic features is due to an inadequate taxonomy that was used to identify the specimens whose sequences are in GenBank, due to the use of different species concepts, and/or due to misidentified sequences in GenBank. Clearly, more research using additional genes is urgently needed.

3.3 Species delimitation based on trees

Tree-based techniques for clustering sequences into species are very popular in the theoretical literature (Wiens & Penkrot 2002) and are also increasingly used in empirical studies (Monaghan *et al.* 2005, Pons & Vogler 2005). The basic observation is that all sequences belonging to the same species tend to form more or less unresolved 'bushes' while the interspecific branches tend to be longer and better resolved. This observation led some authors to propose using the different branching patterns for species delimitation (Vogler & Monaghan 2007). However, drawing the line between interspecific and intraspecific variability on a tree can be difficult and often multiple evolutionary assumptions have to be invoked (e.g., Pons & Vogler 2005). Furthermore, in our test data set consisting of COI sequences many species are not monophyletic on NJ trees (49.2%). Even fewer species are supported by bootstrap values above 70% and it is likely that the situation would worsen if a parsimony search were to be conducted on the same data. It is likely that this is partially due to the use of only one gene in our study; i.e., delimiting species based on trees estimated from DNA sequences would undoubtedly require multiple genes. These genes would have to be sequenced for many specimens because a revision à la DNA taxonomy will be based on sequences instead of the morphology of the study specimens.

4. Cost of a Molecular Taxonomy for Diptera

We had just argued that for a tree-based DNA taxonomy it would be necessary to sequence multiple genes. This raises the question of how expensive DNA taxonomy and/or DNA barcoding really are. It is difficult to provide a definite answer because the literature only discusses the cost of sequencing. The estimates range widely. Stoeckle (2003) is the cheapest and promises amplification and sequencing for USD 1 per sequence, Hebert & Gregory (2005) estimate USD 2, Ball *et al.* (2005) and Tautz *et al.* (2003) 5 Euro. More detailed estimates have been prepared by Pryce *et al.* (2003), who used DNA sequences for diagnosing fungal diseases in humans. They calculated a cost of AUD 10.12 for reagents and AUD 6.54 for labor. Let us apply these estimations to the DNA Barcode of Life project. Hajibabaei *et al.* (2005) proposed that for the Barcoding Life Initiative, approximately 10 specimens should be sequenced for each species. The number of species in the Diptera remains unknown but is likely to reach 1 million. At 10 million specimens and Pryce *et al.*'s estimates (including manpower), the cost would be 124 million US dollars. This would be the molecular cost for DNA barcoding. The cost for a DNA taxonomy treatment of Diptera would be much higher because all specimens would have to be sequenced for multiple genes. For example, Larsen & Meier (2004) revised the 30 species of Asilidae in Denmark. The revision included 4,391 specimens and the cost would have been more than 150,000 USD if all specimens had been sequenced for three genes.

    Unfortunately, all these estimates would only be accurate if we were to already have a cryo-collection that contains 10 specimens for each of the 1 million species of Diptera (preferably from different localities across the species' range). Estimating the cost for establishing such a facility is impossible because the equation involves too many variables. But it is instructive to recall some of the major steps needed in building such a collection. A consultation of any major taxonomic monograph in Diptera reveals that a large proportion of Diptera species are rare, only known from a few localities, and/or highly seasonal (e.g., Grimaldi & Nguyen 1999). Obtaining molecular-grade specimens for all species thus requires sampling using multiple collecting techniques at a fine geographic scale, over a long period of time, and at a global scale. Afterwards, the samples will have to be processed; i.e., labeled, sorted, and identified to species. The vast majority of the species discovered during the sorting will be undescribed and/or unidentifiable because they belong to taxa without adequate identification tools. The new species will have to be described and the identification tools will have to be created. All this assumes that there are no political and permit problems with sampling at a global scale. Clearly, the expensive part of any molecular taxonomy initiative for Diptera is obtaining specimens and not the cost for sequencing.

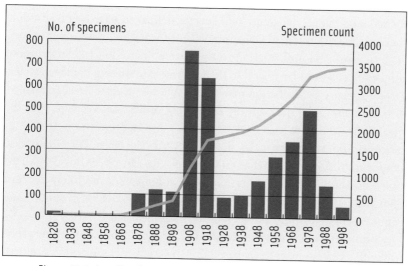

Figure 13.3. Age distribution of museum specimens for Danish Asilidae.

5. The Need for an Integrative Taxonomy

We had earlier argued that there was no need for discussing that DNA sequences are a useful component of any modern taxonomy. However, we believe that there is a need to discuss what would happen if the traditional techniques were to be neglected and DNA taxonomy, i.e., the delimitation and description of species based on DNA sequences, were to be embraced. One of the main concerns is the fate of older museum specimens. We can again use Larsen & Meier's (2004) Danish Asilidae data set as an example (Fig. 13.3). A large proportion of the specimens are relatively old (Fig. 13.3: 20% >100 years, 68% >50 years) and have a small chance for DNA sequence recovery. For example, Hajibabaei *et al.* (2005) reported a sequencing success of only 31% for 'archival moths' (average age: 21 years, 33% of specimens were less than 5 years old). Yet, 98% of all Danish Asilidae in museum collection are more than 21 years old. Such specimens would at best be available for the amplification of short markers (Hajibabaei *et al.* 2006c). Once species delimitations and identification tools are entirely relying on DNA sequences, these specimens and millions others in the natural history collections would become worthless, because they can no longer be (re)identified because the species diagnoses are now based on DNA sequences. The older literature will suffer the same fate because the information in these papers is tied to specimens that may not be suitable

Diptera Diversity: Status, Challenges and Tools
(eds T. Pape, D. Bickel & R. Meier). © 2009 Koninklijke Brill NV.

for sequencing; i.e., when an 'old' species delimited based on traditional methods is split based on DNA sequence information, it becomes impossible to assign the information to the correct 'new' species. Such loss of information would be tolerable if there was convincing evidence that a DNA-sequence based taxonomy is desirable for scientific reasons. However, our empirical data appear to suggest otherwise. We find that the diagnostic value of at least COI sequences is not impressive and that a significant proportion of species share identical consensus DNA barcodes. It appears unlikely to us that all these species have been misdiagnosed based on traditional characters. Instead, it appears more likely and not unexpected from an evolutionary point of view that DNA sequences alone cannot recognize, for example, relatively young species.

Connectivity between the old and the new taxonomy is only one problem with a taxonomy entirely based on DNA sequences. Another problem is that the vast majority of species are found in tropical countries with insufficient funding and infrastructure for large-scale sequencing projects (Seberg 2004). Yet, there is general agreement that taxonomic revisions should be carried out at a global scale and must share at least one common character system across all regions. This is not a problem as long as species descriptions are based on a combination of morphological and DNA sequence characters. But how would one test for the presence of a DNA-based species in a tropical country if it lacks the resources for molecular work (Will & Rubinoff 2004)? Furthermore, how many institutions would be able to identify specimens in the future if species are defined based on only DNA sequences (Dunn 2003, Prendini 2005)? So far, taxonomy had been a relatively egalitarian scientific discipline and even countries with limited resources for scientific research were able to participate. A DNA-sequence based taxonomy would dramatically change this situation.

Fortunately, the problem of DNA-sequence based species is not yet acute. The reality is that most of the species discovered based only on DNA sequences — often called 'cryptic species' — are never described (Ebach & Holdrege 2005). The sequences are submitted to GenBank where they are accumulating, unfortunately often without proper annotation. We have seen submissions under the name of the described species without indication that the sequence may come from a new species (e.g., Frost *et al.* 1998) and with an indication using an 'aff.' (e.g., Feder *et al.* 2003) or 'cf' (e.g., Sharpe *et al.* 2000), or under the genus name with the addition of a 'sp.' (with number: e.g., Smith *et al.* 2005; or without numbering: e.g.,

Hebert *et al.* 2004a). Since every biologist is allowed to submit a sequence under, for example, *Drosophila* sp. 1, 'sp. 1' will be an ambiguous 'name' and we are back to where we were when Linnaeus introduced taxonomic nomenclature in order to avoid having different biologists use the same name for different species. This reluctance to describe and properly name taxa is starting to be a major impediment of all molecular taxonomy initiatives. Sequences not identified to species are quickly accumulating. For example, of the 10,692 sequences with the keyword 'barcode' in GenBank, 34.7% are not identified species (accession 19 May 2007).

6. Conclusions

We would thus conclude that all approaches to taxonomy that rely exclusively or mostly on DNA sequences are undesirable because they suffer from serious connectivity problems: the new will become disconnected from the old and the poor from the rich. At the same time, there is no evidence that a taxonomy relying mostly on DNA sequences will yield a scientifically more satisfactory classification. Claims that such a taxonomy is cheaper and faster are similarly questionable given that specimen acquisition cost and time have been ignored in past estimates for molecular taxonomy initiatives. Under these circumstances, we believe that a better approach is integrative taxonomy. Here, the new remains connected to the old, and taxonomic research will not be restricted to rich countries. Will an integrative taxonomy help overcome the taxonomic impediment? We believe it will. All character systems have their strengths and weaknesses and a new powerful tool in taxonomy — DNA sequences — will surely strengthen the discipline and also speed up several aspects of taxonomic research.

REFERENCES

Armstrong, K.F. & Ball, S.L. (2005) DNA barcodes for biosecurity: invasive species identification. *Philosophical Transactions of the Royal Society B: Biological Sciences* 360: 1813–1823.

Arnqvist, G. (1998) Comparative evidence for the evolution of genitalia by sexual selection. *Nature* 393: 784–785.

Ball, S.L., Hebert, P.D.N., Burian, S.K. & Webb, J.M. (2005) Biological identifications of mayflies (Ephemeroptera) using DNA barcodes. *Journal of the North American Benthological Society* 24: 508–524.

Ballard, J.W.O. (1994) Evidence from 12S ribosomal RNA sequences resolves a morphological conundrum in *Austrosimulium* (Diptera: Simuliidae). *Journal of the Australian Entomological Society* 33: 131–135.

Barrett, R.D.H. & Hebert, P.D.N. (2005) Identifying spiders through DNA barcodes. *Canadian Journal of Zoology* 83: 481–491.

Basso, A., Sonvico, A., Quesada-Allue, L.A. & Manso, F. (2003) Karyotypic and molecular identification of laboratory stocks of the south American fruit fly *Anastrepha fraterculus* (Wied) (Diptera: Tephritidae). *Journal of Economic Entomology* 96: 1237–1244.

Beebe, N.W., Van Den Hurk, A.F., Chapman, H.F., Frances, S.P., Williams, C.R. & Cooper, R.D. (2002) Development and evaluation of a species diagnostic polymerase chain reaction-restriction fragment-length polymorphism procedure for cryptic members of the *Culex sitiens* (Diptera: Culicidae) subgroup in Australia and the Southwest Pacific. *Journal of Medical Entomology* 39: 362–369.

Brower, A.V.Z. (1999) Delimitation of phylogenetic species with DNA sequences: A critique of Davis and Nixon's population aggregation analysis. *Systematic Biology* 48: 199–213.

Cai, J.-F., Liu, M., Ying, B.-W., Dong, J.-G., Deng, Z.-H., Tao, T., Pan, H.-F., Zhang, H.-X., Yan, H.-T. & Liao, Z.-G. (2005) Sequencing of mitochondrial DNA cytochrome oxidase subunit I for identification of sarcosaphagous flies (Diptera) in Chengdu. *Acta Entomologica Sinica* 48: 101–106.

Carew, M.E., Pettigrove, V. & Hoffmann, A.A. (2003) Identifying chironomids (Diptera: Chironomidae) for biological monitoring with PCR-RFLP. *Bulletin of Entomological Research* 93: 483–490.

Carew, M.E., Pettigrove, V. & Hoffmann, A.A. (2005) The utility of DNA markers in classical taxonomy: Using cytochrome oxidase I markers to differentiate Australian *Cladopelma* (Diptera: Chironomidae) midges. *Annals of the Entomological Society of America* 98: 587–594.

Cetre-Sossah, C., Baldet, T., Delecolle, J.-C., Mathieu, B., Perrin, A., Grillet, C. & Albina, E. (2004) Molecular detection of *Culicoides* spp. and *Culicoides imicola*, the principal vector of bluetongue (BT) and African horse sickness (AHS) in Africa and Europe. *Veterinary Research* 35: 325–337.

Chen, W.-Y., Hung, T.-H. & Shiao, S.-F. (2004) Molecular identification of forensically important blow fly species (Diptera: Calliphoridae) in Taiwan. *Journal of Medical Entomology* 41: 47–57.

Clare, E.L., Lim, B.K. Engstrom, M.D., Eger, J.L. & Hebert, P.D.N. (2007) DNA barcoding of Neotropical bats: species identification and discovery within Guyana. *Molecular Ecology Notes* 7: 184–190.

Cockburn, A.F. (1990) A Simple and Rapid Technique for Identification of Large Numbers of Individual Mosquitoes Using DNA Hybridization. *Archives of Insect Biochemistry and Physiology* 14: 191–199.

Cognato, A.I. (2006) Standard percent DNA sequence difference for insects does not predict species boundaries. *Journal of Economic Entomology* 99: 1037–1045.

Cook, S., Diallo, M., Sall, A.A., Cooper, A. & Holmes, E.C. (2005) Mitochondrial markers for molecular identification of *Aedes* mosquitoes (Diptera: Culicidae) involved in transmission of arboviral disease in West Africa. *Journal of Medical Entomology* 42: 19–28.

Cornel, A.J., Porter, C.H. & Collins, F.H. (1996) Polymerase chain reaction species diagnostic assay for *Anopheles quadrimaculatus* cryptic species (Diptera: Culicidae) based on ribosomal DNA ITS2 sequences. *Journal of Medical Entomology* 33: 109–116.

Crabtree, M.B., Savage, H.M. & Miller, B.R. (1995) Development of a species-diagnostic polymerase chain reaction assay for the identification of *Culex* vectors of St. Louis encephalitis virus based on interspecies sequence variation in ribosomal DNA spacers. *American Journal of Tropical Medicine and Hygiene* 53: 105–109.

Craft, J., Donoghue, M., Dragoo, J., Hillis, D. & Yates, T. (2002) *Assembling the tree of life: Harnessing life's history to benefit Science and Society.* National Science Foundation Brochure, Washington, DC.

Cywinska, A., Hunter, F.F. & Hebert, P.D.N. (2006) Identifying Canadian mosquito species through DNA barcodes. *Medical and Veterinary Entomology* 20: 413–424.

Davis, J.I. & Nixon, K.C. (1992. Populations, genetic variation, and the delimitation of phylogenetic species. *Systematic Biology* 41: 421–435.

Depaquit, J., Leger, N. & Robert, V. (2002) First record of *Phlebotomus* from Madagascar (Diptera: Psychodidae). Description of *Phlebotomus* (*Anaphlebotomus*) *fertei* n. sp. and *Phlebotomus* (*Anaphlebotomus*) *huberti* n. sp. *Parasite* 9: 325–331.

Depaquit, J., Leger, N. & Robert, V. (2004) *Phlebotomus* from Madagascar (Diptera — Psychodidae). III — Description of *Phlebotomus* (*Anaphlebotomus*) *fontenillei* n.sp. *Parasite* 11: 261–265.

DeSalle, R., Egan, M.G. & Siddall, M. (2005) The unholy trinity: taxonomy, species delimitation and DNA barcoding. *Philosophical Transactions of the Royal Society B: Biological Sciences* 360: 1905–1916.

Douglas, L.J. & Haymer, D.S. (2001) Ribosomal ITS1 polymorphisms in *Ceratitis capitata* and *Ceratitis rosa* (Diptera: Tephritidae). *Annals of the Entomological Society of America* 94: 726–731.

Dunn, C.P. (2003) Keeping taxonomy based in morphology. *Trends in Ecology & Evolution* 18: 270–271.

Ebach, M.C. & Holdrege, C. (2005) DNA barcoding is no substitute for taxonomy. *Nature* 434: 697.

Eberhard, W.G. (1985) *Sexual Selection and Animal Genitalia*. Harvard University Press: Cambridge, Massachusetts, USA.

Evenhuis, N.L., Pape, T., Pont, A.C. & Thompson, F.C. (eds) (2008) *BioSystematic Database of World Diptera*, Version 10.5. Available at http://www.diptera.org/biosys. htm, accessed 13 February 2008.

Feder, J.L., Berlocher, S.H., Roethele, J.B., Dambroski, H., Smith, J.J., Perry, W.L., Gavrilovic, V., Filchak, K.E., Rull, J. & Aluja, M. (2003) Allopatric genetic origins for sympatric host-plant shifts and race formation in *Rhagoletis*. *Proceedings of the National Academy of Sciences of the United States of America* 100: 10314–10319.

Ferguson, J.W.H. (2002) On the use of genetic divergence for identifying species. *Biological Journal of the Linnean Society* 75: 509–516.

Frost, D.R., Crafts, H.M., Fitzgerald, L.A. & Titus, T.A. (1998) Geographic variation, species recognition, and molecular evolution of cytochrome oxidase I in the *Tropidurus spinulosus* complex (Iguania: Tropiduridae). *Copeia* 1998: 839–851.

Funk, D.J. & Omland, K.E. (2003) Species-Level Paraphyly and Polyphyly: frequency, causes, and consequences, with insights from animal mitochondrial DNA. *Annual Review of Ecology and Systematics* 34: 397–423.

Gleeson, D.M. & Sarre, S. (1997) Mitochondrial DNA variability and geographic origin of the sheep blowfly, *Lucilia cuprina* (Diptera: Calliphoridae), in New Zealand. *Bulletin of Entomological Research* 87: 265–272.

Godfray, H.C.J. (2002) Towards taxonomy's 'glorious revolution': Taxonomy is a triumph of modern science: But its products could still be improved. *Nature* 420: 461.

Grimaldi, D. & Nguyen, T. (1999) Monograph on the spittlebug flies, genus *Cladochaeta* (Diptera: Drosophilidae: Cladochaetini). *Bulletin of the American Museum of Natural History* 241: 1–326.

Gunasekera, M.B., De Silva, B.G.D.N.K., Abeyewickreme, W., Subbarao, S.K., Nandadasa, H.G. & Karunanayake, E.H. (1995) Development of DNA probes for the identification of sibling species A of the *Anopheles culicifacies* (Diptera: Culicidae) complex. *Bulletin of Entomological Research* 85: 345–353.

Hajibabaei, M., deWaard, J.R., Ivanova, N.V., Ratnasingham, S., Dooh, R.T., Kirk, S.L., Mackie, P.M. & Hebert, P.D.N. (2005) Critical factors for assembling a high volume of DNA barcodes. *Philosophical Transactions of the Royal Society B: Biological Sciences* 360: 1959–1967.

Hajibabaei, M., Janzen, D.H., Burns, J.M., Hallwachs, W. & Hebert, P.D.N. (2006a) DNA barcodes distinguish species of tropical Lepidoptera. *Proceedings of the National Academy of Sciences of the United States of America* 103: 968–971.

Hajibabaei, M., Singer, G.A.C. & Hickey, D.A. (2006b) Benchmarking DNA barcodes: Does the DNA barcoding gap exist? *Genome* 49: 851–854.

Hajibabaei, M., Smith, M.A., Janzen, D.H., Rodriguez, J.J., Whitfield, J.B. & Hebert, P.D.N. (2006c) A minimalist barcode can identify a specimen whose DNA is degraded. *Molecular Ecology Notes* 6: 959–964.

Harris, D.J. (2003) Can you bank on GenBank? *Trends in Ecology & Evolution* 18: 317–319.

Harvey, M.L., Dadour, I.R. & Gaudieri, S. (2003) Mitochondrial DNA cytochrome oxidase I gene: Potential for distinction between immature stages of some forensically important fly species (Diptera) in western Australia. *Forensic Science International* 131: 134–139.

Hebert, P.D.N., Cywinska, A., Ball, S.L. & deWaard, J.R. (2003a) Biological identifications through DNA barcodes. *Proceedings of the Royal Society Biological Sciences Series B* 270: 313–321.

Hebert, P.D.N., Ratnasingham, S. & deWaard, J.R. (2003b) Barcoding animal life: Cytochrome c oxidase subunit 1 divergences among closely related species. *Proceedings of the Royal Society Biological Sciences Series B* 270: S96–S99.

Hebert, P.D.N. & Gregory, T.R. (2005) The promise of DNA barcoding for taxonomy. *Systematic Biology* 54.

Hebert, P.D.N., Penton, E.H. Burns, J.M., Janzen, D.H. & Hallwachs, W. (2004a) Ten species in one: DNA barcoding reveals cryptic species in the Neotropical skipper butterfly *Astraptes fulgerator*. *Proceedings of the National Academy of Sciences of the United States of America* 101: 14812–14817.

Hebert, P.D.N., Stoeckle, M.Y., Zemlak, T.S. & Francis, C.M. (2004b) Identification of birds through DNA barcodes. *PLoS Biology* 2: 1657–1663.

Higazi, T.B., Boakye, D.A., Wilson, M.D., Mahmoud, B.M., Baraka, O.Z., Mukhtar, M.M. & Unnasch, T.R. (2000) Cytotaxonomic and molecular analysis of *Simulium (Edwardsellum) damnosum* sensu lato (Diptera: Simuliidae) from Abu Hamed, Sudan. *Journal of Medical Entomology* 37: 547–553.

Hill, S.M., Urwin, R., Knapp, T.F. & Crampton, J.M. (1991) Synthetic DNA probes for the identification of sibling species in the *Anopheles-gambiae* complex. *Medical and Veterinary Entomology* 5: 455–464.

Hogg, I.D. & Hebert, P.D.N. (2004) Biological identification of springtails (Hexapoda: Collembola) from the Canadian Arctic, using mitochondrial DNA barcodes. *Canadian Journal of Zoology* 82: 749–754.

Hwang, U.W., Yong, T.S. & Ree, H.-I. (2004) Molecular evidence for synonymy of *Anopheles yatsushiroensis* and *An. pullus*. *Journal of the American Mosquito Control Association* 20: 99–104.

Janzen, D.H., Hajibabaei, M., Burns, J.M., Hallwachs, W., Remigio, E. & Hebert, P.D.N. (2005) Wedding biodiversity inventory of a large and complex Lepidoptera fauna with DNA barcoding. *Philosophical Transactions of the Royal Society B: Biological Sciences* 360: 1835–1845.

Krzywinski, J. & Besansky, N.J. (2003) Molecular systematics of *Anopheles*: from subgenera to subpopulations. *Annual Review of Entomology* 48: 111–139.

Laamanen, T.R., Petersen, F.T. & Meier, R. (2003) Kelp flies and species concepts: The case of *Coelopa frigida* (Fabricius, 1805) and *C. nebularum* Aldrich, 1929 (Diptera: Coelopidae). *Journal of Zoological Systematics and Evolutionary Research* 41: 127–136.

Lambert, D.M., Baker, A., Huynen, L., Haddrath, O., Hebert, P.D.N. & Millar, C.D. (2005) Is a large-scale DNA-based inventory of ancient life possible? *Journal of Heredity* 96: 279–284.

Larsen, N.M. & Meier, R. (2004) The species diversity, distribution, and conservation status of the Asilidae (Diptera: Insecta) in Denmark. *Steenstrupia* 28: 177–241.

Lefebure, T., Douady, C.J., Gouy, M. & Gibert, J. (2006) Relationship between morphological taxonomy and molecular divergence within Crustacea: Proposal of a molecular threshold to help species delimitation. *Molecular Phylogenetics and Evolution* 40: 435–447.

Li, G.Q., Hu, Y.L., Kanu, S. & Zhu, X.Q. (2003) PCR amplification and sequencing of ITS1 rDNA of *Culicoides arakawae*. *Veterinary Parasitology* 112: 101–108.

Linton, Y.-M., Harbach, R.E., Seng, C.M., C.M. Anthony, C.M. & Matusop, A. (2001) Morphological and molecular identity of *Anopheles* (*Cellia*) *sundaicus* (Diptera: Culicidae), the nominotypical member of a malaria vector species complex in Southeast Asia. *Systematic Entomology* 26: 357–366.

Linton, Y.-M., Smith, L., Koliopoulos, G., Samanidou-Voyadjoglou, A., Zounos, A.K. & Harbach, R.E. (2003) Morphological and molecular characterization of *Anopheles* (*Anopheles*) *maculipennis* Meigen, type species of the genus and nominotypical member of the *maculipennis* complex. *Systematic Entomology* 28: 39–55.

Litjens, P., Lessinger, A.C. & de Azeredo-Espin, A.M.L. (2001) Characterization of the screwworm flies *Cochliomyia hominivorax* and *Cochliomyia macellaria* by PCR-RFLP of mitochondrial DNA. *Medical and Veterinary Entomology* 15: 183–188.

Little, D.P. & Stevenson, D.W. (2007) A comparison of algorithms for the identification of specimens using DNA barcodes: examples from gymnosperms. *Cladistics* 23: 1–21.

Lorenz, J.G., Jackson, W.E., Beck, J.C. & Hanner, R. (2005) The problems and promise of DNA barcodes for species diagnosis of primate biomaterials. *Philosophical Transactions of the Royal Society B: Biological Sciences* 360: 1869–1877.

Lounibos, L.P., Wilkerson, R.C., Conn, J.E., Hribar, L.J., Fritz, G.N. & Danoff-Burg, J.A. (1998) Morphological, molecular, and chromosomal discrimination of cryptic *Anopheles* (*Nyssorhynchus*) (Diptera: Culicidae) from South America. *Journal of Medical Entomology* 35: 830–838.

Malafronte, R.S., Marrelli, M.T., Carreri-Bruno, G.C., Urbinatti, P.R. & Marinotti, O. (1997) Polymorphism in the second internal transcribed spacer (ITS2) of *Anopheles* (*Kerteszia*) *cruzi* (Diptera: Culicidae) from the state of São Paulo, Brazil. *Memórias do Instituto Oswaldo Cruz* 92: 306.

Malafronte, R.S., Marrelli, M.T. & Marinotti, O. (1998) Comparison of *Anopheles darlingi* populations based on ITS2 DNA sequences. *Memórias do Instituto Oswaldo Cruz* 93: 328: 651–655.

Malafronte, R.S., Marrelli, M.T. & Marinotti, O. (1999) Analysis of ITS2 DNA sequences from Brazilian *Anopheles darlingi* (Diptera: Culicidae). *Journal of Medical Entomology* 36: 631–634.

Mallet, J., Isaac, N.J.B. & Mace, G.M. (2005) Response to Harris and Froufe, and Knapp et al.: Taxonomic inflation. *Trends in Ecology & Evolution* 20: 8–9.

Marrelli, M.T., Malafronte, R.S., Flores-Mendonza, C., Kloetzel, J.K. & Marinotti, O. (1998) Polymorphism in the second internal transcribed spacer (ITS2) of ribosomal DNA among populations of *Anopheles oswaldoi*. *Memórias do Instituto Oswaldo Cruz* 93: 206.

Marrelli, M.T., Malafronte, R.S., Flores-Mendoza, C., Lourenco-De-Oliveira, R., Kloetzel, J.K. & Marinotti, O. (1999) Sequence analysis of the second internal transcribed spacer of ribosomal DNA in *Anopheles oswaldoi* (Diptera: Culicidae). *Journal of Medical Entomology* 36: 679–684.

McClellan, D.A. & Woolley, S. (2004) *AlignmentHelper*, version 1.0. Brigham Young University.

Meier, R. (2005) The Role of Dipterology in Phylogenetic Systematics: The Insight of Willi Hennig. Pages 45–62 *in*: Yeates, D.K. & Wiegmann, B.M. (eds), *The Evolutionary Biology of Flies*. Columbia University Press, New York.

Meier, R. & Dikow, T. (2004) Significance of specimen databases from taxonomic revisions for estimating and mapping the global species diversity of invertebrates and repatriating reliable and complete specimen data. *Conservation Biology* 18: 478–488.

Meier, R., Kwong, S., Vaidya, G. & Ng, P.K.L. (2006) DNA Barcoding and Taxonomy in Diptera: a Tale of High Intraspecific Variability and Low Identification Success. *Systematic Biology* 55: 715–728.

Meier, R., Zhang, G. & Ali, F. (2008) The use of mean instead of smallest interspecific distances exaggerates the size of the "Barcoding Gap" and leads to misidentification. *Systematic Biology* 57: 809–813.

Meyer, C.P. & Paulay, G. (2005) DNA barcoding: Error rates based on comprehensive sampling. *PLoS Biology* 3: 2229–2238.

Milankov, V., Stamenković, J., Ludoški, J., Ståhls, G. & Vujić, A. (2005) Diagnostic molecular markers and the genetic relationships among three species of the *Cheilosia canicularis* group (Diptera: Syrphidae). *European Journal of Entomology* 102: 125–131.

Monaghan, M.T., Balke, M., Gregory, T.R. & Vogler, A.P. (2005) DNA-based species delineation in tropical beetles using mitochondrial and nuclear markers. *Philosophical Transactions of the Royal Society B: Biological Sciences* 360: 1925–1933.

Moritz, C. & Cicero, C. (2004) DNA barcoding: promise and pitfalls. *PLoS Biology* 2: 1529–1531.

Muraji, M. & Nakahara, S. (2002) Discrimination among pest species of *Bactrocera* (Diptera: Tephritidae) based on PCR-RFLP of the mitochondrial DNA. *Applied Entomology and Zoology* 37: 437–446.

Nakahara, S., Kato, H., Kaneda, M., Sugimoto, T. & Muraji, M. (2002) Identification of the *Bactrocera dorsalis* complex (Diptera: Tephritidae) by PCR-RFLP analysis: III. Discrimination between *B. philippinensis* and *B. occipitalis*. *Research Bulletin of the Plant Protection Service Japan* 38: 73–80.

Noel, S., Tessier, N., Angers, B., Wood, D.M. & Lapointe, F.J. (2004) Molecular identification of two species of myiasis-causing *Cuterebra* by multiplex PCR and RFLP. *Medical and Veterinary Entomology* 18: 161–166.

Otranto, D. & Traversa, D. (2004) Molecular evidence indicating that *Przhevalskiana silenus*, *P. aegagri* and *P. crossii* (Diptera, Oestridae) are one species. *Acta Parasitologica* 49: 173–176.

Otranto, D., Traversa, D. & Giangaspero, A. (2004) Serological and molecular approaches to the diagnosis of myiasis causing Oestridae. *Parassitologia* 46: 169–172.

Otranto, D., Traversa, D., Milillo, P., De Luca, F. & Stevens, J. (2005) Utility of mitochondrial and ribosomal genes for differentiation and phylogenesis of species of gastrointestinal bot flies. *Journal of Economic Entomology* 98: 2235–2245.

Pages, N. & Monteys, V.S.I. (2005) Differentiation of *Culicoides obsoletus* and *Culicoides scoticus* (Diptera: Ceratopogonidae) based on mitochondrial cytochrome oxidase subunit I. *Journal of Medical Entomology* 42: 1026–1034.

Patsoula, E., Samanidou-Voyadjoglou, A., Spanakos, G., Kremastinou, J., Nasioulas, G. & Vakalis, N.C. (2006) Molecular and morphological characterization of *Aedes albopictus* in northwestern Greece and differentiation from *Aedes cretinus* and *Aedes aegypti*. *Journal of Medical Entomology* 43: 40–54.

Pérez-Bañón, C., Rojo, S., Ståhls, G. & Marcos-García, M.A. (2003) Taxonomy of European *Eristalinus* (Diptera: Syrphidae) based on larval morphology and molecular data. *European Journal of Entomology* 100: 417–428.

Petersen, F.T., Damgaard, J. & Meier, R. (2007) DNA Taxonomy: How many DNA sequences are needed for solving a taxonomic problem? The case of two parapatric species of louse Flies (Diptera: Hippoboscidae: *Ornithomya* (Latreille, 1802). *Arthropod Systematics and Phylogeny* 65: 111–117.

Pons, J. & Vogler, A.P. (2005) Complex pattern of coalescence and fast evolution of a mitochondrial rRNA pseudogene in a recent radiation of tiger beetles. *Molecular Biology and Evolution* 22: 991–1000.

Porter, C.H. & Collins, F.H. (1991) Species-diagnostic differences in a ribosomal DNA internal transcribed spacer from the sibling species *Anopheles freeborni* and *Anopheles hermsi* (Diptera: Culicidae). *American Journal of Tropical Medicine and Hygiene* 45: 271–279.

Powers, T. (2004) Nematode molecular diagnostics: From bands to barcodes. *Annual Review of Phytopathology* 42: 367–383.

Prendini, L. (2005) Comment on "Identifying spiders through DNA barcodes". *Canadian Journal of Zoology* 83: 498–504.

Pryce, T.M., Palladino, S., Kay, I.D. & Coombs, G.W. (2003) Rapid identification of fungi by sequencing the ITSI and ITS2 regions using an automated capillary electrophoresis system. *Medical Mycology* 41: 369–381.

Ratnasingham, S. & Hebert, P.D.N. (2007) BOLD: The Barcode of Life Data System (http://www.barcodinglife.org). *Molecular Ecology Notes* 7: 355–364.

Ready, P.D., Day, J.C., Souza, A.A.D., Rangel, E.F. & Davies, C.R. (1997) Mitochondrial DNA characterization of populations of *Lutzomyia whitmani* (Diptera: Psychodidae) incriminated in the peri-domestic and silvatic transmission of Leishmania species in Brazil. *Bulletin of Entomological Research* 87: 187–195.

Ritchie, A., Blackwell, A., Malloch, G. & Fenton, B. (2004) Heterogeneity of ITS1 sequences in the biting midge *Culicoides impunctatus* (Goetghebuer) suggests a population in Argyll, Scotland, may be genetically distinct. *Genome* 47: 546–558.

Roe, A.D. & Sperling, F.A.H. (2007) Patterns of evolution of mitochondrial cytochrome c oxidase I and II DNA and implications for DNA barcoding. *Molecular Phylogenetics and Evolution* 44: 325–345.

Rubinoff, D. (2006) Utility of Mitochondrial DNA Barcodes in Species Conservation. *Conservation Biology* 20: 1026–1033.

Saigusa, K., Takamiya, M. & Aoki, Y. (2005) Species identification of the forensically important flies in Iwate prefecture, Japan based on mitochondrial cytochrome oxidase gene subunit I (COI) sequences. *Legal Medicine* 7: 175–178.

Saunders, G.W. (2005) Applying DNA barcoding to red macroalgae: a preliminary appraisal holds promise for future applications. *Philosophical Transactions of the Royal Society of London B Biological Sciences* 360: 1879–1888.

Sawabe, K., Takagi, M., Tsuda, Y. & Tuno, N. (2003) Molecular variation and phylogeny of the *Anopheles minimus* complex (Diptera: Culicidae) inhabiting southeast Asian countries, based on ribosomal DNA internal transcribed spacers, ITS 1 and 2, and the 28S D3 sequences. *Southeast Asian Journal of Tropical Medicine and Public Health* 34: 771–780.

Scheffer, S.J., Lewis, M.L. & Joshi, R.C. (2006) DNA barcoding applied to invasive leafminers (Diptera: Agromyzidae) in the Philippines. *Annals of the Entomological Society of America* 99: 204–210.

Scheffer, S. J., Wijesekara, A., Visser, D. & Hallett, R.H. (2001) Polymerase chain reaction-restriction fragment-length polymorphism method to distinguish *Liriomyza huidobrensis* from *L. langei* (Diptera: Agromyzidae) applied to three recent leafminer invasions. *Journal of Economic Entomology* 94: 1177–1182.

Schindel, D.E. & Miller, S.E. (2005) DNA barcoding a useful tool for taxonomists. *Nature* 435: 17.

Schlee, D. (1978) In memoriam Willi Hennig 1913–1976. Eine biographische Skizze. *Entomologica Germanica* 4: 377–391.

Schroeder, H., Klotzbach, H., Elias, S., Augustin, C. & Pueschel, K. (2003) Use of PCR-RFLP for differentiation of calliphorid larvae (Diptera, Calliphoridae) on human corpses. *Forensic Science International* 132: 76–81.

Seberg, O. (2004) The future of systematics: assembling the Tree of Life. *The Systematist — Newsletter of the Systematics Association* 23: 2–8.

Sedaghat, M.M., Linton, Y.-M., Nicolescu, G., Smith, L., Koliopoulos, G., Zounos, A.K., Oshaghi, M.A., Vatandoost, H. & Harbach, R.E. (2003) Morphological and molecular characterization of *Anopheles* (*Anopheles*) *sacharovi* Favre, a primary vector of malaria in the Middle East. *Systematic Entomology* 28: 241–256.

Seifert, K.A., Samson, R.A., deWaard, J.R, Houbraken, J., Levesque, C.A., Moncalvo, J.-M., Louis-Seize, G. & Hebert, P.D.N. (2007) Prospects for fungus identification using CO1 DNA barcodes, with *Penicillium* as a test case. *Proceedings of the National Academy of Sciences of the United States of America* 104: 3901–3906.

Sharley, D.J., Pettigrove, V. & Parsons, Y.M. (2004) Molecular identification of *Chironomus* spp. (Diptera) for biomonitoring of aquatic ecosystems. *Australian Journal of Entomology* 43: 359–365.

Sharpe, R.G., Harbach, R.E. & Butlin, R.K. (2000) Molecular variation and phylogeny of members of the *minimus* Group of *Anopheles* subgenus *Cellia* (Diptera: Culicidae). *Systematic Entomology* 25: 263–272.

Smith, M.A., Fisher, B.L. & Hebert, P.D.N. (2005) DNA barcoding for effective biodiversity assessment of a hyperdiverse arthropod group: the ants of Madagascar. *Philosophical Transactions of the Royal Society B: Biological Sciences* 360: 1825–1834.

Smith, M.A., Woodley, N.E, Janzen, D.H., Hallwachs, W. & Hebert, P.D.N. (2006) DNA barcodes reveal cryptic host-specificity within the presumed polyphagous members of a genus of parasitoid flies (Diptera: Tachinidae). *Proceedings of the National Academy of Sciences of the United States of America* 103: 3657–3662.

Sperling, F. (2003) DNA barcoding. Deus et machina. *Newsletter of the Biological. Survey of Canada (Terrestrial Arthropods)* 22: 50–53.

Sperling, F.A.H., Anderson, G.S. & Hickey, D.A. (1994) A DNA-based approach to the identification of insect species used for postmortem interval estimation. *Journal of Forensic Sciences* 39: 418–427.

Steinke, D., Vences, M., Salzburger, W. & Meyer, A. (2005) TaxI: a software tool for DNA barcoding using distance methods. *Philosophical Transactions of the Royal Society B: Biological Sciences* 360: 1975–1980.

Stoeckle, M. (2003) Taxonomy, DNA, and the Bar Code of Life. *BioScience* 53: 796–797.

Tang, J., Pruess, K. & Unnasch, T.R. (1996. Genotyping North American black flies by means of mitochondrial ribosomal RNA sequences. *Canadian Journal of Zoology* 74: 39–46.

Tang, J., Toe, L., Back, C., Zimmerman, P.A., Pruess, K. & Unnasch, T.R. (1995) The *Simulium damnosum* species complex: Phylogenetic analysis and molecular identification based upon mitochondrially encoded gene sequences. *Insect Molecular Biology* 4: 79–88.

Tautz, D., Arctander, P., Minelli, A., Thomas, R.H. & Vogler, A.P. (2003) A plea for DNA taxonomy. *Trends in Ecology & Evolution* 18: 70–74.

Tenorio, F.M., Olson, J.K. & Coates, C.J. (2003) Identification of three forensically important blow fly (Diptera: Calliphoridae) species in central Texas using mitochondrial DNA. *Southwestern Entomologist* 28: 267–272.

Testa, J.M., Montoya-Lerma, J., Cadena, H., Oviedo, M. & Ready, P.D. (2002) Molecular identification of vectors of *Leishmania* in Colombia: Mitochondrial introgression in the *Lutzomyia townsendi* series. *Acta Tropica* 84: 205–218.

Thompson, J.D., Higgins, D.G. & Gibson, T.J. (1994) CLUSTAL W: improving the sensitivity of progressive multiple sequence alignment through sequence weighting, position-specific gap penalties and weight matrix choice. *Nucleic Acids Research* 22: 4673–4680.

Thyssen, P.J., Lessinger, A.C., Azeredo-Espin, A.M.L. & Linhares, A.X. (2005) The value of PCR-RFLP molecular markers for the differentiation of immature stages of two necrophagous flies (Diptera: Calliphoridae) of potential forensic importance. *Neotropical Entomology* 34: 777–783.

Torres, E.P., Foley, D.H. & Saul, A. (2000) Ribosomal DNA sequence markers differentiate two species of the *Anopheles maculatus* (Diptera: Culicidae) complex in the Philippines. *Journal of Medical Entomology* 37: 933–937.

Vences, M., Thomas, M., Bonett, R.M. & Vieites, D.R. (2005a) Deciphering amphibian diversity through DNA barcoding: chances and challenges. *Philosophical Transactions of the Royal Society B: Biological Sciences* 360: 1859–1868.

Vences, M., Thomas, M., van der Meijden, A., Chiari, Y. & Vieites, D.R. (2005b) Comparative performance of the 16S rRNA gene in DNA barcoding of amphibians. *Frontiers in Zoology* 2: 5.

Vilgalys, R. (2003) Taxonomic misidentification in public DNA databases. *New Phytologist* 160: 4–5.

Vogler, A.P. & Monaghan, M.T. (2007) Recent advances in DNA taxonomy. *Journal of Zoological Systematics and Evolutionary Research* 45: 1–10.

Wallman, J.F. & Donnellan, S.C. (2001) The utility of mitochondrial DNA sequences for the identification of forensically important blowflies (Diptera: Calliphoridae) in southeastern Australia. *Forensic Science International* 120: 60–67.

Ward, R.D., Zemlak, T.S., Innes, B.H., Last, P.R. & Hebert, P.D.N. (2005) DNA barcoding Australia's fish species. *Philosophical Transactions of the Royal Society of London B Biological Sciences* 360: 1847–1857.

Wells, J.D. & Williams, D.W. (2007) Validation of a DNA-based method for identifying Chrysomyinae (Diptera: Calliphoridae) used in a death investigation. *International Journal of Legal Medicine* 121: 1–8.

Wheeler, Q.D. (2004) Taxonomic triage and the poverty of phylogeny. *Philosophical Transactions of the Royal Society of London B Biological Sciences* 359: 571–583.

Wheeler, Q.D. & Meier, R. (eds) (2000) *Species Concepts and Phylogenetic Theory. A Debate*. Columbia University Press, New York; 230 pp.

Wheeler, Q.D. & Platnick, N.I. (2000) The Phylogenetic Species Concept sensu Wheeler and Platnick. Pages 55–69 *in*: Wheeler, Q.D. & Meier, R. (eds), *Species Concepts and Phylogenetic Theory. A Debate*. Columbia University Press, New York.

Wiens, J.J. & Penkrot, T.A. (2002) Delimiting species using DNA and morphological variation and discordant species limits in spiny lizards (*Sceloporus*). *Systematic Biology* 51: 69–91.

Wilkerson, R.C., Li, C., Rueda, L.M., Kim, H.-C., Klein, T.A., Song, G.-H. & Strickman, D. (2003) Molecular confirmation of *Anopheles* (*Anopheles*) *lesteri* from the Republic of South Korea and its genetic identity with *An.* (*Ano.*) *anthropophagus* from China (Diptera: Culicidae). *Zootaxa* 378: 1–14.

Will, K.W., Mishler, B.D. & Wheeler, Q.D. (2005) The perils of DNA barcoding and the need for integrative taxonomy. *Systematic Biology* 54: 844–851.

Will, K.W. & Rubinoff, D. (2004) Myth of the molecule: DNA barcodes for species cannot replace morphology for identification and classification. *Cladistics* 20: 47–55.

Willmann, R. & Meier, R. (2000) A critique from the Hennigian species perspective. Pages 101–118 *in*: Wheeler, Q.D. & Meier, R. (eds), *Species Concepts and Phylogenetic Theory. A Debate.* Columbia University Press, New York.

Xiong, B. & Kocher, T.D. (1991) Comparison of mitochondrial DNA sequences of seven morphospecies of black flies Diptera Simuliidae. *Genome* 34: 306–311.

Xu, J.-N. & Qu, F.-Y. (1997) Ribosomal DNA difference between species A and D of the *Anopheles dirus* complex of mosquitoes from China. *Medical and Veterinary Entomology* 11: 134–138.

Yu, D.-j., Chen, Z.-l., Zhang, R.-j. & Yin, W.-y. (2005) Real-time qualitative PCR for the inspection and identification of *Bactrocera philippinensis* and *Bactrocera occipitalis* (Diptera: Tephritidae) using SYBR Green assay. *Raffles Bulletin of Zoology* 53: 73–78.

Yu, D.J., Zhang, G.M., Chen, Z.L., Zhang, R.J. & Yin, W.Y. (2004) Rapid identification of *Bactrocera latifrons* (Dipt., Tephritidae) by real-time PCR using SYBR Green chemistry. *Journal of Applied Entomology* 128: 670–676.

Zehner, R., Amendt, J., Schuett, S., Sauer, J., Krettek, R. & Povolný, D. (2004) Genetic identification of forensically important flesh flies (Diptera: Sarcophagidae). *International Journal of Legal Medicine* 118: 245–247.

CHAPTER FOURTEEN

DIPTERA BIODIVERSITY INFORMATICS

Shaun L. Winterton

Queensland Department of Primary Industries & Fisheries, Indooroopilly, and University of Queensland, St. Lucia, Australia

Introduction

The study of insects has relied traditionally upon a simple experimental process using a select number of tools (microscope, net, etc.) for observation, postulation, experimentation and prediction. With the development of computers, the Internet and digital technologies such as genomic libraries, automated taxonomic tools and distributed databases, tedious analysis of primary data is no longer the dictum. Metadata analyses provide sophisticated analytical tools for study of mega-diverse insect groups like Diptera, including biodiversity hotspot delimitation, ecological, spatial and temporal predictive modelling, interactive diagnostic tools and matrix-based species descriptions. The chapter provides a basic snapshot in time of the ever-expanding wealth of biodiversity informatics tools and methodologies available for the study of true flies.

Worldwide biodiversity of true flies (Diptera) is immense, and only rivalled among insects by other holometabolous orders like Coleoptera, Hymenoptera and Lepidoptera. With many species still undescribed and more remaining undiscovered, the study of dipteran biodiversity is a daunting task. Almost a quarter of a century ago, Gilbert & Hamilton (1983) summarised the diverse information resources available to entomologists at that time. Only eight pages of their book were devoted to digital or computer-based resources while the rest focused on published literature in hard-copy; there is even information on a very basic internet search for literature. Since then, major advances in computing power and information technology, as well as the wide and rapid availability of information via the Internet, provide a tremendous resource for analysis of Diptera biodiversity data. Biodiversity informatics is an emerging field principally reliant on metadata (i.e. data about primary data), encompassing the creation (e.g. databasing), integration and analysis of these data as a tool for understanding biodiversity on a macro-scale. These are exciting times in a dynamic field in constant flux, with significant advances to the fundamental toolkit with which biodiversity research is conducted occurring it seems daily. To even attempt to summarise all the resources at a single point in time is truly a Sisyphean task and I beg the readers pardon if a particular resource has been omitted or subsequently changed. This chapter is not an exhaustive compilation of all of the informatics resources available or currently being used by those studying Diptera. Instead, it is merely a snapshot in time of the important and innovative digital resources being used in the study of Diptera (especially systematics) and some important primers included from other fields that should be adopted by dipterists to ensure that the study of Diptera remains at the forefront of research on biodiversity.

In stark contrast with vertebrate collections, major entomology collections contain large holdings of specimens at various levels of curation, with many measured in the hundreds of thousands or even millions of specimens. Each specimen usually has accompanying it a wealth of metadata in the form of collection label information. Unfortunately, much of this label data is unavailable for metadata analysis because it is mostly not in digital form, is highly inconsistent in the amount and format of data included, and older specimens have usually handwritten labels that require some taxonomic detective-work and expertise to interpret them. This discontinuity between the primary information regarding the fly specimen in the field and the published label data is a major impediment to value-adding those data for biodiversity informatics. Most entomology collections have some level of digitising (i.e. database entry of primary data) effort in place to retrospectively capture this data, but it is time consuming and error prone. Plant and vertebrate collections are very small in comparison to most entomology collections, and so are easily digitised enabling relevant researchers to take full advantage of biodiversity informatics resources and technology. The sheer magnitude of entomology collections means that digitisation of label data is rarely more than 10–20% of specimen holdings but representing significantly more data points for metadata analyses (Johnson 2007) and making them potentially more valuable in a statistical sense. For dipterists and the wider entomological community to be able to fully capitalise on biodiversity informatics, more resources need to be sourced for digitising specimen label data so that this information can be available for metadata analyses. This data may come directly from label information on specimens, or as suggested by Meier & Dikow (2004), it may be more cost effective and accurate to obtain such label data quoted in taxonomic revisions.

Online information essentially comes in two forms: unstructured content or semantically structured content. Hypertext mark-up language (HTML) is unstructured content that is made up of tags which describe to an internet browser how information is to be presented (e.g. italics, bold, etc.), but does not convey meaning (e.g. identifying a scientific name) (Johnson 2007). Many of the current internet resources available on Diptera are in the form of static web pages. Extensible mark-up language (XML) is designed to help facilitate sharing of information across the Internet by providing a self-descriptive way to communicate the meaning of content through semantic constraints that are user legible.

1. Databases: The Foundation of Informatics

The database is the basis of biodiversity informatics. A wide selection of software is available commercially or open-source for use and range from very basic to complex relational databases. Data or information is searchable in most forms, but the structure of the data determines its ultimate utility and compatibility with other forms of data storage and retrieval. Information may be in a variety of forms varying in organisation: (a) unstructured and un-atomized information (e.g. a word processor document), (b) table of atomised information (e.g. spreadsheet), and (c) atomised and normalized content in a relational database (e.g. Microsoft Access©, FileMaker Pro©, etc.). A relational database essentially stores information in one or more two-dimensional arrays called relations (i.e. tables). Records in each table represent occurrences while fields represent the attributes of that occurrence. Typical relational databases may contain dozens or even hundreds of these interrelated tables from which other tables can source information and present it in completely new tables. Particularly as the size of relational databases increase, normalisation is critical to data integrity as a method to reduce information duplication

Diptera Diversity: Status, Challenges and Tools
(eds T. Pape, D. Bickel & R. Meier). © 2009 Koninklijke Brill NV.

and consequent logic conflicts arising from subsequent non-concerted modification of record duplicates.

Database structure is always highly idiosyncratic, especially across entomology and systematics (including dipterology), and frequently reflects the rather provincial needs of the compiler (e.g. collection) rather than needs of an end-user. The real power and utility of biodiversity informatics comes from combining data about a taxon from multiple sources through integration of information from disparate databases. A distributed database is one physically stored in two or more computer systems. Although geographically dispersed, a distributed database system manages and controls the entire database as a single collection of data. Communication between locations is by a common set of fields and language, with the entire operation under the control of a central management system and single user interface. Integration and sharing of information between disparate databases requires standardisation so that the information is meaningful and uniform in context between databases and requires development of a common language so that it is understood by the data provider(s) and users (Johnson 2007). Various standards are being developed for data transfer depending on the type of data. Development and promotion of a standardised set of data for biodiversity information is a primary mission of *Biodiversity Information Standards* (TDWG) [formerly the *Taxonomic Database Working Group*]. This is a group that develops standards and protocols for sharing biodiversity data, largely using XML schema such as Structure for Descriptive Data (SDD) (Hagedorn *et al.* 2005). The *Global Biodiversity Information Facility* (GBIF) has goals that include facilitation of data mining (metadata analyses), digitisation of collections data, electronic catalogue of names, species information pages, outreach, and digitisation of literature. Table 14.1 contains links to these and various other websites considered useful to dipterists and mentioned in the following text.

1.1 Taxonomic databases

Taxonomic names are the logical identifier for searches for information (e.g. articles, specimen metadata, images, gene sequences, text, etc.) about a taxon. Unfortunately, this identifier is not fixed or unique, since taxonomy changes, a taxonomic binominal may refer to more than one taxon (homonyms) or a single taxon may have more than one binomen (synonyms). There are also confounding issues for database information retrieval regarding lexical variants and gender. Without a unique identifier this creates problems with information retrieval in databases. Taxonomic databases are designed with a purpose of organising names and classification data into a standard format so that each taxonomic entity is uniquely identifiable and the relationships between the names understood. There are a number of independent and collaborative efforts aimed at providing a comprehensive index of organism identifiers. Some examples of the taxonomic databases include *Index to Organism Names* (ION), *International Plant Name Index* (IPNI), *Integrated Taxonomic Information System* (ITIS) (Blum 2000) and the *uBio* project. *uBio* was developed by the scientific library community as a joint initiative called to contribute to international efforts to create a comprehensive catalogue of scientific names of living and extinct organisms. An important taxonomic website for dipterists is *The Diptera Site: The BioSystematic Database of World Diptera* (Evenhuis et al. 2008). This website comprises an extensive taxonomic names database for all Diptera (extant as well as extinct) and enables users to check the validity of taxonomic names as well as supplementary information such as type, family placement and source reference. A literature database provides a comprehensive list of published papers in which Diptera have been described. *ZooBank* is another online index, which is being developed to provide an official registry of zoological nomenclature, including information on publication of nomenclatural acts, authors of names, and type specimen information (Polaszek *et al.* 2005, http://zoobank. org/). On a regional scale, *Fauna Europaea* (http://www.faunaeur.org/) will contribute to the *European Community Biodiversity Strategy* as an exhaustive database of scientific names and distribution records of all European species (Pape 2008).

1.2 Genetic sequence databases

Large sequence databases are searchable online for access to genetic sequence data of numerous organisms for use in a wide variety of fields but with emphasis on human health and disease research. Databases of the *National Center for Biotechnology Information* (NCBI) (i.e. GenBank), the *European Molecular Biology Laboratory Nucleotide Sequence Database* (EMBL) and the *DNA Data Bank of Japan* (DDBJ) exchange data on a daily basis to make available genetic information on more than ¼ million described species (Benson *et al.* 2008). Genetic sequences are submitted by the scientific community, often as a prerequisite for publication

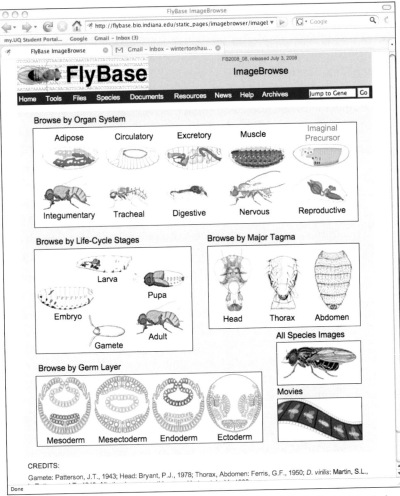

Figure 14.1. Image database of *FlyBase* covering all morphological, physiological and genetic systems of *Drosophila* (Drosophilidae).

by many journals where the results of associated research are published. Besides nucleotide sequence information, additional databases on litera-ture, protein sequence and genomic data are provided, along with tools for data mining, sequence analysis, protein structural modelling, genetic maps, etc. The *FlyBase* project is carried out by a consortium of *Drosophila* researchers and computer scientists at Harvard University, University of Cambridge (United Kingdom) and Indiana University. The *FlyBase* web-site provides an exhaustive database of genes and genomes of *Drosophila* spp. for use by researchers utilising species of this genus as model organisms for the study of developmental biology, genetics, molecular biology, nucleotide and protein evolution, etc. Detailed information on each gene from the *Drosophila* genome is searchable via the GBrowse database. Flybase also provides extensive image databases of anatomical and physiological systems of all life stages of *Drosophila* spp. (Fig. 14.1) (Grumbling *et al.* 2006; Wilson *et al.* 2008).

1.3 Phylogeny databases

Publication of studies of organismal evolution and phylogeny often involves making available online supplementary information such as phylogenetic tree files, DNA sequence alignments and matrices of molecular sequences and morphological scoring in nexus format. *TreeBASE* provides an online storage and retrieval mechanism for phylogenetic information associated with publications on phylogeny of all biotic groups (Sanderson *et al.* 1993; 1994).

1.4 EOL/TOL

The number of described species on Earth is around 1.8 million, and the total including yet undiscovered species is estimated at between 8-9 million (Chapman 2006). Disseminating information on documented species via the Internet in a comprehensive format available to the wider society is of paramount importance, with a number of initiatives in progress.

Encyclopedia of Life (EOL) is an initiative involving massive international collaboration, which aims to produce a comprehensive online reference resource comprising dynamically produced fact sheets on all of the species of named organisms on Earth. Dynamic pages are generated, even personalised based on the end user requirements, with information sourced from many data stores (e.g. mining scientific literature or character databases) throughout the Internet. In a similar initiative, the *Tree of Life Web Project* (TOL) (Maddison *et al.* 2007) is an online collection of web pages about biodiversity compiled collaboratively and arranged phylogenetically through the hierarchy of life. Accounting for the considerable overlap in the goals of both EOL and TOL Web Project, these initiatives have recently begun to coordinate their activities to avoid duplication through improved sharing of content and complementary activities. TOL Web Project will focus more on developing higher-level phylogenetic content while EOL will focus more on species-level content.

DIPTERA DIVERSITY: STATUS, CHALLENGES AND TOOLS
(EDS T. PAPE, D. BICKEL & R. MEIER). © 2009 KONINKLIJKE BRILL NV.

Another initiative is *Wikispecies*, a multilingual, web-based, directory of species that is collectively authored by volunteers from all around the world. Author access is relatively unlimited and articles can be edited by anyone. This type of information resource has shown a capacity to grow enormously in content since its inception in 2001, and now covers most subjects imaginable, including topics relating to dipterology. An obvious drawback is the relative lack of rigorous academic/scientific peer review, resulting in credibility issues of content.

1.5 Digitised texts

Seminal works that are in frequent use have recently started to appear on the Internet to enable wider availability and use. In an effort to make significant older literature on biodiversity available in a searchable digitised format, ten major natural history museum libraries, botanical libraries, and research institutions have joined to form the *Biodiversity Heritage Library Project*. This group is developing a strategy and operational plan to digitize the published literature of biodiversity held in their respective collections. This literature will be available through a global Biodiversity Commons (Moritz 2002). Examples of significant works being made available online in digital form included, for example, *Nomenclator Zoologicus* and *Biologia Centrali-Americana*. *Nomenclator Zoologicus* (Neave 1939-1996) is an essential taxonomic reference for original descriptions of new genera and is an exhaustive list of all published generic names of animals. An online, version of *Nomenclator Zoologicus* is now available as an online database and is continually being updated as new genera are described (Remsen *et al.* 2006). *Biologia Centrali-Americana* (BCA) was originally published as a series of 63 volumes forming the fundamental knowledge-base of the neotropical flora and fauna as was known at the end of the 19th Century until the last volume was published in 1915. This important work has now become *electronic Biologia Centrali-Americana* (eBCA) as all of the pages have been reproduced online as PDF and JPEG images with XML mark-up.

An unfortunate trend recently is that many scientific journals are no longer willing to publish taxonomic papers where readership is likely to be limited to a small group of specialist taxonomists. Fortunately, one scientific journal is filling this void by providing a rapid means of publication of taxonomic descriptions of new species of animals. *Zootaxa* was first published in 2001, and has increased near exponentially in the number of pages published, and as of August 2008, has published descriptions of 8,361 new species (including 1,469 Diptera) in 1,842 papers. In 2008 the journal *ZooKeys* began publication of peer-reviewed, rapidly produced systematics papers on biodiversity similar to *Zootaxa*, but as open access to support the free exchange of ideas and information in response to the recent and rapid developments in the way taxonomic research is undertaken (Penev *et al.* 2008).

1.6 Image databases

Rapid and wide dissemination of biological images is becoming routine in scientific research, often as supplementary material associated with published research in scientific journals where publication of such information is prohibitively expensive. Two recent initiatives for open web repositories of images are *Morphbank* (Morphbank 2007) and *Morphobank* (O'Leary & Kaufman 2007). *Morphbank* serves as a permanent archive of digital images used in specimen-based research allowing them to be shared with other researchers (Morphbank 2007). *Morphobank* facilitates collaborative research on phylogenetics by providing a means to share and annotate character information through dynamic matrices with labelled images (O'Leary & Kaufman 2007). These image databases make available large libraries of images for other researchers (a search for 'Diptera' thus returned 33,153 images) and use search tools such as Web Services and Life Science Identifiers (LSIDs) to link to other online databases. Recently there have been moves to combine *Morphobank* and *Morphbank* into a single database, thus eliminating any redundancy between the two (Wheeler 2007).

1.7 Specimen-level databases

Specimen-level databases comprise information principally focused around the collecting event for an individual specimen. These databases typically are enormous in size and deal with numerous aspects of the individual specimen (often as a relational database), including label data, georeferencing, loan information, taxonomy, nomenclature, bibliography, specimen images and so on. Most entomological collections have some sort of specimen-level digitisation project underway, but like all database projects there is little uniformity in software used (see Berendsohn 2005), and database structure between them (Johnson 2007). Combining metadata from multiple collections via distributed specimen-level databases will ultimately provide a substantial synthetic tool for research on broader questions on organismal ecology, biogeography and evolution, including development of more complex models.

2. Digital Taxonomy

The highly structured nature of taxonomic description makes it highly suitable for presentation in a matrix and therefore for digitisation. Using a matrix, taxa can be scored for the presence or absence of an enumerable range of characters and character states. Although characters are frequently scored as binary, the number of character states is also unlimited. This matrix may form the basis of a character database, it may be used to generate interactive diagnostic tools (see below), or the output used for created highly ordered and standardised Natural Language Descriptions. A previously widely used format is *Descriptive Language for Taxonomy* (DELTA) (Dallwitz 1980; Dallwitz & Paine 2005; Dallwitz *et al.* 1993), which has been primarily used for natural language taxonomic descriptions (e.g. Kim 1994) and interactive key production despite having a rather steep learning curve for new users and is no longer being actively developed or supported. More recently, later versions of *Lucid* (version 3.1 onwards) are able to generate sophisticated natural language descriptions using XML and XML transformations. Considering the chronic and far reaching implications of the taxonomic impediment, with diminishing resources and dwindling expertise for taxonomy in the face of overwhelming undescribed biological diversity, the power of this tool has not been widely utilised for improving the efficiency of species description. Describing species is a time-consuming, careful process requiring specialised expertise and knowledge about a specific group of organisms. Some of the steps in this process are amenable to increased efficiency in data handling by scoring characters in morphological matrices and using these data in digital format to formulate descriptions of new species directly, along with generating interactive keys from the same data. This enables taxonomists to move away from hand-crafted and tediously composed species descriptions in word processors, towards digitised character data where description involves a process of checking appropriate character states followed by subsequent transformation into species descriptions in XML. Deans & Kawada (2008) have recently integrated internet resources within a taxonomic treatment of a new genus of Evaniidae (Hymenoptera), including links to images in *Morphbank*, and species pages on a taxon specific website and the *Tree of Life Web Project*, links to interactive keys and registration of taxon names in *ZooBank* with assignment of LSIDs for each name. The paper is also marked up in *TaxonX*, an XML schema for encoding taxonomic literature structured to facilitate extraction of textual components (scientific names, localities, morphological characters, etc.) by external web applications and use in distributed web resources.

To understand the complexities of adult fly anatomy and morphology, a recent contribution is the online *Anatomical Atlas of Flies* (Yeates *et al.* 2005). This interactive atlas was produced using high-resolution images to accompany interactive keys to flies (e.g. Winterton *et al.* 2005) or as an aid in teaching Diptera morphology. Individual anatomical structures can be identified by clicking on them or by clicking on a term to locate it on the fly (Fig. 14.2). Representatives of four major fly groups (Lower Diptera, Lower Brachycera, Acalyptratae and Calyptratae) are included in dorsal and lateral view, and a dragable magnifying lens is provided to see greater detail.

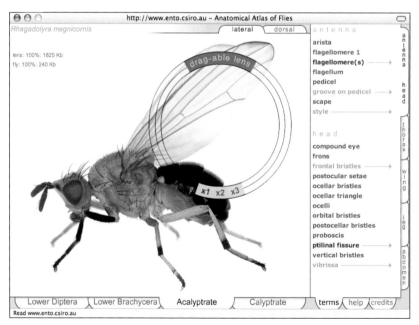

Figure 14.2. Screenshot of *Anatomical Atlas of Flies* showing dragable
magnifying glass and terminology
(from Yeates et al. 2005).

Diptera Diversity: Status, Challenges and Tools
(eds T. Pape, D. Bickel & R. Meier). © 2009 Koninklijke Brill NV.

3. Character Data and Matrices: Interactive Identification Tools

Traditional methods for identification rely on the use of dichotomous keys that usually accompany a taxonomic treatment of a particular group (e.g. revision of a genus), and are produced as a hard copy in a journal publication. Dichotomous keys provide contrasting statements (i.e. leads) of characteristics (usually two) that comprise a couplet. These then lead to either a taxon representing those characteristics, or another couplet. The user wishing to identify a specimen follows the leads through individual couplets until a taxonomic entity is reached. By nature, dichotomous keys are usually full of taxonomic jargon, very few illustrations and are frequently criticised as only being usable by taxonomists familiar with the group already and thereby do not need them (Walter & Winterton 2007). Nevertheless dichotomous keys are used successfully on the Internet, and while the interface and format may vary, the basic premise of a pathway key remains constant. Dichotomous keys are frequently available on the Internet as simple text-based keys included on web pages as they would have been included in a journal publication. Through HTML links and anchors, the interface of this 'flat' key may be improved by representing a couplet as a separate web page and pointers to the next couplet as links [other pages] or anchors [couplets on the same web-page]. This enables the author to populate the key with supplementary images and text that would be otherwise unavailable in the printed publication. The most sophisticated manifestation of the multimedia dichotomous key is *Lucid Phoenix*. This software package is designed to either newly construct or import existing digitised (e.g. via optical character recognition (OCR)) dichotomous keys and populate them with multimedia that is then compiled into an interface that resembles a matrix-based interactive key (see below). This program is particularly useful for making historically important dichotomous keys available on the Internet with supplementary multimedia (Walter & Winterton 2007)

A fatal flaw of dichotomous, or pathway keys, is the unanswerable couplet. This occurs when characters in a lead are not observable for any reason (e.g. damaged specimen, different life stage, opposite sex) or the user is unable to interpret what the author means. The immediate result is termination of key utility at that couplet, or even worse, the user makes a guess, which can lead to incorrect identification (i.e. false-positive). Matrix-based keys have the advantage of giving the user a choice to simply

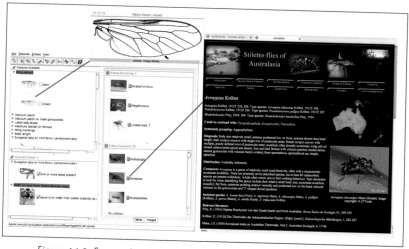

Figure 14.3. Screenshots of Lucid matrix key showing browser interface,
character state images and fact sheet
(*from Winterton et al. 2005*).

choose characters in any order and thus are also known as multi-access or random-access keys. matrix-based keys are made up of a database of character (or feature) states scored for taxa in the key, a program to query the database and a user-interface (usually a set of windows displaying character state selection and taxon remaining or eliminated) (Fig. 14.3) (Dallwitz 2006; Walter & Winterton 2007). The user selects character states by which the program discards those taxa not exhibiting the particular state, thereby an identification is made via a process of elimination. The problem of the afore-mentioned unanswerable couplet problem is avoided by enabling the user to simply choose another character if one character is not observable or confusing. To ensure the most efficient pathway to an identification, such programs can also have functions like 'Best' (generally preferring characters that parse remaining taxa equally) or 'Bingo' (list of unique characters that will derive a single remaining taxon). Matrix-based interactive keys to various insect groups, including Diptera (Table 14.1), are available on either CD or more increasingly on the Internet where the interface is commonly operating in an internet browser. There are numerous interactive key software products available (freely or commercially), but commonly used ones are *20q* (Discover Life), *Intkey* (Dallwitz 1980), *Lucid* (Lucidentral) and *Linnaeus II* (ETI BioInformatics).

Diptera Diversity: Status, Challenges and Tools
(eds T. Pape, D. Bickel & R. Meier). © 2009 Koninklijke Brill NV.

There are various advantages to using matrix-based interactive keys. Besides the problem of the unanswerable couplet, another advantage includes the practically unrestricted use of multimedia (e.g. images, audio, video, search engines and hypertext links). Being matrix-based, users unfamiliar with a character may choose to use another, as the order of character selection is arbitrary. Character parsing algorithms (e.g. 'Best', 'Bingo') enable a highly efficient pathway through a key to make an identification, often reducing the number of required choices (and possible chances for error) exponentially. Error tolerances can also be built into the key by the author using compensatory scoring or weighting schema to anticipate common mistakes (Walter & Winterton 2007). Walter & Winterton (2007) give 10 recommendations for building interactive keys using best practice, all aimed at increasing key efficiency by reducing error and increasing the speed of diagnosis.

Matrix-based interactive keys use character state databases. Therefore, just as distributed databases need to be able to communicate and transfer information between each other, an important consideration when developing interactive keys and key software is to ensure that the key data is available in SDD for future use in distributed databases and integrated online keys. Few interactive key programs currently are SDD-compliant, but those that are include *Lucid* 3 and the *Electronic Field Guide* project (Johnson 2007).

The clear diagnostic advantages of developing interactive keys are overwhelming, with a number of Diptera workers publishing such keys on CD or on the Internet. Unfortunately, key development remains piecemeal at various taxonomic levels and to anyone reading the current taxonomic literature it is obvious that the number published yearly is tiny compared to the number of traditional paper-based keys published. Reasons for this disparity are varied and difficult to quantify, but likely relate to the perceived difficulties of producing such keys, overestimates of the time required to produce the accompanying digital media, and minimal reward from employers for such publications that do not have a citation index.

4. Data Analyses

Specimen-based information served on the Internet can be used in a variety of metadata analyses for the study of Diptera biodiversity. Meier & Dikow (2004) analysed specimen data from a revision of a genus of robber flies (Asilidae: Leptogasterinae: *Euscelidia* Westwood) using nonparametric species richness estimators to estimate global species diversity for that genus and to assess whether biodiversity hotspots established using plant data also reflect significant biodiversity richness for invertebrates. Examples of similar studies of species richness have been conducted on flies by Novotny *et al.* (2005) (Tephritidae), McCreadie *et al.* (2005) (Simuliidae), Roque *et al.* (2006) (Chironomidae), Foley *et al.* (2007) (Culicidae) and Petersen & Meier (2003) (Danish Diptera).

Distribution maps are a commonly used representation of the geographic range of Diptera. The best way to display this is as dot points on the map representing collecting or observation events for a particular taxon. Sometimes distributions are represented as shaded areas on a map based on the researcher's estimation of predicted range from a previous set of distribution points; such predicted distributions are rife with significant and perilous assumptions about homogeneity of occurrence of the taxon throughout the landscape. Information used to plot data points on a distribution map are sourced directly from label data on specimens or from field observations and entered as coordinates (latitude and longitude). Modern field collecting usually involves a global positioning system (GPS) so that collection labels usually now contain GPS coordinates which can be entered directly into GIS software. Older specimens without GPS coordinates require rather tedious georeferencing that is costly in time for manual data entry. Collecting localities on labels can be located using online or networked gazetteers and geographic name servers (Johnson 2007). An important development in GIS on the Internet is *Google Earth*™, an online virtual globe, which represents the earth as a series of superimposed images obtained from high-resolution satellite and aerial photography. While dipterists are slow to use this powerful technology, other fields of entomology have started to utilise it for plotting distribution maps of ant species on a digital globe (Antweb 2007). Predicting distributions based on specimen label data can be estimated using software programs such as Bioclim (Busby 1991). Ecological niche modelling (ENM) or bioclimatic modelling uses georeferenced collection data from specimen labels combined with environmental parameters on distribution maps to compile models of ecological requirements of individual species. Conditions matching those predefined for that species are then represented on digital maps indicating the potential distribution of the species within a range of probabilities (e.g. *Climex* (Creative Research Systems)) (Pearson & Dawson 2003; Peterson & Shaw 2003; Martinez-Meyer 2005). Examples of use in Diptera: *Lutzomyia* Franca (Peterson & Shaw 2003) and Ceratopogonidae (Wittman *et al.* 2001). This type of modelling has particular useful application for pathway analysis for exotic insect incursions, and for climate change effects on distribution. Another software package that can be used for similar analyses of ecological niche modelling is *Lifemapper*, which is an automated equivalent to *DesktopGARP*. *Lifemapper* obtains species distribution data from specimen collection data held in distributed databases that are accessed using DiGIR.

5. Conclusions

Diptera research, especially collection-based systematics studies, is at a crossroads: continue using outdated and time-consuming methodologies to study fly natural history and biodiversity, or embrace new informatics technologies and develop dipteran research in new and innovative directions. Development of databases of information on Diptera and its sharing across distributed resources provides limitless possibilities for metadata analyses, with electronic dissemination of results via the Internet in the fields of systematics, ecology, biodiversity discovery and pest diagnostics, to name but a few.

Initiatives by organisations like GBIF are fostering the development of large distributed databases of specimen-based information accessible via the Internet. This is an extremely powerful analytical tool for studies on Diptera biodiversity. It also allows for unprecedented integration of information in a multimedia format. While it appears to be a panacea of information, this electronic format has its weaknesses. The most common criticism is the apparent ephemeral nature of the content, as websites may disappear or move. Moreover, as touched upon with issues of 'Wiki' type websites, content on the Internet may sometimes lack the rigorous peer review of established print outlets like scientific journals. Issues associated with digital publication include establishing peer review processes, permanency of information and recognition of digital publications by employers and are still being addressed (e.g. recent provision of ISBN numbers for internet publications). These intrinsic problems of electronic media are far outweighed by the advantages gained by using them both as analytical tools and for dissemination of results. Obvious advantages over static print media are that digital media are rapidly delivered, highly integrated with varied information sources, often with multimedia (e.g. video, images, sound) and they are updated as information changes (e.g. as taxonomy changes). This issue of ephemeralness is being addressed through assignment of globally unique identifiers (GUIDs) to biological resources (e.g. specimens, genes, names etc.) and assignment of a digital object identifier (DOI) as a permanent identifier given to electronic documents (PDF, video, software, etc.). Both provide persistent, location-independent identifiers to objects on the Internet regardless of transfer to a different location after initial publication. Images are arguably the most important aspect of diagnostics and studies of biodiversity, and with the Internet and increasing computing power, more information is available faster for contrasting character states, and colour image galleries providing limitless numbers of images of taxa and multiple views of specimens to convey much more information in a comparative way than what could be done in print media. This speed of delivery will improve even more in the future with a move to semantic web (*Web 2.0*), with improved software development which will integrate information sourced from a variety of distributed sources including interactive keys, catalogues (data sourced from other online databases), bibliographies (sourced elsewhere), image galleries, etc.

A fundamental issue of the taxonomic impediment is the immense biodiversity remaining to be discovered and formally documented, and the lack of taxonomic descriptive capacity in the scientific community to even come close to achieving this goal. End users of taxonomic information (i.e. names) typically simply require a name that can be accurately applied to a species, a minimal requirement for communication of concepts and hypotheses associated with it. Systematic neglect of descriptive taxonomy is rapidly resulting in a diminished capability of reliably applying correct names to specimens, thus undermining the accuracy of any information and hypotheses (e.g. ecology, biology, etc.) subsequently generated from them (Meier & Dikow 2004, Bortolus 2008, Wheeler 2007). Traditional methods of taxonomic description are tedious and outdated. Utilisation of digital tools for publication of taxonomic names is an essential step forward for taxonomic description if we are to make any advances in realistic timeframes towards describing the earth's biota. Taxonomic publications need to set new standards for publishing descriptions by incorporating electronic features such as embedded hypertext links, and e-references etc., including links to online databases of specimen collection data, images, nomenclature and DNA sequences. This approach has been embraced by the journal *ZooKeys* (Penev et al. 2008). Natural language output from programs like *DELTA* and *Lucid* will enable rapid taxonomic description coupled with digital media sourced from online resources. An important step in this process is the input of primary descriptive data, in this case into character state matrices, which are then used to output descriptions and generate interactive keys. Recent developments promise to speed up the process through iterative parsing of published taxonomic descriptions (e.g. XML mark-up) and development of character state lists using rule sets and libraries. Rapid digital tools for describing biodiversity have particular advantages considering the recent depletion of traditional taxonomy in both active taxonomists and limited allocation of resources to undertake such endeavours.

Bioinformatics resources for dipterists are immense, and promise to become even more diverse in the near future, presenting opportunities for dipterists to use them both basic and applied research, especially in web-based applications. As a discipline studying one of the most biologically diverse groups of organisms on Earth we cannot ignore the potential benefits of embracing this technology for improving our understanding of these insects and delivering the products of our research in a rapid and meaningful way to both the scientific and wider communities. We need more efficient means to deliver systematic and diagnostic identification services, as well as better web-based visualisation and analysis tools for tracking biodiversity change (Guralnick *et al.* 2007). A lesson that could be learned by the dipterist community is that provided by the *Australian Virtual Herbarium* (Council of Heads of Australian Herbaria 2003). This initiative provides through a single portal, access to the holdings of a network of herbaria throughout Australia, making available to the wider community specimen images, distribution data and maps. A World Virtual Diptera Collection, utilizing information on the holdings of various collections around the world, could eventually provide information on all the species of flies as fact sheets, image galleries, distribution maps, interactive keys, etc. via a single portal such as EOL or GBIF. Such data is also then widely available for metadata analyses for biodiversity research. This utopia may not be far away, as many collections are already actively databasing label data of specimens holdings, as well as systematically posting digital images of types online so that they can reduce the risk of damaging or loosing these precious specimens by sending them in the mail to taxonomists. For this to be realised, the focus though must continue on acquiring the primary label data held in collections, continued discovery and description of new taxa using digital taxonomic tools, and most importantly adoption by the present and future cohorts of taxonomists of these digital tools so that we may come close to documenting the earth's biological diversity within a reasonable timeframe.

DIPTERA DIVERSITY: STATUS, CHALLENGES AND TOOLS
(EDS T. PAPE, D. BICKEL & R. MEIER). © 2009 KONINKLIJKE BRILL NV.

Table 14.1. Examples of general bioinformatics sites and taxon related pages on Diptera found on the Internet or published on CD (accessed July 2008).

The Diptera site	http://www.diptera.org
Bug guide	http://bugguide.net
Dipterist Forum	http://www.dipteristsforum.org.uk/
Hoverfly Recording Scheme	http://www.hoverfly.org.uk/
Diptera.info	http://www.diptera.info
Catalog of the Diptera of the Australasian and Oceanian Regions	http://hbs.bishopmuseum.org/aocat/aocathome.html
Catalogue of the fossil flies of the world (Insecta: Diptera)	http://hbs.bishopmuseum.org/fossilcat/
Tipulidae	http://ip30.eti.uva.nl/ccw/
	http://iz.carnegiemnh.org/cranefly
Dolichopodidae	http://www.fortunecity.com/greenfield/porton/875/
	http://www.cdfa.ca.gov/phpps/ppd/Lucid/Therevidae/Austherevid/
Therevidae	key/Austherevid/Media/Html/opening_page.html
	http://www.inhs.uiuc.edu/cee/therevid/
Morphbank	http://www.morphbank.net/
Morphobank	http://morphobank.org/
ZooKeys	http://pensoftonline.net/zookeys
Zootaxa	http://www.mapress.com/zootaxa/
eBCA	http://www.sil.si.edu/digitalcollections/bca/
Nomenclator Zoologicus	http://www.ubio.org/NomenclatorZoologicus/
TreeBASE	http://www.treebase.org/
FlyBase	http://flybase.bio.indiana.edu/
NCBI (Genbank)	http://www.ncbi.nlm.nih.gov/

Table 14.1. Examples of general bioinformatics sites and taxon related pages on Diptera found on the Internet or published on CD (accessed July 2008).

UBio	http://www.ubio.org/
ION	http://www.organismnames.com/
GBIF	http://www.gbif.org/
TDWG	http://www.tdwg.org/
Wikispecies	http://species.wikimedia.org/
Biodiversity Heritage Library	http://www.biodiversitylibrary.org/
Lucid Phoenix	http://www.lucidcentral.org/
Google Earth	http://earth.google.com/
ITIS	http://www.itis.usda.gov
IPNI	http://www.ipni.org
Tree of Life Web Project	http://www.tolweb.org/
Lucid Central	http://www.lucidcentral.org/
Encyclopedia of Life	http://www.eol.org/
Electronic field guide	http://efg.cs.umb.edu/efg/
Mydidae & Apioceridae	http://www.mydidae.tdvia.de/index.html
Asilidae	http://www.geller-grimm.de
	http://nlbif.eti.uva.nl/bis/euscelidia.php?menuentry=inleiding
Lifemapper	http://www.specifysoftware.org/Informatics/informaticslifemapper/
Lucid Phoenix	http://www.lucidcentral.com/phoenix/
DesktopGarp	http://www.nhm.ku.edu/desktopgarp/index.html
Creative Research Systems: Climex	http://www.climatemodel.com/climex.htm
ZooBank	http://zoobank.org/

Table 14.1. Examples of general bioinformatics sites and taxon related pages on Diptera found on the Internet or published on CD (accessed July 2008).

Sciaroidea		http://www.sciaroidea.info
Ceratopogonidae		http://campus.belmont.edu/cienews/cie.html
Simuliidae		http://www.Blackfly.org.uk
		http://www.syrphidae.com/
Syrphidae		http://www.syrphidae.de
		http://homepage2.nifty.com/syrphidae
	Mosquitoes of New Zealand,	http://www.lucidcentral.org/
Culicidae	CulicID: Mosquitos of Queensland,	http://www.lucidcentral.org/
	Walter Reed Biosystematics Unit	http://www.wrbu.org/
Discover life		http://www.discoverlife.org
Australian Diptera Families		http://www.csiro.au/resources/ps236.html
Drosophilidae	Drosophila del Ecuador	http://www.lucidcentral.org/
	The Fruitfly project	http://projects.bebif.be/fruitfly/index.html
	Pest Fruit Flies of the World,	http://www.sel.barc.usda.gov/diptera/tephriti/tephriti.htm
Tephritidae	Dacine Fruit Flies,	http://delta-intkey.com/ffl/www/_wintro.htm
	DORSALIS,	Lawson et al. (2003)
	Fauna Malesiana	White & Hancock (2004)
Agromyzidae		http://ip30.eti.uva.nl/bis/agromyzidae.php?menuentry=inleiding
Carnidae		http://www.sel.barc.usda.gov/Diptera/carnid/ca-home.html
Milichiidae		http://www.sel.barc.usda.gov/Diptera/milichid/mi-home.html
Tachinidae		http://www.nadsdiptera.org/Tach/TTimes/TThome.htm
Sarcophagidae		http://www.zmuc.dk/entoweb/sarcweb/sarcweb/Sarc_web.htm

References

AntWeb (2008) Available at http://www.antweb.org/, accessed July 2008.

Benson, D.A., Karsch-Mizrachi, I., Lipman, D.J., Ostell, J. & Wheeler, D.L. (2008) Genbank. *Nucleic Acids Research* 36: 25–30.

Berendsohn, W., (2005) Standards, information models, and data dictionaries for biological collections. Available at http://www.bgbm.org/TDWG/acc/Referenc. htm, accessed July 2008.

Blum, S.D. (2000) An overview of Biodiversity Informatics. Available at http://www. calacademy.org/RESEARCH/informatics/sblum/pub/biodiv_informatics.html, accessed July 2008.

Bortolus, A. (2008) Error cascades in the biological sciences: the unwanted consequences of using bad taxonomy in ecology. *Ambio* 37: 114–118.

Busby, J.R. (1991) BIOCLIM - A Bioclimatic Analysis and Prediction System. Pages 64–68 *in*: Margules, C.R. & M.P. Austin (eds), *Nature Conservation: Cost Effective Biological Surveys and Data Analysis.* CSIRO, Canberra.

Chapman, A.D. (2006) Numbers *of Living Species in Australia and the World.* A Report for the Department of the Environment and Heritage, September 2005. Australian Biodiversity Information Services, Toowoomba, 61 pp.

Council of Heads of Australian Herbaria (2003) Australia's virtual herbarium. Available at http://www.chah.gov.au/avh/, accessed July 2008.

Dallwitz, M.J. (1980) A general system for coding taxonomic descriptions. *Taxon* 29: 41–46.

Dallwitz, M.J. (2006) Programs for interactive identification and information retrieval. Available at http://delta-intkey.com/www/idprogs.htm, accessed July 2008.

Dallwitz, M.J. & Paine, T.A. (2005) Definition of the DELTA format. Available at http://www.deltaintkey.com/www/standard.pdf, accessed July 2008.

Dallwitz, M.J., Paine, T.A. & Zurcher, E.J. (1993) User's guide to the DELTA system: a general system for processing taxonomic descriptions, fourth edition. Available at http://www.delta-intkey.com/, accessed July 2008.

Deans, A.R. & Kawada, R. (2008) *Alobevania*, a new genus of Neotropical ensign wasps (Hymenoptera: Evaniidae), with three new species: integrating taxonomy with the World Wide Web. *Zootaxa* 1787: 28–44.

Evenhuis, N.L., Pape, T., Pont, A.C. & Thompson, F.C. (eds) (2008) *BioSystematic Database of World Diptera*, Version 10.5. Available at http://www.diptera.org/biosys. htm, accessed July 2008.

Foley, D.H., Bryan, J.H. & Wilkerson, R.C. (2007) Species-richness of the *Anopheles annulipes* complex (Diptera: Culicidae) revealed by tree and model-based allozyme clustering analyses. *Biological Journal of the Linnean Society* 91: 523–539.

Gilbert, P. & Hamilton, C.J. (1983) *Entomology, A Guide to Information Sources.* Mansell Publishing Ltd., London, 237 pp.

Grumbling, G., Strelets, V. & The FlyBase Consortium, (2006). FlyBase: anatomical data, images and queries. *Nucleic Acids Research*, 34: D484–D488 (doi:10.1093/nar/gkj068).

Guralnick, R.P., Hill, A.W. & Lane, M. (2007) Towards a collaborative, global infrastructure for biodiversity assessment. *Ecology Letters* 10: 663–672 (doi: 10.1111/j.1461-0248.2007.01063.x).

Hagedorn, G., Thiele, K., Morris, R. & Heidorn, P.B. (2005) The Structured Descriptive Data (SDD) w3c-xml-schema, version 1.0. Available at http://www.tdwg.org/standards/116/, accessed July 2008.

Johnson, N.F. (2007) Biodiversity Informatics. *Annual Review of Entomology* 52: 421–438.

Kim, S.P. (1994) *Australian Lauxaniid Flies. Revision of the Australian Species of* Homoneura *Van Der Wulp,* Trypetisoma *Malloch and Allied Genera (Diptera: Lauxaniidae).* CSIRO Publishing, Canberra, 645 pp.

Lawson, A.E., McGuire, D.J., Yeates, D.K., Drew, R.A.I. & Clarke, A.R. (2003) DOR-SALIS: An interactive identification tool for fruit flies of the *Bactrocera dorsalis* complex. [Multimedia CD-ROM.]

Maddison, D.R., Schulz, K.S. & Maddison, W.P. (2007) The Tree of Life Web Project. Pages 19–40 *in*: Zhang, Z.-Q. & Shear, W.A. (eds), *Linnaeus Tercentenary: Progress in Invertebrate Taxonomy. Zootaxa* 1668: 1–766.

Martinez-Meyer, E. (2005) Climate Change and Biodiversity: some considerations in forecasting shifts in species' potential distributions. *Biodiversity Informatics* 2: 42–55.

McCreadie, J.W., Adler, P.H. & Hamada, N. (2005) Patterns of species richness for blackflies (Diptera: Simuliidae) in the Nearctic and Neotropical regions. *Ecological Entomology* 30: 201–209.

Meier, R. & Dikow, T. (2004) Significance of specimen databases from taxonomic revisions for estimating and mapping the global species diversity of invertebrates and repatriating reliable specimen data. *Conservation Biology* 18: 478–488.

Moritz, T. (2002) Building the Biodiversity Commons. *D-Lib Magazine* 8(6). Available at http://www.dlib.org/dlib/june02/moritz/06moritz.html, accessed September 2008.

Morphbank (2007). Florida State University, School of Computational Science, Tallahassee, FL 32306-4026 USA. Available at http://www.morphbank.net, accessed July 2008.

Neave, S.A. (1939–1996) *Nomenclator Zoologicus.* Volumes 1–9. Zoological Society of London, London. [Vol. 9 edited by Edwards, M.A, Manley, P. & Tobias M.A.]

Novotny, V., Clarke, A.R., Drew, R.A.I., Balagawi, S. & Clifford, B. (2005) Host specialization and species richness of fruit flies (Diptera: Tephritidae) in a New Guinea rain forest. *Journal of Tropical Ecology* 21: 67–77.

O'Leary, M.A. & Kaufman, S.G. (2007) MorphoBank 2.5: Web application for morphological phylogenetics and taxonomy. Available at http://www.morphobank.org, accessed July 2008.

Pearson, R.G. & Dawson, T.P. (2003) Predicting the impacts of climate change on the distribution of species: Are bioclimate envelope models useful? *Global Ecology and Biogeography* 12: 361–371.

Penev, L., Erwin, T., Thompson, F.C., Sues, H.-D., Engel, M.S., Agosti, D., Pyle, R., Ivie, M., Assmann, T., Henry, T., Miller, J., Ananjeva, N.B., Casale, A., Lourenço, W., Golovatch, S., Fagerholm, H.-P., Taiti, S. & Alonso-Zarazaga, M. (2008) Zoo-Keys, unlocking Earth's incredible biodiversity and building a sustainable bridge into the public domain: From "print-based" to "web-based" taxonomy, systematics, and natural history. ZooKeys Editorial Opening Paper. *ZooKeys* 1: 1-7 (doi: 10.3897/zookeys.1.11).

Peterson, A.T. & Shaw, J. (2003) *Lutzomyia* vectors for cutaneous leishmaniasis in Southern Brazil: Ecological niche models, predicted geographic distributions, and climate change effects. *International Journal for Parasitology* 33: 919–931.

Petersen, F.T. & Meier, R. (2003) Testing species-richness estimation methods on single-sample collection data using the Danish Diptera. *Biodiversity and Conservation* 12: 667–686.

Polaszek, A., Agosti, D., Alonso-Zarazaga, M.A., Beccaloni, G., Bjorn, P. de Place, Bouchet, P., Brothers, D.J., Earl of Cranbrook, Evenhuis, N.L., Godfray, H.C.J., Johnson, N.F., Krell, F.-T., Lipscomb, D., Lyal, C.H.C., Mace, G.M., Mawatari, S.F., Miller, S.E., Minelli, A., Morris, S. Ng, P.K.L., Patterson, D.J., Pyle, R.L., Robinson, N., Rogo, L., Taverne, J., Thompson, F.C., Tol, J. van, Wheeler, Q.D. & Wilson, E.O. 2005. Commentary: A universal register for animal names. *Nature* 437: 477.

Remsen, D.P., Norton, K. & Patterson, D.J. (2006) Taxonomic Informatics Tools for the Electronic *Nomenclator Zoologicus*. *Biological Bulletin* 210: 18–24.

Roque, F.O., Trivinho-Strixino, S., Milan, L. & Leite, J.G. (2006) Chironomid species richness in low-order streams in the Brazilian Atlantic Forest: a first approximation through a Bayesian approach. *Journal of the North American Benthological Society* 26: 221–231.

Sanderson, M.J., Baldwin, B.G., Bharathan, G., Campbell, C.S., Ferguson, D., Porter, J.M., Von Dohlen C.,Wojciechowski M.F. & Donoghue, M.J. (1993) The growth of phylogenetic information and the need for a phylogenetic database. *Systematic Biology* 42: 562–568.

Sanderson, M.J., Donoghue, M.J., Piel, W. & Eriksson, T. (1994) TreeBASE: a prototype database of phylogenetic analyses and an interactive tool for browsing the phylogeny of life. *American Journal of Botany* 81: 183.

Walter D.E. & Winterton S.L. (2007) Keys and the Crisis in Taxonomy: Extinction or Reinvention? *Annual Review of Entomology* 52: 193–208.

Wheeler, Q.D. (2007) *Digital Innovation and Taxonomy's Finest Hour*. Pages 9–23 *in*: MacLeod, N. (ed.), *Automated Taxon Identification in Systematics: Theory, Approaches and Applications*. The Systematics Association, London, Series 74: xvi + 1–339.

White, I.M. & Hancock, D.L. (2004) *Fauna Malesiana — Interactive key for dacine fruit flies (Diptera: Tephritidae: Dacini)*. Version 1.0. ETI Bioinformatics, Amsterdam.

Wilson, R.J., Goodman, J.L., Strelets, V.B. & The FlyBase Consortium (2008) FlyBase: integration and improvements to query tools. *Nucleic Acids Research* 36: 588–593 [doi:10.1093/nar/gkm930].

Winterton, S.L., Skevington, J.H. & Lambkin, C.L. (2005) *Stiletto Flies of Australasia*. Online Lucid key and information website. Available at http://www.lucidcentral.com/keys/viewKeyDetails.aspx?id=254, accessed July 2008.

Wittmann, E.J., Mellor, P.S. & Baylis, M. (2001) Using climate data to map the potential distribution of *Culicoides imicola* (Diptera: Ceratopogonidae) in Europe. *Revue Scientifique et Technique de l'Office International des Epizooties* 20: 731–740.

Yeates, D.K., Hastings, A., Hamilton, J.R., Colless, D.H., Lambkin, C.L., Bickel, D., McAlpine, D.K., Schneider, M.A., Daniels, G. & Cranston, P. (2005) *Anatomical Atlas of Flies*. Available at http://www.ento.csiro.au/biology/fly/fly.html, accessed July 2008.

CHAPTER FIFTEEN

MEETING THE INTERRELATED CHALLENGES OF TRACKING SPECIMEN, NOMENCLATURE, AND LITERATURE DATA IN *MANDALA*

GAIL E. KAMPMEIER & MICHAEL E. IRWIN

University of Illinois at Urbana-Champaign, Illinois, USA

PREFACE

Storing, manipulating, and accessing biological and nomenclatural data has received considerable attention over the past decade or so. Here we describe a program that efficiently deals with specimens and their associated data, their nomenclature and classification, and the literature pertaining to the taxa involved. Further, this program allows for the association of ecological and biological information and can manage and track large field-collected samples from the bulk sample through to identification of individual specimens. It is a powerful and reliable tool for taxonomists and ecologists, is easily operated and customizable, includes many preformatted search options or allows the experienced user to formulate specialized complex searches, facilitates import and export of data to and from other applications, and uses the popular commercially available cross-platform (Apple Mac® OS and Microsoft Windows®) database engine, FileMaker® Pro.

INTRODUCTION

Systems to store data are probably as old as humans themselves. At first, information was transferred orally from parent to child: what was safe to eat; where to find this animal, this plant, what are its uses; family histories; noteworthy events. These databases of the mind were supplanted in part by the written word, which codified traditions and learning. Scientists and other observers transcribed their observations to paper or to note cards that could be shuffled and reorganized to highlight specific aspects of the data and that might prove useful in formulating and testing various hypotheses. The passing of these facts, thoughts, and traditions was then no longer limited to specific personal contacts but could be savored by people removed in time and space from the originator.

A leap from note cards, which organize single threads of thought, was the advent of punch cards, which could trace multiple threads by matching holes in or along the edges of cards. Now with more user-friendly interfaces of electronic databases, we no longer have to personally remember an oral history, sift through countless stacks of cards, or even go to a library to find the written observations that may be limited to a single field notebook. More stored data than ever before dreamed can now be made accessible with the click of a few keys. Graduate students dream of pushing a button to generate their dissertations. Scientists conjure a multitude of hypotheses that will clearly resolve themselves when all the entered data are appropriately manipulated and presented. But even as we did when oral histories were our only means of databasing, we all still have different ways of working, organizing our thoughts, identifying different pieces of the puzzle that are important to us, and accessing different resources for maintaining, distributing, and analyzing data.

As little as a decade ago, much discussion within the systematics community centered on whether to store data in individual databases or in a centralized repository. Only within the past several years has it been possible to create systems that can access the data over the internet from distributed sources, freeing users to choose from more than one option. Many database choices exist that were not in existence 10 years ago, and most are able to communicate or exchange data with one another.

At the individual scientist or team level, the strength of either creating one's own database or modifying an available desktop or server database to suit a program's needs confers more control over the scope of data that are documented, making possible provisions for working names and those manuscript names awaiting publication, unverified data, and controlling the timeframe during which the data are released to public scrutiny. The downsides to custom designed databases are several, including upkeep once projects have finished or designers have moved on, limited access by others to the data, and the difficulty of integrating data from other databases for field names and content with different standards. The strengths of a centralized database into which all data are fed and manipulated, address many of the downsides of customized data logging; however, the questions of data ownership and credit then arise, and users are restricted to a minimum set of data that can be agreed upon across taxa. Such unified databases exist (e.g., ITIS [Integrated Taxonomic Information System; http://www.itis.gov/], NCBI [National Center for Biotechnology Information; http://www.ncbi.nlm.nih.gov/], etc.) but they are by nature considered authoritatively as an end product or compilation of current knowledge for public consumption, akin to publication rather than the working databases of biologists or systematists that often incorporate unpublished and unverified content. Therefore, foremost in the minds of those creating their own database systems that will eventually feed into such centralized data repositories or be accessed dynamically by data portals such as GBIF: http://data.gbif.org/ or uBio: http://www.ubio.org/ should be the ease and flexibility of manipulating and exporting the data.

0.1 *History of* Mandala[1]

In 1995, with the debut of the US National Science Foundation's Partnerships for Enhancing Expertise in Taxonomy (PEET) grant program,

---

1. *System Requirements. Mandala* requires the commercial application, File-Maker® Pro. FileMaker Pro is available for the Windows and Macintosh computing environments and *Mandala* was designed to be fully cross-platform, i.e., operate equally well under both operating systems. To obtain a trial version of FileMaker Pro, go to their website http://www.filemaker.com/. To use *Mandala* in a multiuser/multiplatform environment, FileMaker Server (FMS) http://www.filemaker.com/products/fms/ on a dedicated computer with a stable IP address on a TCP/IP network is highly recommended. This can allow access by authorized users worldwide. Development of a web-based interface that does not require the user to own FileMaker Pro is also possible and would require the administrator of the database to run FileMaker Server (for PHP queries) or FMS Advanced http://www.filemaker.com/products/fmsa/ for other connections.

File size limitations. Individual file size is limited by the version of FileMaker Pro, currently for FileMaker v.7-9 to 8 TB or the limits of your hard disk. Currently, our largest file is less than 185 MB with thousands of small illustrations, and our specimen-tracking file is only 134 MB with just over 134,000 records.

Disclaimers: Names or links to commercial products are for informational purposes only and do not constitute endorsement. Neither the authors, nor the Illinois Natural History Survey, nor the State of Illinois, nor the University of Illinois, nor the National Science Foundation take any responsibility for lost or damaged data incurred in the use of *Mandala*. A non-exclusive research/internal business use license is now available for *Mandala* software. See http://www.inhs.illinois.edu/research/mandala/ for the latest information and version of *Mandala*.

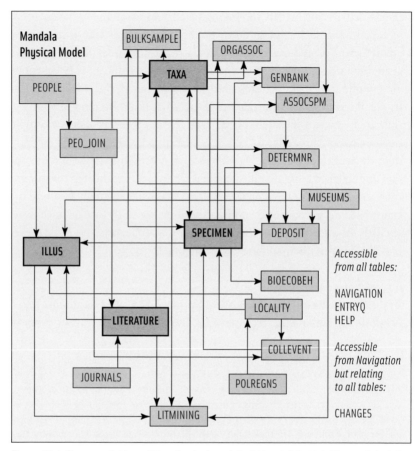

Figure 15.1. Representation of the physical model of *Mandala*'s 23 tables and their interrelationships (see Figs 15.3–15.11 for more specific examples). Primary tables (those 4 in bold surrounded by heavier black borders) can support one to many relationships with secondary tables, e.g., the relationship of a record in SPECIMEN to one or more records in ASSOCSPM, BIOECOBEH, BULKSAMPLE, DEPOSIT, DETERMNR, GENBANK, and ORGASSOC. Most of these underlying tables will only be viewed through the context of SPECIMEN (as portals or views onto these tables related by the SpecimenID) and not on their own, simplifying the user experience. Other tables are used as references (PEOPLE, MUSEUMS, POLREGNS, JOURNALS, LITERATURE, GENBANK) where commonalities may be exploited in economies of scale in a larger setting (e.g., a museum managing many different kinds of collections).

database choices were few for a viable and malleable cross-platform (compatible with both Windows and Macintosh operating systems) system for manipulating data about specimens and documenting appropriate literature and taxonomic nomenclatural histories. *Mandala* (well before it had a name[2]) started out as 5 tables that were quasi-relational by means of lookups from one table to another. Its original design was based on the structure of a then mothballed database designed by Rauch and Irwin (Rauch 1970) to catalogue the approximately 7,000 therevid fly specimens forming the basis of Irwin's dissertation (Irwin 1971). The data were originally entered on 80-entry edge punch cards and later migrated to an IBM mainframe where data entries could be expanded past the punch card limit and stored on magnetic tape. The restored data were provided by Rauch as an electronic text file and were successfully imported into *Mandala*[3]. Ten years later, *Mandala* 7 now boasts 23[4] (from a high of 27) interrelated and interconnected tables, which can be viewed as separate windows (Fig. 15.1). Since the summer of 1995, the therevid PEET team has been continuously entering data about stiletto fly (Diptera: Therevidae) specimens collected by us or obtained on loan from museums and collections around the world, charting the nomenclatural history of fly names in the family Therevidae, and cataloguing literature pertaining to this little known family of flies and its near relatives. During each growth stage of *Mandala*, existing data have been migrated into the structure of every new feature-enhanced version.

0.2 *Philosophy of* Mandala

Contrary to databases developed for commercial distribution, *Mandala* operates under the open source philosophy that allows access to the in-

---

2. *Mandala* (pronounced with all vowels as 'ah', with the accent on the second syllable) derived its name not from the intricate sand paintings of eastern religions, although the interconnected and interrelated nature of these paintings has inspired the splash screen that greets the user. The choice of the name *Mandala* was inspired by the imagery in a detective novel by Jane Dentinger, which described how the electrified glances among people in a room would have formed a 'mandala in the air.'

3. After a certain amount of massaging of the data, with thanks for the generous help in developing automated methods to accomplish the transition from R. Lilienkamp, a FileMaker Pro listserv member who volunteered his expertise.

4. See Appendix for an alphabetical listing of tables and their purpose in *Mandala*

ner workings of the database system. The fully working database system is available without charge, upon request, from its designer, Kampmeier (gkamp@illinois.edu; see also http://www.inhs.illinois.edu/research/mandala/ for additional information and illustrations of *Mandala*'s features). With the open source model, users do not have to be stranded when the database developer leaves the project because the cross-platform, relational backbone of *Mandala* is the commercially supported database engine, FileMaker® Pro (http://www.filemaker.com/). Further, motivated users with unrestricted access privileges can learn how *Mandala* was developed and customize pick (=value) lists with choices relevant to the organisms they are studying. The open source philosophy is not without its downside; it allows users to modify features of the system without the normal safeguards that commercial developers deliberately build in. When present, such safeguards can prevent even administrative user access to and customization of layouts, scripts, and field definitions, and allow navigation only by buttons, whose purposes are defined by the database developer. In *Mandala*, setting multiple levels of password protection can limit user access to design functions; access to fields, layouts, scripts, or value lists; and addition, deletion, or modification of records.

Mandala was originally created to meet our needs for organizing data related to the biology, ecology, nomenclature, and systematics of stiletto flies (Insecta: Diptera: Therevidae), and more recently has been expanded to accommodate and track batches or bulk samples that might be collected during a biotic survey. *Mandala*'s flexible architecture is being used primarily by groups studying taxa bound by the rules of the International Code of Zoological Nomenclature (ICZN[5]) (International Commission on Zoological Nomenclature, 1999). This includes cataloguing specimens from single or multiple collections and organizing taxa to study the biodiversity of a taxonomic group both within and across geographically bound areas. *Mandala* can be used to manage loans; to catalogue illustrations and images associated with specimens, taxa, localities, and literature; to link to resources on the internet such as GenBank and Morphbank records, webpages of museums and people, and geographic and taxonomic names

---

5. Although *Mandala* was created with the rules of the ICZN in mind, it can be and is used to document taxonomic names governed by other international codes of nomenclature. Further modifications would need to be made to accommodate other codes, but they can be done, requiring primarily the knowledge and collaboration of interested parties.

servers; to serve *Mandala* data to the web, establishing a web presence for accumulating data; and to mine through the literature (e.g., allow users to search for all references to immatures, to species with distribution maps, and to phenologies). Like most software applications, *Mandala* is rich with features, not all of which need to be deployed, explored, or exploited by each user. In addition, tasks can be broken up according to expertise. For example, we employ students with a minimum amount of training to catalogue specimen-related data and to input basic taxonomic name information and literature citations. With further training in the use of on-line gazetteers, the links for which are supplied in *Mandala*, data entry workers determine georeferencing coordinates (latitude/longitude) if not previously stated on the label, extrapolate place references to higher divisions of political localities that may have been lacking on the label (e.g., county, country, state or province), and parse out and verify locality information. Additional training is also required to fully catalogue illustrations and to process loans. Specialist knowledge and experience are necessary to completely mine the literature and to detail the often complex nomenclatural history of many taxonomic names. *Mandala* enables the user to trace the subtleties of the nomenclatural history more completely than most printed catalogues. Specific examples are given below illustrating how *Mandala* is being used by our therevid PEET project, but it should be remembered that other groups are using *Mandala* to catalogue their own data, and not all of them are working with arthropods.

1. Getting Started with *Mandala*

Although *Mandala* is currently comprised of 23 tables, the user is always routed through NAVIGATION[6] to sign in so that records created or modified by each user may be date (and often time) stamped and tracked. This pivotal table allows the choice of entering data or finding specific records; creating/viewing reports; finding a glossary of icons and terms, help, or web resources; or troubleshooting problems.

---

6. The names of individual files or tables in *Mandala* will be listed in upper case letters, without the file type suffix that indicates the versions of FileMaker Pro with which they are compatible.

2. Enter Specimen Information

The ability to thoroughly catalogue information about individual specimens and bulk samples in SPECIMEN is one of *Mandala*'s strengths.
Here the focus is on specimen-level data. An in-depth discussion of bulk
samples and how they are handled is reserved for a future paper that
concentrates on bioinventory management. Each specimen or sample,
whether on loan from a museum collection or recently collected, needs
an identifier[7] unique to the database system, which can be compatible and
flexible enough to comply with evolving standards (see http://wiki.tdwg.
org/) for global unique identifiers (institution code, collection code, and
catalog number). That identifier may be pre-designated or it may be created at the time of data entry. Bar codes in use by many museums are
examples of unique identifiers that accompany individual specimens or
bulk samples. If the identity of the taxon is known, its corresponding numeric code (automatically generated in TAXA, see below) is first entered
into the specimen record. The determination that is used at the time of
data entry may merely be to the family or genus level, or it could be to the
species level; this determination can be updated at any time. That taxon
code represents all that has been recorded about a particular taxonomic
name and its history. If changes or additions are made to that taxon record, they are automatically updated everywhere that taxon code appears.
Second, a field is provided to capture verbatim label information from the
specimen or sample, particularly in the case of retrospective data capture
from existing label(s). Verbatim recording of the data from such label(s)
prior to categorizing that information enables easier proof reading, allows
rapid input of label data, eliminates the simultaneous need to interpret it
(e.g., a label designating the locality as 'Red River' could refer to a river, to
a municipality, or to both), and less training of data entry personnel. For
prospective data entry, where locality and collecting event labels will be
generated from the database for specimens or bulk samples (see below),
there is no need to type in a verbatim label (Fig. 15.2).

Third, supplementary data may be entered about the specimen: sex; life
stage collected, as well as the life stage(s) now preserved in the collection;

---

7. Although *Mandala* was originally created as a specimen database of uniquely
identified individuals, it now (as of version 6.6) also tracks bulk samples of specimens. The bulk sample system is integrated with loan tracking features.

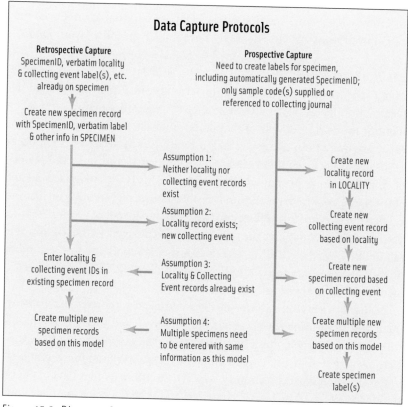

Figure 15.2. Diagram of workflow distinguishing retrospective from prospective data capture for specimens. One distinguishing characteristic may be the need to produce labels in prospective data capture and the likelihood of needing to produce multiple consecutively numbered specimen records with the same data.

pupation and eclosion dates (or appropriate rearing dates noted); curation specifics; condition of the specimen; dissections; preservation method for specimens or parts of specimens being reserved for molecular analysis; and space for a second or alternate identification code.

Fourth, additional tabs to related tables are provided in SPECIMEN to detail a specimen's determination history, track loan activity, and link observations of taxon association, specimen association, and behavioral or ecological observations. Fifth, there is a field for related specimen information that is not specifically pigeonholed elsewhere. Database users may decide that in the interests of time or for other reasons, some of the data associated with specimens will only be recorded for specific kinds of specimens, e.g., name-bearing types. Fields or tabs that are not used may be removed from view in FileMaker's graphically friendly layout mode.

2.1 Detailing the determination history of a specimen

Many specimens, with the exception of types, may not have affixed determination labels bearing the name of the specialist making the taxon identification and the year of that determination, while other specimens may have more than one such label, providing each determiner's concept of the taxon at a given time. *Mandala* permits the recording of this history of determinations for individual specimens. This is the only place outside of the literature where invalid/unavailable names, representing an historical perspective, are not adjusted to the current concept of a taxon's name. To record determinations, the user has only to navigate to the 'Determine' tab in SPECIMEN. Note that the determiner is represented by a number (identifier or key, see below) in the authority file, PEOPLE, and the taxon identifier (ID) is keyed to TAXA (see below).

Mandala's PEOPLE table is ideally a complete authority file representing a wealth of information about the persons listed there and dynamically linked, by their numeric identifier, to the activities in which they have participated. In reality, it is almost always a work in progress, at least in the entomological community, where true authority files have yet to be standardized for those people contributing to the science. Complete contact information, activity or birth/death years, and information about the person's specialty and level of expertise may be recorded. Users can also designate the best representation of a person's name as a senior synonym, and other forms as junior synonyms, helping to resolve multiple entries of a person's name in slightly different formats (e.g., without initials, with a single initial, with more than one initial, first name spelled out, even misspellings). This provides the flexibility of scripting a uniform presentation of the person's name (as the senior synonym) or the representation as it appeared on the work involved (as a junior synonym). PEOPLE are linked by their identifier as determiners of specimens, borrowers, lenders, illustrators, and copyright holders of illustrations. The bridging (join) table, PEO_JOIN, connects PEOPLE in a specified order as collectors in COLLEVENT, authors (in LITERATURE), and authorities of taxonomic names (in TAXA). Although not a table designed for view, PEO_JOIN is essential when multiple people are responsible for a single action in a prescribed order and flexibility in the way the name is presented is important in the output (e.g., this allows flexibility in the display of author names from a literature citation, permitting the first author's initials to appear after the surname and all subsequent authors to have their initials before their surname).

2.2 Recording loan history of a specimen

If a specimen is on loan from an individual or a collection, a loan label should be associated with the specimen indicating the loan number, collection or museum loaning the specimen, the borrower, and the date the specimen was borrowed. Unique 3- or 4-letter codens are used to represent the loaning and borrowing entities in these transactions. The full contact information represented by a coden about a collection is contained in MUSEUMS, which is based on Arnett *et al.* (1993); periodically updated coden information of the Insect and Spider Collections of the World is available from the Bishop Museum's searchable website http://hbs.bishopmuseum.org/codens/. Collections of other sorts of organisms may use different codens, which can be added or modified to suit the user as long as they remain unique (but we suggest selecting standardized codens when they are both available and unique). The navigation tab 'Loan & Deposit' in SPECIMEN then allows the user to record the current loan information (specified in the portal[8] to DEPOSIT) or update the current status of a loan. Except where a single specimen is broken into more than one part and scattered between two or more collections, only a single record designated as current should exist, e.g., where a specimen is physically located at any point in time. When a specimen is returned or sent elsewhere, the information in DEPOSIT should always be updated to reflect the current location of a specimen (we find this is a step that can easily be neglected; if locality transfers are not logged into the database, specimen data can easily get lost in the shuffle).

2.3 Specimen sequences in GenBank

Certain specimens or parts of specimens may be selected to have their genetic structure analyzed and reported through the National Center for Biotechnology's GenBank database (http://www.ncbi.nlm.nih.gov/).

---

8. A portal is a window of data, usually showing multiple records, which are related by a common key such as the unique specimen identifier, which here is common to both SPECIMEN and DEPOSIT.

Mandala's GENBANK table can provide links to the GenBank accession number, the gene sequenced, and the URL to the NCBI online record. Each specimen, whether the primary (the specimen sequenced) or secondary (part of the same collection series or considered conspecific to the specimen sequenced) voucher is linked via its unique specimen identifier to the full record in SPECIMEN. Any number of records linking specimens to GenBank (or other repository) may be created.

2.4 Documenting additional observations about a specimen

Mandala features the ability to record observations about the biology, ecology, and behavior of a specimen: information that is less commonly found on insect labels accompanying the specimen than in field notebooks that can get separated from the specimens or, as often happens, lost or destroyed as the worker retires or switches to a new research focus. Although entering such ancillary data is not a requisite, the ability to collect and compare such observations can be invaluable as the dataset matures (Fig. 15.3). Accessed from tabbed layouts in SPECIMEN, *Mandala* can link (via ASSOCSPM) specimens with unique identifiers that are in some way related to one another at the time of collection, e.g., mating pairs; mimic-model; or through subsequent rearing, (e.g., host–parasite/parasitoid relationships). Where a specimen cannot be linked to another collected and recorded specimen, but only to an identifiable taxon (with a taxonID in TAXA), these observations can be linked to the specimen under the ORGASSOC tabbed layout (e.g., host plant–herbivore relationships). Observations on the biology, ecology, or behavior of specimens at the time of collection should be noted in the BIOECOBEH table, represented under the tab labeled 'Bioassoc & AHC' (AHC = associations, habitats, and conditions) in their raw form, often from the label. Users are encouraged to drag and drop this information from the verbatim label into the field provided to form a basis from which the information may be drawn and parsed into phrases with orthographic consistency and a reduced set of searching criteria for specific activities, adjectives, and associations. A series of controlled language popup lists is provided, including an action, a linkage type, an adjective, and an object (e.g., feeding on yellow flower; resting under dead shrub) with the specimen as the *de facto* subject. Taken individually, such observations are little more than anecdotal; however, repeated observations form patterns that may be used in

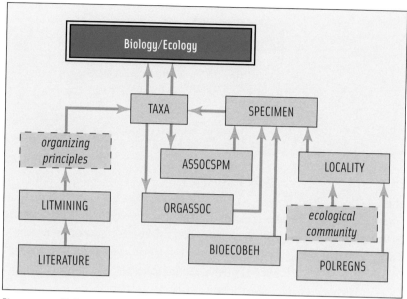

Figure 15.3. Biology and ecology of various taxa are derived from observations made during the collection of specimens or as recorded in the literature. This diagram shows the tables (names capitalized) and their interactions. The biology or ecology may be viewed from the context of a taxon (in TAXA), of a single specimen (in SPECIMEN with supporting observations made through ASSOCSPM, ORGASSOC, and/or BIOECOBEH), through the choice of specified 'organizing principles' (dotted line around field concept) in LITMINING from observations gleaned from LITERATURE, or 'ecological communities' specified in LOCALITY.

predicting or determining biological peculiarities of the taxon involved, or the likelihood of finding live specimens for future studies.

3. Interpret and Standardize Locality and Collecting Event Data

Although knowing the locality and associated collecting event information that was recorded on specimen labels is essential for record verification and historical accuracy, there is little standardization of the way the information appears, how it may be abbreviated, and what is assumed as common knowledge by the collector. In addition, until handheld GPS (Global Positioning System) units became popular, few labels specified georeferenced coordinates, making the task of creating distribution maps (Fig. 15.4) extremely laborious and often somewhat arbitrary. Now, not

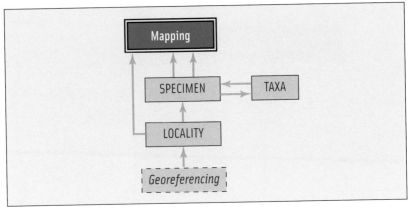

Figure 15.4. Mapping is generally thought of as being specimen-driven, often organized by taxon, but just as important may be places where specimens or taxa were not found in a particular locality. Mapping is derived from georeferencing data (dotted line around field concept) recorded in LOCALITY (table names capitalized) and linked to mapping output either directly or through SPECIMEN.

only do modern collectors record such information on their collecting labels, retrospective capture of mapping coordinates for older specimens has been made increasingly easy with the availability of such on-line gazetteers as the United States Geological Survey's Geographic Names Information System (USGS GNIS), the National Imagery and Mapping Agency (NIMA), the Getty Thesaurus of Geographic Names, the Global Gazetteer, Google™ Maps and Earth, and a number of country or region-specific gazetteers such as Geoscience Australia, Canadian Geographical Names, Town Search for Central/Eastern Europe, and New Zealand Geographic Place Names. Clickable URLs to open an internet browser and access these on-line gazetteers are part of the resources available in *Mandala*. Internet-independent software application tools include Microsoft's Encarta® World Atlas (available for MS Windows only), and other tools available from libraries and bookstores (e.g., printed maps and bound gazetteers).

Locality and collecting event data recorded from the specimen's label(s) and shown in SPECIMEN, need to be uniformly structured to allow them to be electronically searched and output (e.g., for compiling faunal and specimens examined lists (Fig. 15.5)). Two tables in *Mandala*, LOCALITY for the documentation of the geographic and political boundaries related

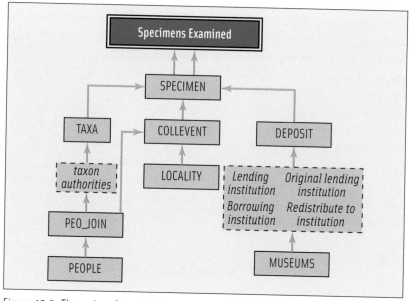

Figure 15.5. The order of text in the output of a specimens examined list is generally determined by the journal in which it is being published. The pieces that figure in such listings are ultimately coordinated by taxon (in TAXA, table names capitalized) listed in SPECIMEN for specific specimen identifiers and sex; associating information about each locality and collecting event, and where a specimen ultimately resides (from DEPOSIT, if one has been diligent, from the collection specified as maintaining ownership, or else from whence came the specimen to the specialist). The output order of specimens examined text is governed by a calculated field, several iterations of which exist in *Mandala* and may be modified by a database administrator for exporting purposes corresponding to individual needs.

to the collecting site, and COLLEVENT for collecting event information, are used to structure these data. The LOCALITY table allows parsing of the hierarchy of political boundary descriptions (e.g., city, county, state, or country) and those of named geographic features (e.g., stream, river, or mountain range; features that may defy political boundaries) by relative size (small, medium, and large), standardizing spellings; adding political divisions missing from the label such as counties or countries; facilitating translation of miles to kilometers for distance displacements and feet to meters for elevation or depth; automatically determining the biogeographic region (based on the related reference table, POLREGNS); and adding georeferencing coordinates and calculating their displacements

by distance and direction from a fixed point. While a locality record
documents the 'where' of a collection, each associated collecting event for
that locality details the 'who, when, how,' and other temporally-relevant
specifics, e.g., collector names (see earlier discussion of PEOPLE and
PEO_JOIN), a date or range of collecting dates, collecting method, trap
number or site specific information, and abiotic conditions at the time
of collection. Such conditions might include temperature, with a built-
in converter from Fahrenheit to Celsius and specification of the medium
measured (air, water, soil), sky, light, wind, humidity, time of day, and
substrate moisture. Each locality may have one or more collecting events
associated with it and a new collecting event is created whenever any information in the record changes.

4. Trace Nomenclatural History

All ranks of taxonomic names are handled within the single table, TAXA,
with the classification hierarchy of valid or unregulated (working, manuscript, or in press) names built in. A layout is provided to aid in importing
taxonomic data from other sources.

4.1 Essentials of documenting a taxonomic name

Although entering a new taxonomic name without knowing and documenting its entire nomenclatural history is a relatively simple operation,
records for taxa with complex histories will need to be completed by a systematist. The essential information associated with any new name record
includes a unique[9], automatically generated 'taxonID' (identifier); the rank
of the name being entered chosen from a popup list, ranging from infrasu-
bspecies to kingdom; the taxon name itself, and if it is a name in combination, specification of its genus or subgenus identifier; and its status (valid,
invalid, unregulated). In the case of valid or unregulated names, the taxonID of the parent taxon (in the case of a species, this is often the genus) is
needed to generate the classification hierarchy. When the status of a name

---

9. An identifier uniquely generated for this database, but not necessarily unique
in the world. Globally unique identifiers and other forms of unique identification
of life science objects, literature, taxa, people, etc. are still very much under discussion at this point in time. For the latest information, see the Biodiversity Information
Standards website http://www.tdwg.org/.

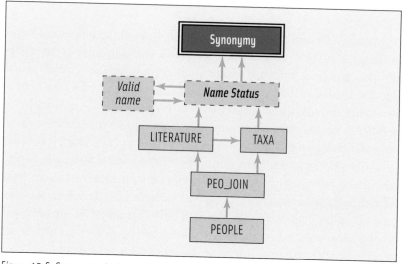

Figure 15.6. Synonymy is derived from status of relationships of a taxon or name in combination with its valid name. These are listed in TAXA (table names capitalized), and are derived from the scientific literature (in LITERATURE) to which it is linked for priority in the year in which synonymic changes have been recorded. Both LITERATURE and TAXA have links to PEOPLE through PEO_JOIN for authors and authorities of taxa, respectively. These are particularly important in distinguishing homonyms and their priority.

is valid or invalid, a valid taxonID needs to be specified. For valid names, this is the same as the taxonID; for invalid names, the taxonID for the accepted valid name should be indicated. The act of designating the valid name of a taxon begins the dynamic process of building a synonymic list (Fig. 15.6), which can be seen through an automatically generated portal in TAXA and is built into a static indented list for potential display or to be exported. Although these few pieces of information constitute the bare minimum of what is required, additional documentation of a taxonomic name involves attaching the authorities of the name and indicating whether, in the case of a species, that taxon name has been moved to a different genus from the one in which it was originally described (thus requiring parentheses around the author name), along with the year and page of publication and a notation of the literature citation number from LITERATURE in which the name first appeared. The reasons for classifying the status of a name (following Thompson 1997) and other taxonomic name-related information can be documented as described below.

4.2 Common names

Common names are generally not governed by anything more than popular usage, and their value in most taxonomic works, at least for many arthropod groups, is minimal except as cited from collecting labels or the popular media. Although there is a field to record common names, it is assumed that most users working with arthropods will not be troubled to validate it further than confirming its association with a scientific name on one of the taxonomic names servers (=TNS, e.g., ITIS, or various other specialty TNS for specific kingdoms; the Entomological Society of America provides a standardized, juried list of common names of insects and related organisms primarily of North America http://www.entsoc.org/pubs/common_names/) on the internet. URLs to many TNS are provided among *Mandala*'s web resources. Clicking on the globe icon appearing next to the URL will open a browser and take the user to the website.

4.3 Name changes and conflicts

Beyond taxon synonymies, name changes such as the preceding and subsequent combination of a name and the replacement name for and replaced-by names may be detailed along with references to the literature that established these name changes. Clicking on the magnifier icon appearing next to the taxon number transports the user to the related taxon record and history of the changes to the taxonomic name. Name conflicts with another taxon such as homonymy, unjustified new name, unjustified emendation, incorrect original spelling, improper formation, published in synonymy, misspelling, misidentification, and subsequent usage may also be specified along with the relevant literature reference.

4.4 Documenting the type

The navigation tab leading to the 'type' layout in TAXA is context sensitive by taxon rank, allowing entry of the unique identifier of a type specimen for a species-level name along with a separate designation of the kind of name-bearing type (which may be different from the current information with the specimen), the original name for the taxon, and a field for recording the type label as it was cited in the original publication. If no exemplar name-bearing type has been designated, specimens retrieved from an automatically generated list of those specimens designated as syntypes or cotypes for that nominal species may be examined and a suitable specimen designated as the exemplar name-bearing type.

For a taxonomic name of a rank where a type specimen is not delineated (i.e., names above the species-group), the type may be designated as a taxonomic name of a lower rank (e.g., a type genus of a family), whose appropriate type designation may be indicated from an editable list (original designation, species-group name indication, monotypy, absolute tautonomy, Linnaean tautonomy, subsequent designation, or ICZN decision[10]). In addition, for each type layout there is a field for listing all of the biogeographic regions from which the taxon is known, and links to the literature citing the first designation of the type.

5. Track Loans

Loan management depends on perspective and context (Figs 15.7–15.9) and is ultimately handled in *Mandala* by DEPOSIT. A lending entity (museum or established collection referenced by MUSEUMS and/or specific lender or curator found in PEOPLE) sending out a loan may base the loan on numbers of taxa (listed in BULKSAMPLE, Fig. 15.7) shipped from its collection to an individual (borrower listed in PEOPLE, who may be at an institution referenced in MUSEUMS). When the borrower receives the loan, it is broken out and tagged (if not already done so) by a unique specimen identifier (in SPECIMEN, Fig. 15.8) and linked back to the loan (in DEPOSIT). Loans both sent and received can be complex (e.g., the loan might be broken apart with some specimens being retained while others are returned at different times). *Mandala* tracks the fate of loans by specimen and taxon, and can keep a running tally of those returned and those kept, along with the date of each transaction and additional comments.

A third perspective of loan management is the specific case of subsample management (Fig. 15.9) from a bulk sample, e.g., Malaise trap sample (logged in SPECIMEN with a common locality and collecting event) rough sorted to order or family (in BULKSAMPLE from a portal in SPECIMEN) to be dispersed, often as loans (linked in BULKSAMPLE and DEPOSIT), to specialists (as borrowers in PEOPLE) and by loan number (recorded in both BULKSAMPLE and DEPOSIT) for further study and identification. This context differs from the straight loan tracking described above in being from the perspective of the specialist, who may accumulate fractions

---

10. The appropriateness of this list may depend on the code of nomenclature, and as with all such lists, may be modified to suit the taxa of focus.

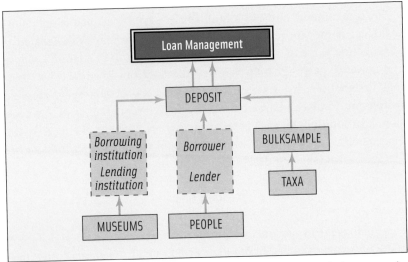

Figure 15.7. Loan management based on numbers of specimens of particular taxa can be from the perspective of an entity (dotted line around field concept of 'loaning institution' in MUSEUMS, table names capitalized) and/or person (in PEOPLE) lending or from the perspective of the person or entity receiving the loan. Both should be recorded in DEPOSIT under a single identifier in BULKSAMPLE (data entry takes place in DEPOSIT). Partial loan returns may be managed here, although it may be easier to do this by specimen (in SPECIMEN, see Fig. 8) if taxonomic names have changed substantially through synonymy.

(all the spiders, for example) of many Malaise trap samples in a single loan.

Boilerplate letters are also provided for collections personnel managing loans, allowing electronic management via email with loan reports as PDF attachments (portable document format of Adobe® Acrobat®). Loan tracking includes fields for documenting shipping details, including the type of loan, how and when a loan was shipped, tracking and packing information, insured value of each specimen and of the total loan, number of packages in the shipment, and any additional comments.

6. Illustrations

A large part of many taxonomic tomes includes rendering of distribution maps and illustrations for descriptions of new and revised taxa, to convey what would take voluminous pages of text to adequately describe. For an

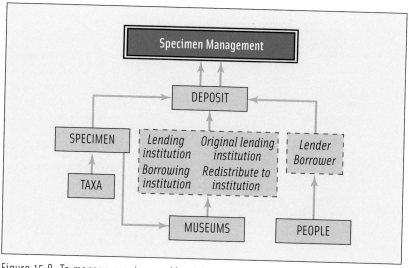

Figure 15.8. To manage specimens, *Mandala* can distinguish the provenance, current location, originating lending institution, and ultimate destination (dotted line around fields representing these concepts) by codens from MUSEUMS (table names capitalized) of individual specimens (from SPECIMEN) in DEPOSIT.

image to be displayed in *Mandala*, it can be linked via a file path reference to the digital image residing on a separate server or the digital image may be imbedded in the ILLUS[11] table. Aside from the more obvious linkages of illustrations with specific taxa (in TAXA, illustrating internal or external morphology, or type labels), specimens (in SPECIMEN, often reserved for types, including their labels), literature (in LITERATURE, linking to its citation, and in LITMINING, providing specifics of page and classification of the type of illustration for more organized retrieval), people (in PEOPLE for illustrators, photographers, or copyright holders), collecting locality (in LOCALITY for specific habitats or geography), and collecting event (in COLLEVENT, illustrating perhaps collectors or abiotic conditions at the time of collection) records may be tied to an illustra-

---

11. Imbedding large image files or other documents, rather than linking to them in *Mandala*, generally leads to an engorged file size that threatens easy backup procedures onto removable media, and hampers the speed at which the database functions, but may be the solution in a FileMaker Server environment where links referencing images must reside on a server external to the databases being served.

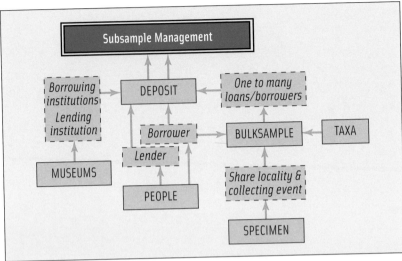

Figure 15.9. Subsample management, the most complex of the loan management types, is elegant in its ability to draw together and detail subsamples from many samples destined for a single specialist in a single loan shipment. Although data may be viewed from DEPOSIT (table names capitalized), and loan forms generated there, data are entered and managed from layouts in SPECIMEN that collect and distribute information on the number of taxa in subsamples (listed in BULKSAMPLE) going to particular specialists, each with his/her own loan number, of a sample (listed in SPECIMEN on the bulk sample layout) that shares a common locality and collecting event. As with earlier management strategies (Figs 15.7–15.8), both MUSEUMS and PEOPLE play their roles as lenders and borrowers, but unlike those strategies, the only required concept is that of the 'borrower' (concepts of fields surrounded by a dotted line). The borrower and loan number are the key to collecting one to many subsamples as a loan to a specialist.

tion (Fig. 15.10). Details (referred to as 'controlled language' in Fig. 15.10) of the medium, method, background, view, subject, life stage, and gender may be recorded from editable lists, or a more detailed description may be added. By specifying a subject for the illustration, comparisons among other illustrations with the same subject may be viewed. Finally, archiving details of both physical and electronically based illustrations may be fully documented, including multiple file types (.jpg, .psd, .bmp, .tif, .png, etc.), greatest resolution (dpi), software used, and information about original, working files, URLs for web renderings, and details of physical locations of storage media and artwork.

7. Record & Dissect Taxonomic Literature

Pertinent references may be documented in LITERATURE. Identified by unique[9] accession numbers, literature citations include the expected fields allowing output flexibility of literature cited, including title of a smaller work (article or chapter) within a larger work (journal or book); authors (linked to PEOPLE and PEO_JOIN to establish author order and provide flexibility in the formation of citations); and full publication data. Additional fields include original language, title translation (of large and smaller works), type of publication being catalogued (which determines how a citation is formatted), curation details of the literature, and space for a linkable URL, which might go directly to a PDF or other representation of the literature online. Periodicals (serials or journals) also have their own reference table (JOURNALS) to minimize typographical errors and provide flexibility of citing the full title of a journal or an accepted abbreviation.

Each literature citation may be dissected in LITMINING, linking citations and page numbers to over 30 categories of information relating to specimens (for types; specimens examined), illustrations (see above), and taxa (detailing the status of a name and repository of a type; information about immatures, distribution, phenology, behavior, ecology; taxonomic keys and lists; and various description types). Although tackling the job of mining through the literature is not generally the first priority in most projects, and never a small job to be undertaken, particularly by the inexperienced, in a thorough treatment of any group of organisms it would be invaluable to have the answer at one's fingertips to questions like 'where can I find all of the literature detailing information about immatures of [a specific taxon]?'

8. Querying *Mandala*

Only when significant portions of data are recorded in *Mandala* does the real value of having spent all that time entering and checking the accuracy of input begin to pay substantial dividends. The beauty of a database versus a word processing file, file cards, spreadsheets, or relying on human memory, is to easily and accurately combine data in novel ways, to look up and display specific pieces of the data without extraneous information, to ease the task of assembling information for monographs or faunal

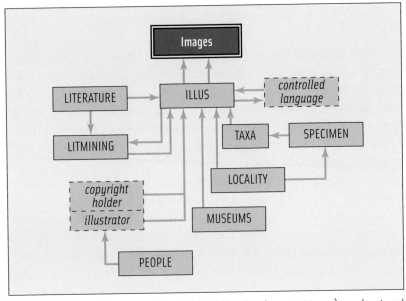

Figure 15.10. Images (and now with FileMaker 7/8, other document types) may be stored in or linked to ILLUS (table names capitalized) via URLs or file paths. Circumscription of the document takes place in ILLUS via controlled language (dotted line around concept) editable popup lists that not only describe the subject of the image but the medium, background, and other details pertaining to it. Links are established to TAXA, SPECIMEN, LOCALITY, and LITERATURE, with subsequent links to the page number of published images in LITMINING. Curatorial details such as acknowledgements of the artist and copyright owner (from PEOPLE), where a physical rendition of the image may reside (from MUSEUMS), and details specific to digitization, archiving, and representation on the web (as a URL) may also be profiled in ILLUS.

lists, to make it possible for others to access the data via the web, and to plan targeted expeditions for collecting new species or gathering fresh specimens for genetic analysis by identifying key collecting localities, times of the year, taxa found previously, and sex and numbers of specimens already collected. *Mandala*'s reporting features, accessed through NAVIGATION's 'Reports' tab, help users accomplish these goals through specialized find layouts or automatically generated views of selected data via portals (windows onto other tables, filtered through the relationship of the record being viewed). Examples of such automatically generated reports include the synonymic listing of a valid name and its junior synonyms, other occurrences of a taxonomic name for quick identification of duplicate entries and possible conflicts, and image comparisons based on subject. Specialized search layouts guide the user through the task of finding, sorting, displaying, printing, and exporting data. Although experienced users may use FileMaker's robust search engine tools to create additional ways of querying any field in *Mandala* to yield the desired results without necessarily using the custom layouts provided, *Mandala*'s specialized reporting features can help organize specimens examined lists (Fig. 15.5), export data to plot phenologies (Fig. 15.11) in charting applications such as Microsoft® Excel®, and provide coordinates to display distributions on maps (Fig. 15.4) in iMap: http://www.biovolution.com/, ESRI™: http://www.esri.com/ mapping software, Google™ Earth http://earth.google.com/, Discover Life: http://www.discoverlife.org/, or elsewhere. Collections managers can also use *Mandala* to find borrowers with overdue loans and prepare associated letters that can be printed or made into a PDF for emailing to the borrower. *Mandala* also aids in compilation of faunal lists to view, print, or export, based on political locality descriptions (e.g., counties, states, or provinces, named conservation areas), named geographic features, or biogeographic regions.

9. Special Features

Tabs in NAVIGATION are devoted to additional web resources, help, glossary, and troubleshooting. Users can access *Mandala*'s integrated online help system (HELP), which features general help tips, and contextual help for both individual fields and generic level information about tables. Individual help records or the entire contents may be printed. Additional documentation is also available in abundantly illustrated PowerPoint® slide shows (included with the demo, contact the first author at gkamp@illinois.edu), which help novice users become acquainted with the database system's many features and get started with the process of data entry. *Mandala* also features record flagged electronic tracking of user questions and problems and their resolution (ENTRYQ). For the developers and those customizing *Mandala* for their own use, there is also a table (CHANGES) to catalogue changes made to the structure of the database system, such as adding customized fields, changing field definitions, adding or changing scripts, layouts, etc. The troubleshooting layout in NAVIGATION leads database administrators to specialized layouts so that they can deal with duplicate entries, verify records, recover from a crash or just

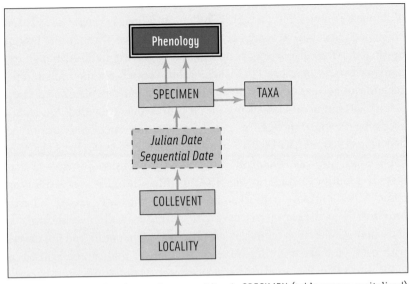

Figure 15.11. A specialized query layout residing in SPECIMEN (table names capitalized) guides the user through searches to describe the phenology or days of the year when taxa have been found in a particular locality. It makes use of the automatic calculation of the Julian date (day of any year) from the initial (and sometimes only) collecting date recorded in COLLEVENT. A phenology of a particular snapshot in time (where the year matters, or where time crosses the boundary between years, such as collecting in the southern hemisphere) may be circumscribed using a range of sequential dates (calculator for any date provided on the query layout in SPECIMEN). The results of these searches may be sorted and plotted in your favorite graphing application, such as Microsoft Excel.

open all tables, and deal with the ramifications of changing or deleting record keys (unique identifiers) that may have dependent (child) records attached that would become orphaned and difficult to resolve once changes were enacted.

10. Future Plans for *Mandala*

Although feature-rich, often dauntingly so to the first time user, there is no module to catalogue specimen character data for exporting to various other specialized applications to analyze and construct phylogenetic trees. Such a feature is not beyond the bounds of being included in the future should users request and help develop it. Future versions will see a consolidation of the current 23 tables into a fewer number of files in *Mandala* as FileMaker 7 and beyond now allows multiple tables per file. Because *Mandala* includes overhead (e.g., complex relationships, unindexed calculated fields, scripts, etc.) that would unnecessarily slow the processing of data queries via the web, a separate database file with matching DarwinCore (http://www.tdwg.org/activities/darwincore/) field names will optimize the speed with which information may be accessed via PHP, an essential step to fully sharing biodiversity information with others such as GBIF (Global Biodiversity Information Facility: http://data.gbif.org/) and Discover Life: http://www.discoverlife.org/. Support for web-based data entry is more complex, and may be supported in limited instances such as in updating specimen identifications. See http://www.inhs.illinois.edu/research/mandala/ for the latest developments.

Acknowledgements

The authors wish to thank Kristin Algmin, Amanda Buck, Robin Carlson, Neal Evenhuis, Stephen Gaimari, Martin Hauser, John Pickering, Kevin Holston, Christine Lambkin, J. Marie Metz, Mark Metz, Donald Webb, Brian Wiegmann, David Yeates, and the members of the therevid PEET database entry team for their input and suggestions that have contributed to the growth and evolution of *Mandala*. We would also like to especially thank F. Christian Thompson, USDA-Systematic Entomology Laboratory, for the inspiration and collaboration on the nomenclatural aspects of this project. We wish to acknowledge the support of grants from the Schlinger Foundation and the National Science Foundation (DEB 95-21925, 99-77958, 07-31528, and EF 03-34448), USDA Hatch Act ILLU 875-380, and support from the Illinois Natural History Survey and the Department of Natural Resources & Environmental Sciences at the University of Illinois at Urbana-Champaign. Any opinions, findings, and conclusions or recommendations expressed in this material are those of the authors and do not necessarily reflect the views of the National Science Foundation nor of any of the institutions supporting this research.

References

Arnett, R.H., Jr., Samuelson, G.A. & Nishida, G.M. (1993) The insect and spider collections of the world. 2nd ed. *Flora & Fauna Handbook*, no. 11.

International Commission on Zoological Nomenclature (1999) *International Code of Zoological Nomenclature*. 4th Ed. International Trust for Zoological Nomenclature, London, 306 pp.

Irwin, M.E. (1971) *Ecology and biosystematics of the pherocerine Therevidae (Diptera)*. Ph.D. Dissertation in Entomology, University of California, Riverside, 256 pp. + 3 appendices (181 pp.).

Rauch, P.A. (1970) *Electronic data processing for entomological museums, an economical approach to an expensive problem*. Ph.D. Dissertation in Entomology, University of California, Riverside, 78 pp.

Thompson, F.C. (1997) Data dictionary and standards. BioSystematic Database of World Diptera: Family Tephritidae, 15 pp. Available at http://www.diptera.org/names/BDWDstan.pdf, accessed September 2008.

Appendix

1. Alphabetical Listing of Tables & their Purpose in *Mandala*

ASSOCSP

Purpose: Makes linkages of one specimen to one or more other specimens all with unique Specimen identifiers for the purpose of solidifying the association among mating pairs, mimic/model, predator/prey, host/parasite relationships. Accessed from layout in SPECIMEN.

BIOECOBEH

Purpose: Provides opportunity to enter controlled language phrases from editable popup lists about the biology, ecology, and behavior observed during the collection of a specimen. Accessed from layout in SPECIMEN.

BULKSAMPLE

Purpose: Used to track single loans when accessed from DEPOSIT, and track taxa and loans by borrower of subsamples when accessed from SPECIMEN.

CHANGES

Purpose: Administrative utility table to track changes made or needed to the structure of *Mandala*. Primary access is via NAVIGATION.

COLLEVENT

Purpose: Records the collecting event information associated with a locality, including collecting method, date or range of dates of collection, collectors, and abiotic conditions at the time of collection. Create new records here by starting with a LOCALITY.

DEPOSIT

Purpose: Manages loans of multiple specimens/taxa (through BULKSAMPLE), specimen-based loans (through SPECIMEN), and subsample loans to specialists (SPECIMEN and BULKSAMPLE). Tracks physical parameters of loan shipments, loans by borrower, overdue loans, and can generate loan forms.

DETERMNR

Purpose: Provides a place to record the history of determinations of a specimen, including determiner, taxon identification, and year.

ENTRYQ

Purpose: User utility of electronic sticky notes to record problems or notations specific to records in *Mandala*, along with their resolution.

GENBANK

Purpose: Links specimens and taxa that have be analyzed for deposit in GenBank or other repository of molecular or genetic information. Will link directly to record in such on-line databases.

HELP

Purpose: Context sensitive table and field-level help available to users from all tables. Important scripted button and layout functions may also be included. Includes a section on General help tips.

ILLUS

Purpose: Provides means to catalog, archive, and compare images according to illustrator, subject, life stage, medium, method, view, image background, etc.

JOURNALS

Purpose: Authority file of journal titles and accepted abbreviations that provides orthographic consistency and flexibility in reporting when constructing citations in LITERATURE.

LITERATURE

Purpose: Vital table cataloging all types of literature. Citations are linked to taxonomic names in nearly a dozen ways, and to published images.

LITMINING

Purpose: Provides organized means of dissecting relevant literature with references to page numbers in citations for over thirty topics.

LOCALITY

Purpose: Provides an interpretation of a verbatim locality label, filling in missing information from gazetteers or maps, including georeferencing. A locality consists of a political boundary description and may include named geographic features that cross political boundaries. A locality may be linked to many collecting events.

MUSEUMS

Purpose: An authority table of collections and institutions housing collections. Identified by unique codens (3- or 4-letter codes), these are primarily linked to loans and the physical archiving of documents, files, or images in ILLUS.

NAVIGATION

Purpose: Clearinghouse for navigating to data entry, searching, and reporting features; contains glossary to icons and terms; and troubleshooting.

ORGASSOC

Purpose: Links one or more taxa (from TAXA) that may be associated with a specimen but not represented as a collected specimen. Accessed as layout in SPECIMEN.

PEO_JOIN

Purpose: A bridging or join file linking PEOPLE in a specified numerical order with a taxonID for authorities of taxonomic names; a LocCollEventID for collectors of specimens or bulk samples; or a literature citationID for its authors.

PEOPLE

Purpose: An authority table of names of people as collectors, authors, authorities of taxonomic names, determiners, illustrators, copyright holders, borrowers, and lenders. Contains contact and demographic information.

POLREGNS

Purpose: Authority file of biogeographic regions and the political boundaries contained within them, including names of countries no longer in use. Most are at the country level, but some countries, such as Mexico and China, lie in more than one biogeographic region and must further match state subdivisions for correct placement. Matched within country and state/province fields to LOCALITY.

SPECIMEN

Purpose: Linchpin table to *Mandala*, where information on all specimens is recorded and accessed by a unique specimen identifier.

TAXA

Purpose: Essential table in *Mandala* recording taxonomic names, their history, rank, status, and classification. Links in one way or another to nearly all other tables.

SPECIES OF DIPTERA PER FAMILY FOR ALL REGIONS

Table A.1. Numbers of described species for all families of Diptera, globally as well as for the biogeographic regions.

AF = Afrotropical

AU = Australasian/Oceanian

NE = Nearctic

NT = Neotropical

OR = Oriental

PA = Palaearctic

Data from Evenhuis, N.L., Pape, T., Pont, A.C. & Thompson, F.C. (eds) (2007) *Biosystematic Database of World Diptera*, Version 10; available at http://www.diptera.org/biosys.htm.

ALL FAMILIES	Total	AF	AU	NE	NT	OR	PA
Acartophthalmidae	4			2			4
Acroceridae	394	67	52	62	94	40	84
Agromyzidae	3,013	279	287	763	383	325	1,274
Anisopodidae	158	17	16	9	64	25	34
Anthomyiidae	1,896	68	20	691	106	108	1,158
Anthomyzidae	94	23	3	21	8	5	42
Apioceridae	169	4	68	64	33	1	
Apsilocephalidae	3		2	1	1		
Apystomyiidae	1				1		
Asilidae	7,413	1,686	579	1,073	1,485	1,017	1,673
Asteiidae	132	15	29	17	32	11	34
Atelestidae	10			2	2		6
Athericidae	122	22	16	4	28	27	27
Aulacigastridae	18	4		3	4	3	5
Australimyzidae	9		9				
Austroleptidae	8		3		5		
Axymyiidae	6			1			5
Bibionidae	754	74	96	86	192	140	184
Blephariceridae	322	28	43	33	76	37	106
Bolitophilidae	59			20		2	40
Bombyliidae	5,030	1,437	448	988	717	317	1,370
Brachystomatidae	145	11	46	17	32	9	30
Braulidae	7	3	1	1	2	4	4
Calliphoridae	1,524	208	300	103	152	227	585
Camillidae	40	23		4	1	1	14
Canacidae	119	25	32	12	23	18	19
Canthyloscelidae	16		5	3	4		4
Carnidae	90	5		20	4	4	67
Cecidomyiidae	6,051	215	321	1,247	561	569	3,275
Celyphidae	116	12	3			90	15
Ceratopogonidae	5,621	916	839	614	1,084	876	1,537
Chamaemyiidae	349	14	40	80	54	18	175
Chaoboridae	55	8	7	13	11	7	13
Chironomidae	6,951	569	530	1,111	790	816	3,579
Chloropidae	2,863	404	529	302	428	560	765
Chyromyidae	106	10	5	9	3	3	82

ALL FAMILIES	Total	AF	AU	NE	NT	OR	PA
Clusiidae	349	12	59	41	148	59	39
Coelopidae	35	5	21	5		2	6
Conopidae	783	161	66	74	203	123	205
Corethrellidae	66	3	6	5	45	3	5
Cryptochetidae	33	12	4	1	1	11	9
Ctenostylidae	10	1			5	4	
Culicidae	3,616	810	627	182	952	975	268
Curtonotidae	61	25	1	1	20	14	3
Cylindrotomidae	67		8	8	1	30	22
Cypselosomatidae	34		10	3	7	12	3
Deuterophlebiidae	14			6		5	4
Diadocidiidae	19		3	3		6	9
Diastatidae	48	12		8	4	12	18
Diopsidae	183	140	3	2		38	3
Ditomyiidae	93		37	6	31	2	17
Dixidae	185	8	18	45	28	27	62
Dolichopodidae	7,118	766	1,186	1,382	1,189	1,057	1,716
Drosophilidae	3,925	460	1,141	248	876	1,019	416
Dryomyzidae	25		1	8	1	4	15
Empididae	2,911	214	188	468	377	249	1,425
Ephydridae	1,977	341	211	484	395	166	582
Eurychoromyiidae	1				1		
Evocoidae	1				1		
Fanniidae	319	13	14	111	75	31	150
Fergusoninidae	29		28	1		1	
Glossinidae	23	23					
Gobryidae	5		2			3	
Helcomyzidae	12		4	1	5		2
Heleomyzidae	717		105	153	88	21	337
Helosciomyzidae	23	62	21		2		
Hesperinidae	6			1	1		4
Heterocheilidae	2			1			1
Hilarimorphidae	32			27		1	4
Hippoboscidae	786	130	143	43	239	216	96
Homalocnemiidae	7	1	4		2		
Huttoninidae	8		8				

ALL FAMILIES	Total	AF	AU	NE	NT	OR	PA
Hybotidae	1,882	143	114	316	289	319	729
Inbiomyiidae	10				10		
Ironomyiidae	1		1				
Iteaphila group	27		1	18			10
Keroplatidae	907	170	151	85	186	110	207
Lauxaniidae	1,893	91	472	157	369	408	430
Limoniidae	10,334	1,027	1,917	926	2,648	2,323	1,625
Lonchaeidae	480	64	54	136	94	47	118
Lonchopteridae	58	6	1	5	1	21	29
Lygistorrhinidae	30	7	2	1	6	13	1
Marginidae	3	3					
Megamerinidae	15				1	8	6
Micropezidae	578	65	81	37	284	69	55
Milichiidae	276	66	31	43	80	48	60
Mormotomyiidae	1	1					
Muscidae	5,153	992	743	631	882	845	1,502
Mycetophilidae	4,105	258	319	673	1,056	370	1,549
Mydidae	463	200	43	75	88	11	51
Mystacinobiidae	1		1				
Mythicomyiidae	346	56	4	186	19	8	82
Nannodastiidae	5		1		2	1	3
Natalimyzidae	1	1					
Nemestrinidae	275	51	55	8	58	22	87
Neminidae	14	7	7				
Neriidae	111	20	22	2	41	31	2
Neurochaetidae	20	12	4			4	
Nothybidae	8					8	
Nymphomyiidae	7			2		2	3
Odiniidae	62	8		11	23	3	23
Oestridae	192	37	7	49	58	6	66
Opetiidae	5						5
Opomyzidae	61	5		11		1	48
Oreogetonidae	36		4	8	19		5
Oreoleptidae	1			1			
Pachyneuridae	5			1		1	4
Pallopteridae	66		10	9	6	1	40

ALL FAMILIES	Total	AF	AU	NE	NT	OR	PA
Pantophthalmidae	20				20		
Pediciidae	494	10	5	149	10	127	210
Periscelididae	84		14	7	27	13	15
Perissommatidae	5		4		1		
Phaeomyiidae	3		1				3
Phoridae	4,022	452	291	422	1,487	587	906
Piophilidae	82	7	9	37	11	5	39
Pipunculidae	1,381	154	158	158	235	193	507
Platypezidae	252	42	26	77	15	21	78
Platystomatidae	1,162	286	495	42	24	273	68
Psilidae	321	52	2	32	10	73	156
Psychodidae	2,886	302	448	123	929	397	763
Ptychopteridae	74	9		18	1	17	29
Pyrgotidae	351	143	73	11	54	47	24
Rangomaramidae	39	4	16		17	1	1
Rhagionidae	707	54	90	105	108	122	233
Rhiniidae	363	160	22	1		127	78
Rhinophoridae	167	28	16	5	16	4	101
Richardiidae	174			8	166		
Ropalomeridae	33			2	31		
Sarcophagidae	3,071	429	183	451	872	249	1,013
Scathophagidae	392	4	3	151	6	7	273
Scatopsidae	323	43	72	77	46	41	61
Scenopinidae	414	68	91	148	32	12	125
Sciaridae	2,224	75	251	176	273	492	988
Sciomyzidae	604	64	43	196	90	27	233
Sepsidae	375	149	31	35	55	66	89
Simuliidae	2,080	213	213	242	390	315	729
Somatiidae	7				7		
Spaniidae	43		7	13	1	1	21
Sphaeroceridae	1,580	325	148	285	387	169	436
Stratiomyidae	2,666	391	410	312	935	355	400
Strongylophthalmyiidae	47		4	1	1	36	8
Syringogastridae	10			1	9		
Syrphidae	5,935	591	417	818	1,516	878	2,058
Tabanidae	4,387	816	468	394	1,168	819	778

ALL FAMILIES	Total	AF	AU	NE	NT	OR	PA
Tachinidae	9,629	1,032	864	1,439	2,729	835	3,051
Tachiniscidae	3	2			1		
Tanyderidae	38	1	21	4	3	3	6
Tanypezidae	21			2	19		1
Tephritidae	4,625	995	829	372	785	1,051	891
Teratomyzidae	8		5		3		
Tethinidae	193	24	53	28	34	15	60
Thaumaleidae	173	2	34	25	7	18	87
Therevidae	1,125	164	335	165	162	46	272
Tipulidae	4,324	375	394	620	779	911	1,303
Trichoceridae	160		27	30	10	27	77
Ulidiidae	672	24	24	139	285	15	248
Valeseguyidae	1		1				
Vermileonidae	59	29		3	4	4	20
Xenasteiidae	13	2	5			3	3
Xylomyidae	134	6	10	12	9	60	43
Xylophagidae	136		10	28	12	50	38
TOTAL	**152,715**	**20,163**	**18,920**	**21,449**	**31,088**	**22,543**	**44,894**

INDEX